Current Topics in Behavioral Neurosciences

Volume 44

Current Topics in Behavioral Neurosciences provides critical and comprehensive discussions of the most significant areas of behavioral neuroscience research, written by leading international authorities. Each volume in the series represents the most informative and contemporary account of its subject available, making it an unrivalled reference source. Each volume will be made available in both print and electronic form.

With the development of new methodologies for brain imaging, genetic and genomic analyses, molecular engineering of mutant animals, novel routes for drug delivery, and sophisticated cross-species behavioral assessments, it is now possible to study behavior relevant to psychiatric and neurological diseases and disorders on the physiological level. The Behavioral Neurosciences series focuses on translational medicine and cutting-edge technologies. Preclinical and clinical trials for the development of new diagnostics and therapeutics as well as prevention efforts are covered whenever possible. Special attention is also drawn on epigenetical aspects, especially in psychiatric disorders.

CTBN series is indexed in PubMed and Scopus.

More information about this series at http://www.springer.com/series/7854

Golam M. Khandaker • Urs Meyer • Peter B. Jones
Editors

Neuroinflammation and Schizophrenia

Springer

Editors
Golam M. Khandaker
Department of Psychiatry
University of Cambridge School of Clinical Medicine
Cambridge, UK

Urs Meyer
Verhaltensneurobiologie
ETH Zürich
Schwerzenbach, Switzerland

Peter B. Jones
Department of Psychiatry
University of Cambridge School of Clinical Medicine
Cambridge, UK

ISSN 1866-3370 ISSN 1866-3389 (electronic)
Current Topics in Behavioral Neurosciences
ISBN 978-3-030-39140-9 ISBN 978-3-030-39141-6 (eBook)
https://doi.org/10.1007/978-3-030-39141-6

This Springer imprint is published by the registered company Springer Nature Switzerland AG.
The registered company address is: Gewerbestrasse 11, 6330 Cham, Switzerland

Preface

Over recent years, there has been an exponential growth in research into the possible role of the immune system in major mental illness, but research into immune links to schizophrenia and related psychotic disorders has a long history (Khandaker et al. 2015). For example, in 1926 Karl Menninger published 200 cases of postinfluenzal psychosis, one-third of which were reported to resemble dementia praecox (conceptual predecessor of schizophrenia). In 1927, the Nobel Prize for medicine was awarded to Julius Wanger Jurug for his work on medical inoculation of malarial parasites as a treatment for syphilitic psychosis. Ten years later, in 1937, Lehman Facius described autoantibodies against brain structures in the cerebrospinal fluid of patients with schizophrenia, paving the way to one of the many neuroimmune concepts of schizophrenia and related disorders.

More recently, a large number of epidemiological, genetic and clinical studies have pointed to links between schizophrenia and infection, inflammation, atopy and autoimmunity. Contrary to the traditional view that the brain is an immunologically privileged site shielded behind the blood–brain barrier, studies in the past 20 years have demonstrated complex interactions between the immune system, systemic inflammation and the brain, which can lead to changes in mood, cognition and behaviour. Recent studies provide compelling evidence that the immune system in general, and neuroinflammation in particular, may play a role in the aetiology of schizophrenia and could be a promising therapeutic target. With significant investments from the academia and the industry to further investigate immunological mechanisms and treatment options for major psychiatric disorders, research in this field is set to grow exponentially in the coming years. Yet currently there is no book on the topic of the neuroinflammation and schizophrenia.

This book aims to fill this gap by bringing together an internationally renowned panel of experts to provide a comprehensive and authoritative summary of cutting edge scientific evidence regarding the role of immune system in pathogenesis and treatment of schizophrenia and related psychotic disorders. Comprising 11 chapters, this book tackles a number of key aspects including historical perspectives and basic immunobiology relevant for psychiatric disorders; possible role of infection, inflammatory mediators, microbiota, atopy and autoimmunity in illness pathogenesis; possible mechanisms including role of genes, cognition and microglial activation; and clinical implications particularly as regards to treatment of schizophrenia and comorbidity between psychiatric and physical illness.

Chapter 1 ("A Brief History of Immunological Research into Psychosis and Pathways for Immune Influence of the Brain") by Norbert Müller gives a historical overview of the wax and wane in research interest in this field over the past 150 years. Chapter 2 ("Basic Concept of Microglia Biology and Neuroinflammation in Relation to Psychiatry") by Daniele Mattei and Tina Notter provides an overview of the basic concepts of neuroinflammation and neuroinflammatory processes, including potential role of neuroinflammation and microglia activity in psychiatric disorders. Chapter 3 ("Epidemiological Studies of Prenatal and Childhood Infection and Schizophrenia") by Håkan Karlsson and Christina Dalman provides an up-to-date review of this topic including discussions on mechanistic issues such as sensitive periods for exposure and possible mediating effects of inflammation and cognition. Chapter 4 ("Cytokines, Oxidative Stress and Cellular Markers of Inflammation in Schizophrenia") by Rachel Upthegrove and Golam M. Khandaker reviews current evidence linking immune dysfunction in schizophrenia and related psychotic disorders focusing particularly on circulating cytokines, oxidative stress and cellular markers of inflammation in various stages of illness from drug naïve first episode psychosis to chronic schizophrenia. They discuss evidence of causality, potential reasons for reported trans-diagnostic associations for inflammatory cytokines and clinical implications as regards to new treatment and prediction of treatment response.

Chapter 5 ("From Infection to the Microbiome: An Evolving Role of Microbes in Schizophrenia") by Emily G. Severance and Robert H. Yolken discusses how disruptions in microbiome as the result of pathogen invasion, stress or immune gene deficiency may give rise to localized inflammation, endothelial barrier compromise, translocation of gut-derived products and systemic inflammation leading to neuropsychiatric consequences relevant for schizophrenia. Chapter 6 ("Autoantibodies and psychosis") by Eric Kelleher, Helen Barry, David R. Cotter, Aiden Corvin and Kieran C. Murphy gives a comprehensive overview of psychiatric and other clinical features of anti-NMDA-R autoantibody-mediated encephalitis as well as a number of other autoantibodies against neuronal cell surface proteins. They discuss evidence from clinical studies regarding treatment response to immunotherapy in patients with antibody-mediated psychosis. Chapter 7 ("Herpes Simplex Virus Type-1 Infection: Associations with Inflammation and Cognitive Aging in Relation to Schizophrenia") by Vishwajit L. Nimgaonkar, Triptish Bhatia, Abdelaziz Mansour, Maribeth A. Wesesky and Smita Deshpande presents evidence linking persistent infection with Herpes Simplex Virus Type 1 and cognitive ageing. They discuss potential relevance of these findings for patients with schizophrenia, who are particularly vulnerable due to disorder-related cognitive impairment. Chapter 8 ("Role of Infection, Autoimmunity, Atopic Disorders, and the Immune System in Schizophrenia: Evidence from Epidemiological and Genetic Studies") by Michael E. Benros and Preben B. Mortensen reviews the evidence linking infection and schizophrenia in light of the recent large-scale epidemiological and genetic studies in an attempt to elucidate familial confounding and shared genetic risk for schizophrenia and susceptibility to common infections.

Chapter 9 ("Microglial Activation and Psychotic Disorders: Evidence from Pre-clinical and Clinical Studies") by Tatiana Barichello, Lutiana R. Simoes, Joao Quevedo and Xiang Y. Zhang delves into the mechanisms particularly looking at neuroinflammation and microglial activation based on pre-clinical schizophrenia models. They discuss the possible links between schizophrenia and neuroinflammation in clinical studies and mechanisms by which microglial activation may influence development of schizophrenia. Chapter 10 ("Early-Life Adversity, Systemic Inflammation and Comorbid Physical and Psychiatric Illnesses of Adult Life") by Maria Antonietta Nettis, Carmine M. Pariante and Valeria Mondelli reviews evidence from pre-clinical and clinical studies suggesting that inflammation could be a key mechanism through which childhood maltreatment/abuse increases the risk of chronic psychiatric and physical illnesses of adult life including schizophrenia and cardiometabolic disorders. The final chapter (Chap. 11: "Inflammation, Antipsychotic Drugs, and Evidence for Effectiveness of Anti-inflammatory Agents in Schizophrenia") by Ananda K. Pandurangi and Peter F. Buckley provides an overview of a range of therapeutic options that could be used to reduce inflammation and their potential as treatment for schizophrenia. The authors discuss emerging evidence from studies of biologics and other novel anti-inflammatory drugs and highlights key issues for randomised controlled trials in this field today, notably the need for useful biomarkers for patient selection/stratification.

This book captures the current state of science in this fast moving field, and we are extremely grateful to the authors for their contribution. The chapters provide up-to-date summaries of immunological risk factors for schizophrenia and related psychotic disorders, and underlying mechanisms as informed by neuroimaging, genetic, clinical and experimental studies. In addition, this book illuminates the scope for immunological treatment for schizophrenia. We hope this contribution would be of interest to researchers and clinicians alike.

Cambridge, UK — Golam M. Khandaker
Schwerzenbach, Switzerland — Urs Meyer
Cambridge, UK — Peter B. Jones

Reference

Khandaker GM et al (2015) Inflammation and immunity in schizophrenia: implications for pathophysiology and treatment. Lancet Psychiatry 2:258–270. https://doi.org/10.1016/s2215-0366(14)00122-9

Contents

A Brief History of Immunological Research into Psychosis and Pathways for Immune Influence of the Brain

Norbert Müller

Contents

Abstract Infection and inflammation resulting from a malfunctioning of the immune system have been discussed as pathological factors in psychosis for more than 130 years. The first immune-modulating therapeutic approaches for psychosis were developed more than 100 years ago, but the breakthrough of antipsychotic treatment in the 1950s shifted the emphasis of research to catecholaminergic neurotransmission. In the 1990s, however, the unsatisfactory therapeutic effects of antipsychotics, and the fact that the pathological mechanisms of psychosis were still unknown, reignited the scientific interest in other topics, including inflammation. In parallel, the further development of immunological methods enabled a more sophisticated examination of immunological and inflammatory mechanisms. Psychiatrists' interest in this interdisciplinary field increased as a consequence of encouraging results of psychoneuroimmunological research and broader funding of the field. In the meantime, the benefits of anti-inflammatory treatment in psychosis have been demonstrated in clinical studies and meta-analyses. Future studies are warranted to evaluate the exact immunological mechanisms in the pathophysiology of the disease, optimize the anti-inflammatory treatment approach and develop more targeted, personalized therapies in psychosis.

Keywords History · Inflammation · Kynurenine · Psychoneuroimmunology · Psychosis · Schizophrenia

One of the main pathophysiological mechanisms in schizophrenia relates to abnormal functioning of the dopaminergic system, and all currently available antipsychotic drugs act on dopaminergic transmission. Consequently, over the last six decades,

---

N. Müller (✉)
Department of Psychiatry and Psychotherapy, Ludwig Maximilian University and Marion von Tessin Memory Center, Munich, Germany
e-mail: n.mueller@mvt-zentrum.de

Curr Topics Behav Neurosci (2020) 44: 1–8
DOI 10.1007/7854_2018_82
Published Online: 19 February 2019

research on the neurobiology of schizophrenia has focussed overwhelmingly on disturbances in dopaminergic neurotransmission, with the aim to elucidate the exact role of dopamine in schizophrenia (Carlsson 1988). Despite all the research, however, the precise pathogenic mechanisms underlying schizophrenia remain unclear. Many findings point to processes beyond the dopamine hypothesis, including the fact that the efficacy of dopamine-based antipsychotic drugs is inadequate especially in chronic courses of psychosis. Hence, new concepts are required to elucidate pathogenic processes that go beyond the dopamine hypothesis. Psychoneuroimmunological research is one such new concept, which focuses on a possible immune/inflammatory pathogenesis of schizophrenia in general and its relation to abnormal dopamine functions in particular.

Interestingly, long before the era of dopaminergic research and modern antipsychotics, infections and inflammation were discussed as causes for psychosis, and therapeutic considerations were developed on this basis. It is well known in psychiatry that infectious diseases provoke psychiatric symptoms. As early as 1890 – 9 years before he introduced the modern classification of psychoses into psychiatry – Emil Kraepelin (1856–1926), one of the founders of modern psychiatry, described cases of psychiatric syndromes due to infection. During an influenza epidemic, he reported 11 cases of psychiatric disorders that presented with various syndromes, such as depressed mood, a paranoid and hallucinatory syndrome, involuntary movements, cognitive deterioration and a delirious state (Kraepelin 1890). Kraepelin demonstrated with these cases that:

1. Psychiatric syndromes can be caused by infectious agents.
2. The same infectious agent (in this case influenza virus) can cause completely different psychiatric syndromes.

Kraepelin's groundbreaking insights are still valid today, although the hypotheses of modern research are much more generalized. Nowadays, infection and/or inflammation is hypothesized to play a substantial role in psychiatric disorders, without necessarily being overtly associated with an acute infectious disease.

Three years before Kraepelin's publication, the young Austrian psychiatrist Julius Wagner (1857–1940) (whose full name is Julius Ritter Wagner von Jauregg) had observed that psychiatric patients were more resistant to infectious diseases than a comparison group. Other authors from England, France and Germany described the same phenomenon during epidemic typhus outbreaks in asylums. Wagner-Jauregg published these observations in a scientific review article in 1887 (see Table 1). While only about 17% (the mean of figures provided by different authors) of the mentally ill in the asylums became infected by *Salmonella typhi*, 39% of the guards caught the infection.

Interestingly, Wagner-Jauregg reported that the mentally ill people who recovered from the typhus infection often showed an improvement of their psychiatric symptoms (up to 75% of patients) or even recovered from their psychiatric illness (up to 48%). In this early 'meta-analysis', a mean of 32% of the cases showed an improvement, and 20% of the cases became healthy, i.e. the mental state improved in more than 50% of those who recovered from the typhus infection (see Table 2).

Table 1 Typhus infections of the mentally ill and guards (1887)

Author	Mentally ill (%)	Guards (%)
Gaye	12.4	54
Wille	50	80
Wille	15	42
Campbell	13.8	38
Gottlob	4.7	12
Rath	5.3	6
Mean	17	39

Table 2 Influence of typhus infection on mental illness (1887)

Author	Patients	Unchanged	Healthy	Improved
Gaye	62	58 (94%)	–	4 (6%)
Nasse	21	6 (29%)	10 (48%)	5 (24%)
Campbell	22	16 (72%)	3 (14%)	3 (14%)
Rath	22	7 (32%)	5 (23%)	10 (45%)
Forel	12	8 (67%)	1 (8%)	3 (25%)
Jolly	12	–	3 (25%)	9 (75%)
Mean		48%	20%	32%

These observations were made in the second half of the nineteenth century, a time when almost no therapeutic options were available for psychiatric diseases.

On the basis of these observations, Wagner-Jauregg tried to develop an immune-therapeutic approach. Before the Kraepelinian classification of psychoses into 'dementia praecox' (a concept similar to Eugen Bleuler's 'schizophrenia') and manic-depressive insanity, he successfully treated patients with different vaccines, including attenuated *Mycobacterium tuberculosis*, plasmodia and *Salmonella typhi*. Immunologically, these infectious agents stimulate the type-1 immune response (a contribution of a lack of type-1 activation to an immune imbalance was repeatedly hypothesized in schizophrenia). The most successful of his experiments was the development of the 'malaria therapy' or 'fever therapy' of syphilis, for which he earned the Nobel Prize in 1927. In addition to treating syphilis, he tried to develop additional vaccination therapies for nervous diseases (Wagner-Jauregg 1926). Wagner-Jauregg was, without doubt, a pioneer in the immune treatment of psychiatric disorders. However, although his immune-based 'vaccination' therapy was developed in the first decades of the twentieth century – and looked promising – his work fell into oblivion and was not followed up outside German-speaking countries, particularly after the introduction of electroconvulsive therapy and, later, treatment with antipsychotics.

In parallel, the Russian psychiatrist Alexander Rosenblum (1826–1903) published studies of attenuated infectious agents in his psychotic patients. He induced fever with malaria, typhoid and relapsing fever and reported that 11 of his 22 psychiatric patients were cured after receiving treatment. However, his discovery did not receive widespread attention because his results were published in an obscure journal (Rosenblum and Zakon 1943).

Starting from his serological research on syphilis, a result of infection with *Treponema pallidum*, Felix Plaut (1877–1940) studied the role of neurosyphilis in psychotic processes and the pathogenesis of psychiatric disorders. He argued that this specific infection – around one third of the inpatients in the fast-growing asylums at the end of the nineteenth century and beginning of the twentieth century had neurosyphilis – provoked a variety of reactions in the brain, including psychosis. He described several cases of psychotic patients, including catatonic patients, with a positive Wassermann reaction (a complement-binding reaction used to diagnose syphilis). Together with the bacteriologist August Wassermann, Plaut studied the Wassermann reaction and in particular neurosyphilis by performing research on cerebrospinal fluid. He recognized that the infection could be a trigger of psychosis in those particular cases. At that time, it was already well known that different pathogenic mechanisms might be responsible for psychiatric syndromes. Back in Munich, at the university Department of Psychiatry chaired by Emil Kraepelin, Plaut and his co-workers started animal experiments and clinical studies of the pathogenesis and therapy of neurosyphilis (including the malaria therapy of syphilis). The immunology of neurosyphilis was his central research topic for 30 years. In this context, he was the first to describe the groundbreaking finding of autochthone antibody production in the brain of laboratory animals, i.e. Plaut demonstrated for the first time an immune reaction of the brain. Moreover, although infection was known as a trigger of psychosis long before Plaut, he was one of the first to systematically study how infection with *Treponema pallidum* and the subsequent immune response were associated with psychosis.

The introduction of antibiotics after the Second World War revolutionized the treatment of infections, including syphilis. Effective treatment of syphilis with antibiotics meant that the state of neurolues was seldom reached and, consequently, was no longer a focus of research. Similarly, scientific interest in the tight connection between infection and psychosis subsided and was only broader revitalized in the 1990s (Bechter 2004; Benros et al. 2011).

The discovery of the 'neuroleptic' effect of chlorpromazine and the role of catecholaminergic neurotransmission in psychiatric disorders in the 1950s also caused the focus of psychiatric research to shift. Research concentrated on dopaminergic, noradrenergic and serotonergic neurotransmission as pathological mechanisms in psychiatric disorders, while alternative hypotheses and theories were neglected to a large extent. Because psychoses have a substantial genetic component, interest in genetic research grew in parallel with the development of molecular genetic methods as of 1980s. Interestingly, the strongest genetic signal for schizophrenia was found on chromosome 6p22.1, in a region related to the human leucocyte antigen (HLA) system and other immune functions.

Modern theories about the pathogenesis of psychiatric disorders changed with the emerging evidence suggesting that catecholaminergic functions can be influenced by additional factors, such as the immune system and inflammatory processes. Interactions between the immune system, hormones and neurotransmitters came into the focus of research in biological psychiatry during the 1980s. The work of Robert Ader (1932–2011) on the conditioning of immunity and his overview of the field contributed to growing interest in the mechanisms and the topic, especially among basic researchers (Ader 1981). During the 1980s, only isolated groups of researchers were interested in the field, the methodological resources of immunological research were limited, and the interpretation of the findings is often speculative because of the limited knowledge of the interactions within the immune system. This may have contributed to the fact that in the United States, funding of this research area was viewed critically and only occasionally provided (Rapaport et al. 1989). In Germany, on the other hand, the Volkswagen Foundation launched a large international funding programme for psychoneuroimmunology at the beginning of the 1990s. This funding allowed the first German university chair in psychoneuroimmunology to be established at the University of Marburg: Hugo Besedovsky, a pioneer in research into the interaction between the immune system, glucocorticoid hormone system and neurotransmission, who showed that the immune system influences the corticoid system in the brain, held this chair for around 15 years (Besedovsky et al. 1983). Moreover, this programme contributed both to a lively young scientific scene in the field of psychoneuroimmunology in Germany, which continues to this day, and to the founding of a scientific association called GEBIN (German Endocrine Brain Immune Network), which holds regular biannual congresses.

For the reasons mentioned above, psychoneuroimmunological research was mostly concentrated in Europe during the late 1980s and early 1990s. The research groups of Michael Maes in Antwerp (Belgium) and Maastricht (the Netherlands), which focussed on major depression, and my group in Munich (Germany), which focussed on schizophrenia, contributed actively to the field and showed that pro-inflammatory cytokines and an activated immune state play a role in schizophrenia, major depression and bipolar disorder (Maes et al. 1990, 1992; Muller et al. 1987, 1991, 1993). On the one hand, the evaluation of immunological mechanisms in schizophrenia concentrated on the neurodevelopmental effects of immune activation using animal models (Meyer et al. 2006) and, on the other hand, on the role of the blood-brain barrier (Muller and Ackenheil 1995; Muller et al. 1999), blood-brain axis of pro- and anti-inflammatory cytokines (Besedovsky et al. 1983; Dantzer and Kelley 1989; Muller and Ackenheil 1995; Muller et al. 1999) and the immune effects of antipsychotic treatment. Another aspect of the immune system's involvement in psychoses was introduced by research into the role of tryptophan/kynurenine metabolism. This metabolism pathway is driven by pro- and anti-inflammatory cytokines and has a direct influence on melatonergic, serotonergic and glutamatergic neurotransmission through different neuroactive metabolites, such as kynurenic acid and quinolinic acid (Kim et al. 2009; Linderholm et al. 2012).

Research programmes on schizophrenia and bipolar disorder were funded in the United States and internationally from the mid-1990s by the Theodore and Vada Stanley Research Foundation. Because of the scientific emphasis of the two protagonists of the foundation, Fuller Torrey and Robert Yolken, these programmes had a scientific focus on the roles of infection and immunity. A lot of scientific work in the field of infection, immunity and inflammation in schizophrenia and bipolar disorder was – and still is – enabled by the Stanley Foundation. Similar to the Volkswagen Foundation in Germany, the Stanley Foundation boosted research internationally in the field of psychoneuroimmunology. Moreover, additional focuses and funds were directed towards the elucidation of therapeutic approaches involving immunomodulatory, anti-infective and anti-inflammatory compounds in schizophrenia and bipolar disorder. The explicit therapeutic focus of Stanley programme distinguished it from other funding programmes, the later of which mostly focussed on elucidating the pathophysiological mechanisms of the psycho-immune interaction.

Therapeutic studies on modulating the immune system in psychiatric disorders provided further validation for the psychoneuroimmunological approach, for example, by proving a benefit of anti-inflammatory drugs in patients with schizophrenia. These studies have broadened the therapeutic spectrum in psychiatry and stimulated further pathophysiological and therapeutic studies. For example, the cyclooxygenase-2 (COX-2) inhibitor celecoxib showed a therapeutic benefit in schizophrenia, and, interestingly, COX-2 inhibition also had a positive effect on cognition in schizophrenia (Muller et al. 2005). The efficacy of therapy with a COX-2 inhibitor seems most pronounced in the first years of the schizophrenic disease process (Muller et al. 2010). Other studies demonstrated a beneficial effect of acetylsalicylic acid and other anti-inflammatory compounds in schizophrenia and schizophrenia spectrum disorders (Laan et al. 2010). A meta-analysis of the clinical effects of non-steroidal anti-inflammatory drugs in schizophrenia showed significant effects on the total, positive and negative symptoms of schizophrenia (Zheng et al. 2017), although this effect seems to be dependent on the duration of the disease and most pronounced in first manifestation schizophrenia (Nitta et al. 2013) – findings that are in accordance with an inflammatory origin of the disease. Despite these promising findings, however, further research is needed to elucidate the exact immunological mechanisms in the pathophysiology of schizophrenia and related disorders and to optimize immune-based treatment approaches in order to develop more targeted, personalized therapies in psychosis.

Acknowledgement The author thanks Jacquie Klesing, board-certified editor in the life sciences (ELS), for editing assistance with the manuscript.

Conflict of Interest None to declare.

Funding The work has been supported by the foundation 'Immunität und Seele'.

References

Ader R (ed) (1981) Psychoneuroimmunology. Academic Press, New York

Bechter K (2004) The mild encephalitis-hypothesis – new findings and studies. Psychiatr Prax 31 (Suppl 1):S41–S43. https://doi.org/10.1055/s-2004-828428

Benros ME, Nielsen PR, Nordentoft M et al (2011) Autoimmune diseases and severe infections as risk factors for schizophrenia: a 30-year population-based register study. Am J Psychiatry 168 (12):1303–1310. https://doi.org/10.1176/appi.ajp.2011.11030516

Besedovsky H, del Rey A, Sorkin E et al (1983) The immune response evokes changes in brain noradrenergic neurons. Science 221(4610):564–566

Carlsson A (1988) The current status of the dopamine hypothesis of schizophrenia. Neuropsychopharmacology 1(3):179–186

Dantzer R, Kelley KW (1989) Stress and immunity: an integrated view of relationships between the brain and the immune system. Life Sci 44(26):1995–2008

Kim YK, Myint AM, Verkerk R et al (2009) Cytokine changes and tryptophan metabolites in medication-naive and medication-free schizophrenic patients. Neuropsychobiology 59 (2):123–129. https://doi.org/10.1159/000213565

Kraepelin E (1890) Über Psychosen nach Influenza. Dtsch Med Wochenschr 16:209–212

Laan W, Grobbee DE, Selten JP et al (2010) Adjuvant aspirin therapy reduces symptoms of schizophrenia spectrum disorders: results from a randomized, double-blind, placebo-controlled trial. J Clin Psychiatry 71(5):520–527. https://doi.org/10.4088/JCP.09m05117yel

Linderholm KR, Skogh E, Olsson SK et al (2012) Increased levels of kynurenine and kynurenic acid in the CSF of patients with schizophrenia. Schizophr Bull 38(3):426–432. https://doi.org/10.1093/schbul/sbq086

Maes M, Bosmans E, Suy E et al (1990) Immune disturbances during major depression: upregulated expression of interleukin-2 receptors. Neuropsychobiology 24(3):115–120

Maes M, Lambrechts J, Bosmans E et al (1992) Evidence for a systemic immune activation during depression: results of leukocyte enumeration by flow cytometry in conjunction with monoclonal antibody staining. Psychol Med 22(1):45–53

Meyer U, Schwendener S, Feldon J et al (2006) Prenatal and postnatal maternal contributions in the infection model of schizophrenia. Exp Brain Res 173(2):243–257. https://doi.org/10.1007/s00221-006-0419-5

Muller N, Ackenheil M (1995) Immunoglobulin and albumin content of cerebrospinal fluid in schizophrenic patients: relationship to negative symptomatology. Schizophr Res 14(3):223–228

Muller N, Ackenheil M, Eckstein R et al (1987) Reduced suppressor cell function in psychiatric patients. Ann N Y Acad Sci 496:686–690

Muller N, Ackenheil M, Hofschuster E et al (1991) Cellular immunity in schizophrenic patients before and during neuroleptic treatment. Psychiatry Res 37(2):147–160

Muller N, Hofschuster E, Ackenheil M et al (1993) Investigations of the cellular immunity during depression and the free interval: evidence for an immune activation in affective psychosis. Prog Neuro-Psychopharmacol Biol Psychiatry 17(5):713–730

Muller N, Riedel M, Hadjamu M et al (1999) Increase in expression of adhesion molecule receptors on T helper cells during antipsychotic treatment and relationship to blood-brain barrier permeability in schizophrenia. Am J Psychiatry 156(4):634–636

Muller N, Riedel M, Schwarz MJ et al (2005) Clinical effects of COX-2 inhibitors on cognition in schizophrenia. Eur Arch Psychiatry Clin Neurosci 255(2):149–151. https://doi.org/10.1007/s00406-004-0548-4

Muller N, Krause D, Dehning S et al (2010) Celecoxib treatment in an early stage of schizophrenia: results of a randomized, double-blind, placebo-controlled trial of celecoxib augmentation of amisulpride treatment. Schizophr Res 121(1–3):118–124. https://doi.org/10.1016/j.schres.2010.04.015

Nitta M, Kishimoto T, Muller N et al (2013) Adjunctive use of nonsteroidal anti-inflammatory drugs for schizophrenia: a meta-analytic investigation of randomized controlled trials. Schizophr Bull 39(6):1230–1241. https://doi.org/10.1093/schbul/sbt070

Rapaport MH, McAllister CG, Pickar D et al (1989) Elevated levels of soluble interleukin 2 receptors in schizophrenia. Arch Gen Psychiatry 46(3):291–292

Rosenblum AS, Zakon SJ (trans) (1943) Relation of febrile disease to the psychoses. Arch Derm Syphilol 48:52, Russian. Trudi vrach. Odessk. g. boln. 1876–1877;1872
Wagner-Jauregg J (1926) Fieberbehandlung bei psychosen. Wien Med Wochenschr 76:79–82
Zheng W, Cai DB, Yang XH et al (2017) Adjunctive celecoxib for schizophrenia: a meta-analysis of randomized, double-blind, placebo-controlled trials. J Psychiatr Res 92:139–146. https://doi.org/10.1016/j.jpsychires.2017.04.004

Basic Concept of Microglia Biology and Neuroinflammation in Relation to Psychiatry

Daniele Mattei and Tina Notter

Contents

Abstract The hypothesis that the neuroimmune system plays a role in the pathogenesis of different psychiatric disorders, including schizophrenia, depression, and bipolar disease, has attained increasing interest over the past years. Previously thought to have the sole purpose of protecting the central nervous system (CNS) from harmful stimuli, it is now known that the central immune system is critically involved in regulating physiological processes including neurodevelopment, synaptic plasticity, and circuit maintenance. Hence, alterations in microglia – the main immune cell of the CNS – and/or inflammatory factors do not unequivocally connote ongoing neuroinflammation or neuroinflammatory processes per se but rather might signify changes in brain homoeostasis. Despite this, psychiatric research tends to equate functional changes in microglia or alterations in other immune mediators with neuroinflammation. It is the main impetus of this chapter to overcome some of the current misconceptions and possible oversimplifications with respect to neuroinflammation and microglia activity in psychiatry. In order to do so, we will

D. Mattei
Institute of Pharmacology and Toxicology, University of Zurich-Vetsuisse, Zurich, Switzerland

T. Notter (✉)
Neuroscience and Mental Health Research Institute, Cardiff University, Cardiff, Wales, UK
e-mail: nottert@cardiff.ac.uk

Curr Topics Behav Neurosci (2020) 44: 9–34
DOI 10.1007/7854_2018_83
Published Online: 10 February 2019

first provide an overview of the basic concepts of neuroinflammation and neuroinflammatory processes. We will then focus on microglia with respect to their ontogeny and immunological and non-immunological functions presenting novel insights on how microglia communicate with other cell types of the central nervous system to ensure proper brain functioning. And lastly, we will delineate the non-immunological functions of inflammatory cytokines in order to address the possible misconception of equating alterations in central cytokine levels with ongoing central inflammation. We hereby hope to help unravel the functional relevance of neuroimmune dysfunctions in psychiatric illnesses and provide future research directions in the field of psychoneuroimmunology.

Keywords Cytokines · Microglia · Microglia Sensome · Neuroinflammation · Psychiatry · Schizophrenia

1 Introduction

The neuroimmune hypothesis in schizophrenia has experienced a reappraisal. The possible role of inflammatory processes in psychiatric disorders, which in the context of the central nervous system (CNS) is frequently referred to as neuroinflammation, has attained increasing interest over the past decade (Graeber 2014; Masgrau et al. 2017). In particular, functional abnormalities in microglia – the resident, myeloid immune cell of the CNS – have attained increasing interest in psychiatry in general and schizophrenia in particular (Laskaris et al. 2016). Microglia act as the first line of defence against invading pathogens and play a key role in central infections and central inflammation (Kettenmann et al. 2011). Similar to monocytes/macrophages in the periphery, they constantly survey the CNS and rapidly respond to invading pathogens, changes in the physiological microenvironment, and CNS injury (Gomez-Nicola and Perry 2015; Hanisch and Kettenmann 2007). Upon activation by pathological insults, microglia can rapidly alter their transcriptional profiles and morphological appearance, increase their motility and phagocytic activity, and produce and secrete various factors that are integral for combating pathogens and/or initiating and promoting tissue remodelling and repair (Gomez-Nicola and Perry 2015; Graeber and Streit 1990; Lawson et al. 1992; Ransohoff and Engelhardt 2012; Svahn et al. 2014; Wolf et al. 2017). The functional diversity and dynamics of microglia are enormous, and their activation is heterogeneous and critically depends on the nature of the pathological insult (Gomez-Nicola and Perry 2015; Graeber and Streit 1990; Lawson et al. 1992; Ransohoff and Engelhardt 2012; Svahn et al. 2014; Wolf et al. 2017). However, while microglia play a key role in inflammatory processes, their activation does not equal neuroinflammation per se (Graeber 2014; Masgrau et al. 2017). In fact, microglia can detect, process, and respond to signals in an entirely noninflammatory way (Salter and Beggs 2014). Comparative to peripheral macrophages, microglia support normal tissue function, which in the case of the CNS is neuronal integrity (Hanisch and Kettenmann 2007; Kettenmann et al. 2011; Nimmerjahn et al. 2005; Ransohoff and Perry 2009; Scheffel et al. 2012). We now know that microglia are involved in the regulation of neuronal development, synaptic plasticity, and circuit maintenance (Arnold and Betsholtz 2013; Bilimoria and Stevens 2015; Reemst et al. 2016; Salter and Beggs 2014). As discussed in detail below, there is substantial commonality in the molecular signalling cascades used by microglia to exert its different functions. Thus, alterations in microglia and inflammatory factors do not unequivocally convey ongoing neuroinflammation or neuroinflammatory processes but might instead signify changes in brain homoeostasis (Estes and McAllister 2014; Reemst et al. 2016; Salter and Beggs 2014; Thion and Garel 2017). These novel findings critically challenge the concept that disorders involving changes in microglia or inflammatory mediators are de facto neuroinflammatory disorders (Estes and McAllister 2014; Salter and Beggs 2014). Despite this, psychiatric research tends to equate changes in microglia activity with neuroinflammation (Biesmans et al. 2013; Brites and Fernandes 2015; Doorduin et al. 2009; Gracia-Rubio et al. 2016; Haarman et al. 2014, 2016; Kenk et al. 2015; Monji et al. 2013; Na et al. 2014; Najjar and Pearlman 2015; Nakatomi et al. 2014; Setiawan et al. 2015; Suridjan et al. 2014). Such oversimplifications could obscure the functional complexity of immune cells and molecules in physiological brain processes beyond that of their classically defined roles in inflammation resulting in possible misconceptions of disease aetiology.

In the spirit of John Maynard Keynes: 'The difficulty lies not in the new ideas, but in escaping the old ones'. The main incentive for writing this chapter was to overcome some of the current misconceptions and oversimplifications with respect to neuroinflammation and microglia activity in psychiatry. In the first sections, we will provide an overview of the basic concepts of neuroinflammation and microglia. We will then present key immunological and non-immunological functions of microglia and inflammatory mediators in order to increase the awareness of the complexity and difficulty to interpret changes in microglia and immune mediators in psychiatric disorders.

2 Basic Concept of Neuroinflammation and Neuroinflammatory Processes in Relation to Psychiatry

Associations between psychiatric diseases and immune system dysfunctions have been postulated more than 100 years ago (Kraepelin 1890; Menninger 1919) and have remained a matter of discussion ever since. With the reconceptualization of the 'immune privilege' of the CNS, the field of psychoimmunology has experienced a reappraisal. Advances in the fields of immunology and genetics, as well as the increasing understanding of how immunological processes can influence brain development and functions (Reemst et al. 2016; Thion and Garel 2017), have further contributed to the growing interest and recognition of immune system dysfunction in psychiatry. Indeed, abnormal neuroimmune functions have been implicated in the aetiology and pathophysiology of a number of psychiatric disorders, including depression (Dantzer et al. 2008; Du Preez et al. 2016; Miller and Raison 2016; Muller and Schwarz 2007), schizophrenia (Horvath and Mirnics 2014; Khandaker et al. 2015; Muller et al. 2000; Yolken and Torrey 2008), autism spectrum disorders (Ashwood et al. 2006; Estes and McAllister 2015; Meltzer and Van de Water 2017), and bipolar disorder (Isgren et al. 2017; Wang and Miller 2017; Watkins et al. 2014).

The possible role of aberrant immune functions involving altered inflammatory and neuroinflammatory processes is currently among the timeliest topics in psychiatry. In this field, however, immunological changes that are being revealed in the CNS are frequently (and often misleadingly) referred to as neuroinflammation (Graeber 2014; Masgrau et al. 2017). The extent to which the brain is considered to be ‘inflamed’ is typically evaluated against the background of altered expression of secreted inflammatory mediators (including cytokines and chemokines) together with numerical, morphological, and/or functional abnormalities of astrocytes and microglia (Graeber 2014; Masgrau et al. 2017).

The word inflammation was coined by the ancients and is derived from the Latin word *inflammare* (‘to set on fire’) (Scott et al. 2004). The Roman Celsus is considered the first to have described the four cardinal signs of inflammation more than 2,000 years ago: rubor et tumor cum calore et dolore (redness and swelling with heat and pain) (Rocha e Silva 1978). In the late nineteenth century, the German pathologist Rudolf Virchow added the fifth cardinal sign: loss of function (Scott et al. 2004). This early definition was based on the assumption that inflammation represents a purely pathological process, which was later revised to acknowledge that it encompasses concomitant beneficial effects on tissue healing. Hence, inflammation denotes a complex cascade of concurrent processes that cause both tissue damage and repair (Schwartz and Baruch 2014; Serhan and Savill 2005).

Today, inflammation is considered an integral part of the body’s homoeostatic repair and defence mechanisms and engages physiological interactions between resident and recruited immune cells, soluble factors, and tissue-specific elements (Schwartz and Baruch 2014; Serhan and Savill 2005). Upon initiation and proper orchestration, it limits the spread of infection and/or tissue damage and is typically followed by a resolution phase. The latter ensures that the affected tissues are structurally and functionally restored and that the immunological components attain their original functional state (Schwartz and Baruch 2014; Serhan and Savill 2005).

In general, the processes of classical inflammation can occur in the CNS like in any other organ and show largely the same characteristics on the cellular and molecular level (Denes et al. 2010; Filiou et al. 2014; Graeber 2014; Masgrau et al. 2017; Schwartz and Baruch 2014). Illustrative examples of neurological conditions where this occurs are multiple sclerosis, stroke, traumatic brain injury, and CNS infections (Filiou et al. 2014; Graeber 2014; Masgrau et al. 2017). The immune-driven CNS responses underlying these pathologies have been the cornerstones of defining ‘neuroinflammation’ and involve (1) initiation of a local immune response by CNS-resident immune cells, (2) increased production of pro-inflammatory cytokines and chemokines, (3) additional recruitment of CNS-resident immune cells to the primary site of trauma or infection, (4) blood-brain barrier (BBB) leak and infiltration of blood-derived leucocytes into the brain parenchyma, and (5) resolution of inflammation and tissue remodelling. Hence, the term 'neuroinflammation' was historically well defined and mirrored the hallmarks of classical inflammation in the periphery (Estes and McAllister 2014; Masgrau et al. 2017). Over the past decade, however, the term 'neuroinflammation' has been frequently used to describe isolated aspects of neuroinflammatory processes with no known causative insult or overt changes in the BBB integrity (Graeber 2014; Masgrau et al. 2017). This has led to the oversimplified assumption that a wide range of psychiatric and neurodegenerative disorders underlie neuroinflammatory dysfunctions. The increasing understanding of how microglia and inflammatory mediators exert regulatory functions in brain development and maturation independent of inflammation adds another level of complexity on how to interpret central immune dysfunction in these different disorders (Reemst et al. 2016; Thion and Garel 2017).

As suggested by Estes and McAllister (2014), we therefore propose that the denotation 'neuroinflammation' should only be applied when all five signs of pathological inflammation – increased cytokines and chemokines, activated microglia and astrocytes, disturbance in BBB integrity and blood leucocyte infiltration, degenerative tissue damage, and resolution of inflammation and tissue remodelling – are present. For alterations in isolated inflammatory mediators within the CNS, we suggest to refer to the terms 'changes in neuroinflammatory mediators or processes' or 'changes in microglia activity states' and specify them separately. Clarifying the term 'neuroinflammation' is warranted in order to prevent oversimplifications, which in turn could result in the false assumption that various psychiatric and neurodegenerative diseases involve the same or similar pathologies. Such oversimplifications may, in fact, impede scientific progress regarding the understanding of disease-specific aetiologies and, consequently, developing adequate interventions with maximal therapeutic benefits.

3 Microglia: Historical Perspective

Microglia were first described by the German psychiatrist and neuropathologist Franz Nissl in the late nineteenth century as 'Stäbchenzellen' (rod cells) that represent reactive glial elements with migratory, phagocytic, and proliferative potential (Ginhoux et al. 2013). During the same time, W. Ford Robertson introduced the term 'mesoglia', which attempted to denote phagocytic elements with mesodermal origin. In 1913, Santiago Ramón y Cajal introduced the classification of central elements as 'the first element' (neurons), 'the second element' (neuroglia, a term introduced by Rudolf Virchow, which comprise astrocytes and oligodendrocytes), and 'the third element of the nervous system' (cells with small round nuclei), whereby he too stated that cells of the latter were probable to have a mesodermal origin (Ginhoux et al. 2013; Ransohoff and Cardona 2010). In 1920, the Spanish neuroscientist, and student of Ramón y Cajal, Pio del Rio Hortega, coined the term 'microglia' (Perez-Cerda et al. 2015). del Rio Hortega's early observations and descriptions were of tremendous accuracy. He observed the invasion of amoeboid microglia into the developing brain during early embryonic development and hypothesized that they originated from meningeal macrophages and/or peripheral monocytes penetrating the CNS (Ginhoux et al. 2013; Kettenmann et al. 2011). He also described that microglia change their appearance during brain maturation into ramified cells, with a small round soma and an intricate network of fine ramifications. Furthermore, he reported that in the mature brain, microglia are present throughout the entire brain parenchyma occupying defined, non-overlapping territories (Kettenmann et al. 2011). Upon pathological events, he observed that they were able to retract their processes, become amoeboid, and display migratory and phagocytic functions (Ginhoux et al. 2013; Kettenmann et al. 2011). Astonishingly, most of these early observations and interpretation from Rio Hortega largely hold true until today (Kettenmann et al. 2011).

4 Microglia: Ontogeny and General Facts

Although the ontogeny of microglia has been the subject of debates for decades, their origin – primitive yolk sac (YS) macrophages – was only fully established in 2010 (Ginhoux et al. 2010). By applying an inducible lineage-tracing model using the runt-related transcription factor 1 (Runx1) to label YS progenitors, including YS macrophages, Ginhoux et al. could show that adult microglia arise unequivocally from YS macrophages that invade the developing CNS at embryonic day (E) 9.5 through the bloodstream, where they proliferate in situ and are maintained throughout adulthood (Ginhoux et al. 2010, 2013; Salter and Stevens 2017). The lack of foetal monocyte contribution to the microglia progenitor pool could be explained by the inaccessibility of foetal monocytes to the developing brain, as embryonic tissue colonization of foetal monocytes starts at around E13.5, which coincides with the formation of the BBB (Daneman et al. 2010).

YS macrophages represent an independent lineage and arise before the development of other myeloid cells that differentiate from definitive haematopoietic stem cells (Hoeffel et al. 2015; Orkin and Zon 2008). In contrast to other macrophage populations, they have a unique development in the sense that they can bypass the monocyte stage (Hoeffel et al. 2015; Takahashi et al. 1989). Hence, although microglia may be considered to be similar to tissue-resident macrophages in peripheral tissues, they are the only 'myeloid' cells that are derived solely from yolk sac precursors under 'normal' conditions (Hoeffel et al. 2015; Sheng et al. 2015).

Recent genome-wide chromatin and expression profiling coupled with single-cell transcriptomic analyses throughout development revealed that microglia undergo three distinct developmental stages along with brain development: early, pre-, and adult microglia, which were shown to underlie distinct regulatory circuits (Matcovitch-Natan et al. 2016). Morphologically, microglia undergo maturational changes as well. While microglia during early brain development display an amoeboid cell morphology, they mature into ramified cells with numerous thin processes at around postnatal day 15 (Cunningham et al. 2013; Harry 2013; Salter and Beggs

2014). In a healthy mouse brain, depending on the region analysed, microglia account for 10–15% of all brain parenchymal cells (Gomez-Nicola and Perry 2015; Graeber and Streit 1990; Lawson et al. 1990; Ransohoff and Engelhardt 2012; Svahn et al. 2014; Wolf et al. 2017). In contrast to this, their density shows marked regional differences in the non-diseased adult human brain parenchyma, ranging from approximately 0.5 to 16% of all cells (Mittelbronn et al. 2001).

The unique ontogeny of microglia with no contribution of foetal monocytes suggests that microglia population persist in the brain parenchyma through self-renewal of resident microglia. Previous estimates based on [^{3}H]thymidine, 5-ethynyl-2′-deoxyuridine (EdU), or 5-bromo-2′-deoxyuridine (BrdU) incorporation suggested that 0.05–1.04% of microglia in adult healthy mice of different strains and 2.35% of microglia in the young adult healthy macaque were entering cell cycle each day (Lawson et al. 1992; Shankaran et al. 2007; Tonchev et al. 2003). A recent report in humans using C^{14} retrospective birth dating of microglia isolated from postmortem brains of adults born across six decades estimated that 0.08% of microglia entered cell cycle per day in the healthy human brain, confirming the slow rate in microglia renewal (Reu et al. 2017). However, more recent studies performed in mice challenged the assumption that microglia are long-lived cells with slow proliferation rates. Using a multicolour fluorescence fate mapping system approach, Tay et al. revealed that microglia displayed higher and heterogeneous turnover rates in different brain compartments that occurred in a context-dependent manner (Tay et al. 2017). Another recent report confirmed a high region-dependent turnover rate for murine microglia revealing that proliferation is temporally and spatially coupled to intrinsic apoptosis (Askew et al. 2017). In this study, on average 0.69% of the total microglial cells were estimated to be proliferating, suggesting that the whole population is renewed several times during a lifetime.

In the healthy brain, microglia form a near-regular three-dimensional lattice in which each microglial cell occupies a unique territory. For decades, ramified microglia have been mistakenly denoted as ‘resting’ cells. Recent studies, however, revealed ramified microglia to be the opposite of resting: They constantly scan the brain parenchyma for potential insults (Davalos et al. 2005; Hristovska and Pascual 2015; Nimmerjahn et al. 2005). Estimates suggest that microglia scan the entire brain volume within a few hours (Nimmerjahn et al. 2005). While scanning the brain, the fine microglial processes continuously contact neurons, axons, and dendritic spines (Salter and Stevens 2017; Sierra et al. 2013; Tremblay et al. 2011). Furthermore, process motility was shown to dramatically change in response to adenosine triphosphate (ATP), neuronal activity, and neurotransmitters, whereas the latter is partly indirectly mediated through ATP (Davalos et al. 2005; Dissing-Olesen et al. 2014; Eyo et al. 2014; Fontainhas et al. 2011; Li et al. 2012). Although microglia process motility and interaction with neuronal synaptic elements is an established phenomenon, the functional implications remain to be discovered.

5 Microglia: The CNS Immune Cell

Microglia represent the central immune cell with the potential to sense and initiate active immune defence in the CNS (Gomez-Nicola and Perry 2015; Graeber and Streit 1990; Ransohoff and Engelhardt 2012; Svahn et al. 2014; Wolf et al. 2017). They express numerous cell surface and intracellular receptors including pattern recognition receptors (PRRs) that recognize pathogen-associated molecular patterns (PAMPs) as well as damage-associated molecular patterns (DAMPs) (Kettenmann et al. 2011; Santoni et al. 2015) through which they can virtually sense any pathological event or changes in homoeostatic conditions and respond accordingly (Kettenmann et al. 2011; Ransohoff and Cardona 2010; Ransohoff and Perry 2009).

Upon sensing a pathological insult, microglia rapidly alter their morphological appearance and transcriptional programme in a context-dependent manner. In vivo two-photon microscopy studies revealed that upon brain injury, microglial processes rapidly and autonomously assemble on the site of injury without cell body movement, establishing a potential barrier and thereby protecting the surrounding, healthy tissue (Davalos et al. 2005; Hines et al. 2009; Szalay et al. 2016). This process assembly was shown to be mediated by ATP (released either by damaged cells or in a more regulated manner by astrocytes) and the microglia purinoreceptor $P2Y_{12}$ (Haynes et al. 2006). Subsequent to this immediate barrier formation, microglia are known to retract the processes adopting a more amoeboid-like morphology. These morphological changes were found to be associated with a downregulation of $P2Y_{12}$, a conversion of ATP to adenosine by microglia ectoenzymes CD73 and CD39, and an increase in expression of adenosine A2A receptors (Orr et al. 2009). At the end of the range of morphological changes upon 'activation', microglia display a rounded cell body with an increase in soma size and only sparse processes, which is termed 'amoeboid' (Kettenmann et al. 2011). Similar to peripheral immune responses, the density of microglia increases at the site of an insult in order to provide more immune cells to fight invading pathogens, as well as to assure the protection and restoration of tissue homoeostasis. Microglia can become motile and actively migrate to the sight of insult following chemotactic gradients as well as increase their density through local proliferation (Kettenmann et al. 2011). As tissue macrophages, microglia increase their phagocytic activity to engulf invading pathogens and toxic molecules as well as to promote and regulate tissue remodelling and repair by phagocytizing apoptotic cells and cellular debris (Kettenmann et al. 2011). During neuroinflammation, microglia can also act as antigen-presenting cells (APCs) to activate invading lymphocytes of the adaptive immune system (Kettenmann et al. 2011). Lastly, microglia produce and secrete various immune-mediating factors, including pro- and anti-inflammatory cytokines and chemokines, as well as neurotrophic factors, that are crucial for coordinating the combat against pathogens and/or initiating and promoting tissue remodelling and repair (Gomez-Nicola and Perry 2015; Graeber and Streit 1990; Kettenmann et al. 2011; Ransohoff and Cardona 2010; Ransohoff and Engelhardt 2012; Svahn et al. 2014; Wolf et al. 2017).

Descriptive studies of microglia morphology in a variety of different diseases and animal models suggested that their 'activation' pattern follows a linear range (Perry et al. 2010). This hypothesis, however, has been replaced by the macrophage-derived polarization terminology (Perry et al. 2010). This concept of activation was based on findings in peripheral macrophages where different stimuli could induce different activation states termed M1 (classically activated) and M2 (alternatively activated), whereby M2 phenotypes were further refined into M2a, M2b, and M2c (Martinez and Gordon 2014). The classically activated M1 macrophages are designated to specialize in pathogen elimination, whereas alternatively activated M2 macrophages are involved in tissue remodelling and repair (Geissmann et al. 2010; Mantovani et al. 2005). However, this schema of activation has several limitations (as explained by Martinez and Gordon 2014), which undermines the possibility of applying the M1/M2 framework to microglia (Ransohoff 2016). Despite this, there are numerous publications that employ M1/M2 terminology in order to characterize microglia 'activation' states.

We now know that the functional diversity and dynamics of microglia are enormous, and their activation is heterogeneous and critically dependent on the nature of the pathological insult (Gomez-Nicola and Perry 2015; Graeber and Streit 1990; Ransohoff and Engelhardt 2012; Svahn et al. 2014; Wolf et al. 2017). Thus, identifying the diverse phenotypes and functions they can adopt in response to an insult or in different diseases remains to be a major challenge. Novel technologies including two-photon imaging, whole-genome transcriptomic and epigenomic analysis with complementary bioinformatics and unbiased proteomics, and cytometry by time of flight (CyTOF; Fluidigm) have been able to shed light into the complex world of microglia and microglia 'activation' (Ransohoff 2016). Despite these major advances and exciting new insights, most of the research published today still relies on morphological analyses and measurements of specific cellular markers in order to identify microglia-specific phenotypes. While these measures can, to a certain extent, detect alterations in microglia activation states, they fail to adequately identify all the diverse phenotypes and functions that these cells can adopt (Gomez-Nicola and Perry 2015; Graeber and Streit 1990; Perry et al. 2010; Ransohoff 2016; Ransohoff and Engelhardt 2012; Svahn et al. 2014; Wolf et al. 2017).

6 Microglia Function: Beyond Central Immune Cell

For decades, microglia were regarded as the brain-resident immune cells having the sole purpose to sense and protect the brain from harmful stimuli. In the past years, however, the functional roles of microglia have been extended to non-immunological functions, including the regulation of neurogenesis, myelination, angiogenesis, and synaptic pruning (Arnold and Betsholtz 2013; Bilimoria and Stevens 2015; Reemst et al. 2016; Salter and Beggs 2014; Thion and Garel 2017). A large body of evidence accumulated suggest that microglia are critically involved in neurodevelopmental processes throughout prenatal stages up to postnatal maturation of the CNS (Paolicelli and Ferretti 2017; Reemst et al. 2016).

During embryonic development, microglia were shown to regulate the size of the neural precursor cell (NPC) pool through phagocytosis of NPCs in the subventricular zone (SVZ) of the developing cerebral cortex (Cunningham et al. 2013). Besides controlling the neuronal progenitor cell pool, microglia were implicated in regulating the wiring of forebrain circuits during embryonic development (Squarzoni et al. 2014). In utero perturbations of microglia activity resulted in impaired outgrowth of dopaminergic axons in the forebrain and affected the laminar positioning of subsets of neocortical interneurons (Squarzoni et al. 2014). Furthermore, in utero microglia depletion resulted in defective fasciculation of the axonal tracts in the dorsal corpus callosum (Pont-Lezica et al. 2014).

In postnatal developmental periods, microglia were shown to continue regulating the NPC pool in the SVZ. Besides microglia-dependent regulation of NPC numbers through phagocytosis, microglia were shown to promote neurogenesis and oligodendrogenesis through the production and release of pro-inflammatory cytokines, including IL-1β, IL-6, TNF-α, and IFN-γ (Shigemoto-Mogami et al. 2014). Microglia were further shown to control developmental cell death in the hippocampus during early postnatal development in mice by means of phagocytosis of apoptotic neurons (Wakselman et al. 2008). Moreover, microglia were implicated to support the survival of layer V cortical neurons during postnatal development (Ueno et al. 2013).

Besides regulating survival and death, migration, and positioning, as well as axonal guidance of neurons, microglia are now known to actively participate in synaptic pruning in a complement-dependent manner (Schafer et al. 2012; Wu et al. 2015). The complement system is a major effector of the innate immune system and an adjuvant for the adaptive immunity (Parkin and Cohen 2001). It consists of numerous soluble and cell-surface proteins that can recognize endogenous and foreign materials (Mayilyan et al. 2008; Parkin and Cohen 2001). In the context of CNS circuit refinement, the complement system is critical for the tagging, recognition, and elimination of synapses (Orsini et al. 2014; Presumey et al. 2017; Stephan et al. 2012). As of today, microglia-dependent synaptic pruning has been best studied in the mouse retinogeniculate system. This system has proven to be an excellent model system for studying developmental CNS synapse elimination as it involves the removal of excess synapses in the dorsal lateral geniculate nucleus (dLGN) of the thalamus in an activity-dependent manner (Stephan et al. 2012; Stevens et al. 2007). The findings from these studies suggest that microglia-dependent pruning is mediated by the synaptic deposition of the complement component C1q, which initiates the proteolytic cascade of the complement system ultimately resulting in the deposition of the activated C3 fragment on synapses, which itself is recognized by the complement receptor 3 (CR3) expressed by microglia inducing phagocytosis (Schafer et al. 2012; Stevens et al. 2007). Intriguingly, this process is not random but has been shown to be activity-dependent, as microglia cells engulf weaker, less active synapses only and sparing the strong ones (Schafer et al. 2012).

Besides the complement system, fractalkine (CX3CL1) has been implicated to regulate microglia-dependent synaptic pruning (Arnoux and Audinat 2015; Paolicelli et al. 2011). CX3CL1 is produced by neurons and astrocytes and present in a membrane-bound ('do-not-eat-me' signal) or soluble ('find-me' signal) form and the ligand for the microglia-specific receptor CX3CR1 (Table 1). Findings in CX3CR1 knock-out mice demonstrated that the lack of fractalkine-mediated chemoattraction resulted in delayed recruitment of microglia and impaired synapse formation of pyramidal cells of the CA1 region of the hippocampus (Paolicelli et al. 2011), resulting in long-term alterations of hippocampal functional connectivity (Zhan et al. 2014), and the impaired functional development of thalamocortical synapses of the barrel cortex (Hoshiko et al. 2012).

Glia-dependent pruning of excess or weak synapses has been proposed to contribute to synapse elimination during two distinct phases of postnatal development: the first phase including the first 3 weeks after births in rodents and approximately the first 5 postnatal years in human infants (Johnson 2001; Neniskyte and Gross 2017) and a second phase spanning week 3–8 in rodent development, which represents adolescence in humans (Blakemore 2012; Konrad et al. 2013; Neniskyte and Gross 2017). During the first phase, sensory circuits along with circuits associated with cognition and behaviour are being refined (Johnson 2001; Neniskyte and Gross 2017). The second phase of refinement is crucial for the establishments of circuits involved in goal-directed behaviour, planning, and impulse control in associated brain regions such as the medial prefrontal cortex (mPFC) (Blakemore 2012; Konrad et al. 2013; Neniskyte and Gross 2017). In support of the notion that microglia-dependent synaptic pruning is involved in later postnatal development and maturation are recent findings where transient microglia depletion in mice at postnatal day 19 or 30 altered the development and fine-tuning of synapses necessary for proper learning and memory tasks (Parkhurst et al. 2013).

In the healthy adult brain, microglia have been primarily studied with regard to their role as immune cell of the CNS. However, the few studies assessing the non-immunological function of microglia have shown that microglia continue to participate in the regulation of the NPC pool in regions with adult neurogenesis (Sierra et al. 2010, 2014). The extent to which microglia regulate synaptic pruning and remodelling in the adult brain, on the other hand, remains ill defined. However, recent findings suggest that microglia are involved in regulation of behaviour. Temporal microglia depletion in the hippocampus of adult mice resulted in cognitive deficits and impaired social behaviour (Torres et al. 2016). Furthermore, microglia processes have been shown to make temporary contacts with elements of the neuropil, including dendritic spines and axonal terminals (Salter and Stevens 2017; Sierra et al. 2013; Tremblay 2011). Confocal and immuno-gold electron microscopy studies have identified both pre- and postsynaptic elements within microglia processes following such brief contacts with synapses, suggesting that microglia could actively participate in synaptic remodelling and pruning (Linnartz et al. 2012; Paolicelli et al. 2011; Tremblay et al. 2010).

Against this background, it becomes evident that the neuroimmune system, and microglia in particular, critically regulates proper neuronal development and

Table 1 Overview of the regulatory signals based on their response they evoke in microglia

Ligands/signals	Expression	Microglia receptor	Reference
'Do-not-eat-me'			
CD47 (integrin-associated protein)	Various cell types including neurons and myelin	Signal regulatory protein-alpha (SIRP-a)	Zhang et al. (2015)
Polysialic acid residues	Neuronal glycocalyx	Sialic acid-binding immunoglobulin-like lectins (SIGLECs)	Brown and Neher (2014), Claude et al. (2013), and Wang and Neumann (2010)
Membrane-bound fractalkine ligand (CX3CL1)	Neurons	CX3CR1	Brown and Neher (2014), Paolicelli et al. (2014), and Cardona et al. (2006)
'Find-me'/'help-me'			
Adenosine triphosphate (ATP) ('find-me')	Released by neurons and astrocytes	$P2Y_{12}$	Dissing-Olesen et al. (2014), Haynes et al. (2006), Hristovska and Pascual (2015), and Eyo et al. (2014)
Soluble CX3CL1 ('find-me')	Released by neurons	CX3CR1	Garton et al. (2001), Noda et al. (2011), Maciejewski-Lenoir et al. (1999), Liang et al. (2009), and Zhang et al. (2012)
Interleukin-34 (IL-34) (find-me/help-me)	Released by neurons	Colony-stimulating factor-1 receptor (CSFR-1)	Xing and Lo (2017), Mizuno et al. (2011), and Luo et al. (2013)
Fibroblast growth factor-2 (FGF-2) (find-me/help-me)	Released by neurons	Fibroblast growth factor-3 (FGFR3) (chemotaxis) FGFR1 (restorative microglia phenotype)	Noda et al. (2014) and Xing and Lo (2017)
'Eat-me'			
Phospholipid phosphatidylserine	Exposed on cell surface of neurons	Brain-specific angiogenesis inhibitor-1 (BAI-1)	Brown and Neher (2014), Wakatsuki and Araki (2017), Marker et al. (2012), and Mazaheri et al. (2014)
Opsonin (milk fat globule factor-E8 (MFG-E8)-tagged phospholipid phosphatidylserine	Released by microglia and astrocytes	Vitronectin receptors (VNRs)	Cardona et al. (2006), Fricker et al. (2012), and Neniskyte and Brown (2013)
Membrane debris	Apoptotic cells	Triggering receptor expressed on myeloid cells-2 (TREM2)	Fu et al. (2014) and Takahashi et al. (2005)
Complement component C1q-tagged glycoproteins	Neuronal surface	Complement receptor 3 (CR3)	Schafer et al. (2012), Linnartz et al. (2012), Stephan et al. (2012), and Brown and Neher (2014)

refinement of brain circuitry during embryonic and postnatal development and possibly as well in the adult brain (Arnold and Betsholtz 2013; Bilimoria and Stevens 2015; Reemst et al. 2016; Thion and Garel 2017; Wu et al. 2015). It is therefore conceivable that alterations in the neuroimmune system could impact neurodevelopment and therefore play an important role in the aetiology of neurodevelopmental psychiatric disorders. To what extent dysfunctions in the neuroimmune system and microglia contribute to the pathogenesis of neurodevelopmental disorders, however, warrants further examination.

7 The Microglia 'Sensome'

A crucial prerequisite for the functions of microglia is its proper communication with CNS cells, in particular neurons. Diverse microglia receptors have been identified that recognize and respond to specific neuronal ligands (both soluble and membrane-bound). The set of receptors expressed by microglia in order to enable them to sense brain environment and neuronal states and respond accordingly is highly complex and can be referred to as the 'microglia sensome' (Brown and Neher 2014; Diaz-Aparicio et al. 2016). The microglia sensome is not stable but rather has been shown to adapt to changing brain environments such as present during the development of the CNS (Hickman et al. 2013; Matcovitch-Natan et al. 2016). The regulatory signals (ligands), on the other hand, can be classified based on the response they evoke in microglia (Table 1). These include 'do-not-eat-me' signals presented by healthy neurons to prevent microglial phagocytosis, 'find-me'/ 'help-me' signals from neurons that induce microglial chemotaxis and adhesion to neuronal components (e.g. dendritic spines), and 'eat-me' signals that initiate phagocytosis (Brown and Neher 2014; Sierra et al. 2013).

Phagocytosis is not only important for the physiological maintenance of the CNS, but it is also a crucial mechanism during inflammation to engulf invading pathogens, injured neurons, and cellular debris (Rosales and Uribe-Querol 2017). Although triggered by different signals that induce different intracellular signalling cascades, the phagocytic cascades under noninflammatory or inflammatory conditions both depend on the activation of small GTPases including Rac and Rho, which catalyse cytoskeletal rearrangement in order to enable the formation of a phagocytic cup and eventually (Gumienny et al. 2001; Lee et al. 2007; Patel et al. 2011; Rosales and Uribe-Querol 2017; Sierra et al. 2013; Underhill and Goodridge 2012).

Neurotransmitter receptors expressed by microglia have been suggested to be an integral part of the 'microglia sensome' to mediate the bidirectional communication between neurons and microglia (Liu et al. 2016). Indeed, evidence suggests that neurotransmitter signalling can modulate 'microglia activation', phagocytic clearance, and phenotypic polarization (Liu et al. 2016). For example, microglia express both ionotropic and metabotropic glutamate receptors, which were shown to alter cytokine release (Noda et al. 2000), chemotaxis (Liu et al. 2009), as well as process motility (Fontainhas et al. 2011) in an ATP-dependent and ATP-independent manner. Furthermore, microglia express both ionotropic GABA(A) and metabotropic GABA(B) receptors (Liu et al. 2016), which were both shown to decrease the release of pro-inflammatory cytokines upon an inflammatory stimulus (Kuhn et al. 2004; Lee et al. 2011). Microglia also express both α-1/2 and as β-1/2 adrenergic receptors (Liu et al. 2016). Depending on the receptors expressed, noradrenaline (NA) was shown to regulate the microglia immune profile in response to an inflammatory stimulus (Johnson et al. 2013; Liu et al. 2016), chemotaxis, and phagocytosis (Heneka et al. 2010), as well as ATP-dependent process motility and cell mobility (Gyoneva and Traynelis 2013). Moreover, microglia were shown to express functional serotonin receptors, which promote injury-induced and ATP-mediated microglia process motility and cell mobility, as well as inhibit phagocytosis (Krabbe et al. 2012). Lastly, histamine was also identified as a regulator of microglia motility, migration, and cytokine release (Ferreira et al. 2012), as well as modifying their morphological appearance and immune response in specific brain regions (Frick et al. 2016).

Considering the above, we are only now starting to appreciate the complexity of neuron-microglia interactions and how neuronal activity governs microglia activity and vice versa. Neurotransmitters otherwise designated to regulate our mood, wakefulness, and cognitive processes are now known to directly or indirectly interact with brain-resident immune cells and thereby modulate a broad array of microglia functions including chemotaxis, process motility, phagocytosis, and cytokine release. The latter is of particular interest, as changes in cytokine levels measured in the brain or cerebral spinal fluid (CSF) of psychiatric patients are often interpreted as ongoing inflammatory processes or neuroinflammation. It is, however, only now becoming clear that inflammatory cytokines in the brain are constantly produced at low levels in a region-specific and diurnal manner whereby they exert various physiological tasks independent of immunological processes (Cearley et al. 2003; Krueger et al. 2011).

8 The Role of Central Cytokines Beyond Inflammation

Besides orchestrating and controlling the function of immune cells (Parkin and Cohen 2001), cytokines have been increasingly recognized to be involved in the regulation of various physiological processes of the CNS including sleep, learning, memory, neural plasticity, and neurogenesis (Cearley et al. 2003; Donzis and Tronson 2014; Krueger et al. 2011; Yirmiya and Goshen 2011).

The two prototypical pro-inflammatory cytokines IL1β and TNFα were found to be constitutively expressed in the healthy adult rat brain following a diurnal expression pattern in specific brain regions (Cearley et al. 2003) and were shown to stimulate non-rapid eye movement (NREM) sleep (Krueger 2008). Furthermore, hippocampal IL1β gene expression was shown to regulate contextual fear memory formation (Goshen et al. 2007). Hippocampal IL-1β levels were shown to increase 24 h after contextual fear conditioning and that interfering with IL-1 signalling (excess or blocking the IL-1 signalling pathway) could impede hippocampus-dependent memory formation (Goshen et al. 2007). Intriguingly, sleep deprivation, which is associated with cognitive decline, was shown to cause an increase in central IL-1β and TNF-α levels, which was suggested to contribute to the cognitive deficits evident after excessive lack of sleep (Krueger et al. 2011). The notion that IL-1β is involved in regulating cognitive processes was further strengthened by a study that found hippocampal IL-1β to be increased in an ATP- and microglia-dependent manner after a spatial recognition task (Labrousse et al. 2009). ATP was identified as a key regulator of central IL-1 induction through binding to the microglia-specific purinergic receptor P2X7 (Ferrari et al. 2006; Mingam et al. 2008). Indeed, mice lacking the P2X7 receptor showed no task-dependent IL-1β induction, which was associated with impaired spatial learning (Labrousse et al. 2009). These findings are in line with previous studies showing that impaired IL-1 signalling impeded hippocampus-dependent learning and memory processes, including long-term potentiation (LTP) (Avital et al. 2003; Yirmiya et al. 2002). The chemokine fractalkine (CX3CL1) was also suggested to be involved in learning and memory processing, more specifically to play a role in the protective plasticity process of synaptic scaling (Sheridan et al. 2014). CX3CL1 was shown to be upregulated in the rat hippocampus during a brief temporal window following spatial learning and LTP-inducing stimulation of the dentate gyrus. Furthermore, physiologically relevant levels of CX3CL1 inhibited LTP maintenance and were shown to dampen glutamate-mediated calcium increase in both neurons and microglia (Sheridan et al. 2014). The cytokine TNF-α, on the other hand, was implicated in regulating the NPC pool in adult neurogenesis (Chen and Palmer 2013). NPCs were shown to express TNF receptors (TNFR) 1 and 2, which differentially regulate NPC cell fate, whereby TNFR1 signalling favours proliferation and TNFR2 signalling favours apoptosis (Chen and Palmer 2013).

In light of this book chapter and the presented physiological functions of cytokines described above, there is a need to carefully consider how to interpret alterations in cytokine levels measured between patient groups and controls. Indeed, numerous studies have identified significant changes in cytokine levels both in brain tissue and CSF of psychiatric patients (Miller and Raison 2016; van Kesteren et al. 2017; Wang and Miller 2017). However, although significant, the observed changes are very small in comparison to the neurological conditions that underlie neuroinflammation: For example, a significant increase of CSF IL-6 levels was detected in a subgroup of schizophrenic patients where the levels in healthy controls were found to be at 3 pg/mL and that of patients 4.5 pg/mL (Garver et al. 2003). Also in chronic schizophrenic patients, CSF IL-6 levels were significantly increased, with a mean CSF IL-6 concentration of 1.5 pg/mL in controls and 2.68 pg/mL in patients (Schwieler et al. 2015). Furthermore, significant increased CSF IL-6 levels were measured in recent-onset schizophrenic patients (median 0.85 pg/mL) relative to controls (median 0.52 pg/mL) (Coughlin et al. 2016). Another study found that patients who attempted violent suicide had significantly higher CSF IL6 levels (5.26 pg/mL) as compared to control (0.64 pg/mL) (Lindqvist et al. 2009). In contrast to this, CSF IL-6 levels measured in multiple sclerosis (MS) patients have been found to increase from a mean of 0.87 pg/mL in controls to 13.4 pg/mL in MS patients (Stelmasiak et al. 2000). Furthermore, CSF IL-6 levels measured in patients suffering from meningitis have been found to peak up to 500 pg/mL (Pinto Junior et al. 2011). Similar to IL-6, CSF IL-1β levels have been found to be increased in schizophrenic patients relative to controls, whereby schizophrenic patients displayed a median IL-1β of 4.37 pg/mL and controls 0.78 pg/mL (Soderlund et al. 2009). IL-1β was also found to be elevated in the CSF of patients with acute depression, where the mean level was 1.14 pg/mL in patients as compared to controls who had an average level of 0.14 pg/mL (Levine et al. 1999). In comparison to this, meningitis patients displayed CSF levels of IL-1β that can reach a peak of 1,000 pg/mL (Coutinho et al. 2013). Lastly, in patients with traumatic brain injury, the levels of CSF pro-inflammatory cytokines can increase up to several 100-fold in comparison to controls (Sordillo et al. 2016).

It becomes evident that there is a substantial difference with respect to the measured levels of pro-inflammatory cytokines in patients suffering from conditions or diseases with ongoing neuroinflammation or psychiatric patients. The question arises as to whether these observed alterations in psychiatric patients truly reflect ongoing inflammatory processes or changes in the general physiological state of the brain. To answer this question, future studies are needed to expand our knowledge of the physiological roles of pro-inflammatory cytokines in health and disease.

9 Concluding Remarks

The growing understanding that central immune mediators are functionally involved in regulating physiological processes of the CNS has revolutionized the field of neuroimmunology. Microglia and cytokines have been implicated in the regulation of neurodevelopment, neuronal wiring, and synaptic plasticity. The functional relevance and underlying mechanisms of these non-immunological functions remain, however, largely unknown and await further investigation. It is, however, clear that the reductive conception of microglia as merely central immune cells is too simplistic. Rather, they emerge as a distinct but heterogeneous cell population of the CNS with a high degree of functional diversity and complexity. Unequivocally implying changes in microglia activity profiles and/or inflammatory factors with ongoing neuroinflammation or neuroinflammatory processes may therefore be too simplistic and could result in misconceptions. In contrast, alterations in neuroimmune systems – particularly in neurological and psychiatric diseases where there is no apparent ongoing inflammation that is evident – should be interpreted in relation to the functional complexity of immune cells and molecules in physiological brain processes. This could help unravelling the functional relevance of neuroimmune dysfunctions in psychiatric illnesses and aid defining future research directions in the field of psychoneuroimmunology.

References

Arnold T, Betsholtz C (2013) Correction: The importance of microglia in the development of the vasculature in the central nervous system. Vasc Cell 5:12. https://doi.org/10.1186/2045-824x-5-12

Arnoux I, Audinat E (2015) Fractalkine signaling and microglia functions in the developing brain. Neural Plast 2015:689404. https://doi.org/10.1155/2015/689404

Ashwood P, Wills S, Van de Water J (2006) The immune response in autism: a new frontier for autism research. J Leukoc Biol 80:1–15. https://doi.org/10.1189/jlb.1205707

Askew K et al (2017) Coupled proliferation and apoptosis maintain the rapid turnover of microglia in the adult brain. Cell Rep 18:391–405. https://doi.org/10.1016/j.celrep.2016.12.041

Avital A, Goshen I, Kamsler A, Segal M, Iverfeldt K, Richter-Levin G, Yirmiya R (2003) Impaired interleukin-1 signaling is associated with deficits in hippocampal memory processes and neural plasticity. Hippocampus 13:826–834. https://doi.org/10.1002/hipo.10135

Biesmans S et al (2013) Systemic immune activation leads to neuroinflammation and sickness behavior in mice. Mediat Inflamm 2013:271359. https://doi.org/10.1155/2013/271359

Bilimoria PM, Stevens B (2015) Microglia function during brain development: new insights from animal models. Brain Res 1617:7–17. https://doi.org/10.1016/j.brainres.2014.11.032

Blakemore SJ (2012) Development of the social brain in adolescence. J R Soc Med 105:111–116. https://doi.org/10.1258/jrsm.2011.110221

Brites D, Fernandes A (2015) Neuroinflammation and depression: microglia activation, extracellular microvesicles and microRNA dysregulation. Front Cell Neurosci 9:476. https://doi.org/10.3389/fncel.2015.00476

Brown GC, Neher JJ (2014) Microglial phagocytosis of live neurons. Nat Rev Neurosci 15:209–216. https://doi.org/10.1038/nrn3710

Cardona AE et al (2006) Control of microglial neurotoxicity by the fractalkine receptor. Nat Neurosci 9:917–924. https://doi.org/10.1038/nn1715

Cearley C, Churchill L, Krueger JM (2003) Time of day differences in IL1beta and TNFalpha mRNA levels in specific regions of the rat brain. Neurosci Lett 352:61–63

Chen Z, Palmer TD (2013) Differential roles of TNFR1 and TNFR2 signaling in adult hippocampal neurogenesis. Brain Behav Immun 30:45–53. https://doi.org/10.1016/j.bbi.2013.01.083

Claude J, Linnartz-Gerlach B, Kudin AP, Kunz WS, Neumann H (2013) Microglial CD33-related Siglec-E inhibits neurotoxicity by preventing the phagocytosis-associated oxidative burst. J Neurosci 33:18270–18276. https://doi.org/10.1523/jneurosci.2211-13.2013

Coughlin JM et al (2016) In vivo markers of inflammatory response in recent-onset schizophrenia: a combined study using [(11)C]DPA-713 PET and analysis of CSF and plasma. Transl Psychiatry 6:e777. https://doi.org/10.1038/tp.2016.40

Coutinho LG, Grandgirard D, Leib SL, Agnez-Lima LF (2013) Cerebrospinal-fluid cytokine and chemokine profile in patients with pneumococcal and meningococcal meningitis. BMC Infect Dis 13:326. https://doi.org/10.1186/1471-2334-13-326

Cunningham CL, Martinez-Cerdeno V, Noctor SC (2013) Microglia regulate the number of neural precursor cells in the developing cerebral cortex. J Neurosci 33:4216–4233. https://doi.org/10.1523/jneurosci.3441-12.2013

Daneman R, Zhou L, Kebede AA, Barres BA (2010) Pericytes are required for blood-brain barrier integrity during embryogenesis. Nature 468:562–566. https://doi.org/10.1038/nature09513

Dantzer R, O'Connor JC, Freund GG, Johnson RW, Kelley KW (2008) From inflammation to sickness and depression: when the immune system subjugates the brain. Nat Rev Neurosci 9:46–56. https://doi.org/10.1038/nrn2297

Davalos D et al (2005) ATP mediates rapid microglial response to local brain injury in vivo. Nat Neurosci 8:752–758. https://doi.org/10.1038/nn1472

Denes A, Thornton P, Rothwell NJ, Allan SM (2010) Inflammation and brain injury: acute cerebral ischaemia, peripheral and central inflammation. Brain Behav Immun 24:708–723. https://doi.org/10.1016/j.bbi.2009.09.010

Diaz-Aparicio I, Beccari S, Abiega O, Sierra A (2016) Clearing the corpses: regulatory mechanisms, novel tools, and therapeutic potential of harnessing microglial phagocytosis in the diseased brain. Neural Regen Res 11:1533–1539. https://doi.org/10.4103/1673-5374.193220

Dissing-Olesen L, LeDue JM, Rungta RL, Hefendehl JK, Choi HB, MacVicar BA (2014) Activation of neuronal NMDA receptors triggers transient ATP-mediated microglial process outgrowth. J Neurosci 34:10511–10527. https://doi.org/10.1523/jneurosci.0405-14.2014

Donzis EJ, Tronson NC (2014) Modulation of learning and memory by cytokines: signaling mechanisms and long term consequences. Neurobiol Learn Mem 115:68–77. https://doi.org/10.1016/j.nlm.2014.08.008

Doorduin J, de Vries EF, Willemsen AT, de Groot JC, Dierckx RA, Klein HC (2009) Neuroinflammation in schizophrenia-related psychosis: a PET study. J Nucl Med 50:1801–1807. https://doi.org/10.2967/jnumed.109.066647

Du Preez A, Leveson J, Zunszain PA, Pariante CM (2016) Inflammatory insults and mental health consequences: does timing matter when it comes to depression? Psychol Med 46:2041–2057. https://doi.org/10.1017/s0033291716000672

Estes ML, McAllister AK (2014) Alterations in immune cells and mediators in the brain: it's not always neuroinflammation! Brain Pathol (Zurich, Switzerland) 24:623–630. https://doi.org/10.1111/bpa.12198

Estes ML, McAllister AK (2015) Immune mediators in the brain and peripheral tissues in autism spectrum disorder. Nat Rev Neurosci 16:469–486. https://doi.org/10.1038/nrn3978

Eyo UB, Peng J, Swiatkowski P, Mukherjee A, Bispo A, Wu LJ (2014) Neuronal hyperactivity recruits microglial processes via neuronal NMDA receptors and microglial P2Y12 receptors after status epilepticus. J Neurosci 34:10528–10540. https://doi.org/10.1523/jneurosci.0416-14.2014

Ferrari D et al (2006) The P2X7 receptor: a key player in IL-1 processing and release. J Immunol (Baltimore, MD: 1950) 176:3877–3883

Ferreira R, Santos T, Goncalves J, Baltazar G, Ferreira L, Agasse F, Bernardino L (2012) Histamine modulates microglia function. J Neuroinflammation 9:90. https://doi.org/10.1186/1742-2094-9-90

Filiou MD, Arefin AS, Moscato P, Graeber MB (2014) 'Neuroinflammation' differs categorically from inflammation: transcriptomes of Alzheimer's disease, Parkinson's disease, schizophrenia and inflammatory diseases compared. Neurogenetics 15:201–212. https://doi.org/10.1007/s10048-014-0409-x

Fontainhas AM et al (2011) Microglial morphology and dynamic behavior is regulated by ionotropic glutamatergic and GABAergic neurotransmission. PLoS One 6:e15973. https://doi.org/10.1371/journal.pone.0015973

Frick L, Rapanelli M, Abbasi E, Ohtsu H, Pittenger C (2016) Histamine regulation of microglia: gene-environment interaction in the regulation of central nervous system inflammation. Brain Behav Immun 57:326–337. https://doi.org/10.1016/j.bbi.2016.07.002

Fricker M, Neher JJ, Zhao JW, Thery C, Tolkovsky AM, Brown GC (2012) MFG-E8 mediates primary phagocytosis of viable neurons during neuroinflammation. J Neurosci 32:2657–2666. https://doi.org/10.1523/jneurosci.4837-11.2012

Fu R, Shen Q, Xu P, Luo JJ, Tang Y (2014) Phagocytosis of microglia in the central nervous system diseases. Mol Neurobiol 49:1422–1434. https://doi.org/10.1007/s12035-013-8620-6

Garton KJ, Gough PJ, Blobel CP, Murphy G, Greaves DR, Dempsey PJ, Raines EW (2001) Tumor necrosis factor-alpha-converting enzyme (ADAM17) mediates the cleavage and shedding of fractalkine (CX3CL1). J Biol Chem 276:37993–38001. https://doi.org/10.1074/jbc.M106434200

Garver DL, Tamas RL, Holcomb JA (2003) Elevated interleukin-6 in the cerebrospinal fluid of a previously delineated schizophrenia subtype. Neuropsychopharmacology 28:1515–1520. https://doi.org/10.1038/sj.npp.1300217

Geissmann F, Gordon S, Hume DA, Mowat AM, Randolph GJ (2010) Unravelling mononuclear phagocyte heterogeneity. Nat Rev Immunol 10:453–460. https://doi.org/10.1038/nri2784

Ginhoux F et al (2010) Fate mapping analysis reveals that adult microglia derive from primitive macrophages. Science (New York, NY) 330:841–845. https://doi.org/10.1126/science.1194637
Ginhoux F, Lim S, Hoeffel G, Low D, Huber T (2013) Origin and differentiation of microglia. Front Cell Neurosci 7:45. https://doi.org/10.3389/fncel.2013.00045
Gomez-Nicola D, Perry VH (2015) Microglial dynamics and role in the healthy and diseased brain: a paradigm of functional plasticity. Neuroscientist 21:169–184. https://doi.org/10.1177/1073858414530512
Goshen I et al (2007) A dual role for interleukin-1 in hippocampal-dependent memory processes. Psychoneuroendocrinology 32:1106–1115. https://doi.org/10.1016/j.psyneuen.2007.09.004
Gracia-Rubio I, Moscoso-Castro M, Pozo OJ, Marcos J, Nadal R, Valverde O (2016) Maternal separation induces neuroinflammation and long-lasting emotional alterations in mice. Prog Neuro-Psychopharmacol Biol Psychiatry 65:104–117. https://doi.org/10.1016/j.pnpbp.2015.09.003
Graeber MB (2014) Neuroinflammation: no rose by any other name. Brain Pathol (Zurich, Switzerland) 24:620–622. https://doi.org/10.1111/bpa.12192
Graeber MB, Streit WJ (1990) Microglia: immune network in the CNS. Brain Pathol (Zurich, Switzerland) 1:2–5
Gumienny TL et al (2001) CED-12/ELMO, a novel member of the CrkII/Dock180/Rac pathway, is required for phagocytosis and cell migration. Cell 107:27–41
Gyoneva S, Traynelis SF (2013) Norepinephrine modulates the motility of resting and activated microglia via different adrenergic receptors. J Biol Chem 288:15291–15302. https://doi.org/10.1074/jbc.M113.458901
Haarman BC et al (2014) Neuroinflammation in bipolar disorder - a [(11)C]-(R)-PK11195 positron emission tomography study. Brain Behav Immun 40:219–225. https://doi.org/10.1016/j.bbi.2014.03.016
Haarman BC et al (2016) Volume, metabolites and neuroinflammation of the hippocampus in bipolar disorder - a combined magnetic resonance imaging and positron emission tomography study. Brain Behav Immun 56:21–33. https://doi.org/10.1016/j.bbi.2015.09.004
Hanisch UK, Kettenmann H (2007) Microglia: active sensor and versatile effector cells in the normal and pathologic brain. Nat Neurosci 10:1387–1394. https://doi.org/10.1038/nn1997
Harry GJ (2013) Microglia during development and aging. Pharmacol Ther 139:313–326. https://doi.org/10.1016/j.pharmthera.2013.04.013
Haynes SE, Hollopeter G, Yang G, Kurpius D, Dailey ME, Gan WB, Julius D (2006) The P2Y12 receptor regulates microglial activation by extracellular nucleotides. Nat Neurosci 9:1512–1519. https://doi.org/10.1038/nn1805
Heneka MT et al (2010) Locus ceruleus controls Alzheimer's disease pathology by modulating microglial functions through norepinephrine. Proc Natl Acad Sci U S A 107:6058–6063. https://doi.org/10.1073/pnas.0909586107
Hickman SE, Kingery ND, Ohsumi TK, Borowsky ML, Wang LC, Means TK, El Khoury J (2013) The microglial sensome revealed by direct RNA sequencing. Nat Neurosci 16:1896–1905. https://doi.org/10.1038/nn.3554
Hines DJ, Hines RM, Mulligan SJ, Macvicar BA (2009) Microglia processes block the spread of damage in the brain and require functional chloride channels. Glia 57:1610–1618. https://doi.org/10.1002/glia.20874
Hoeffel G et al (2015) C-Myb(+) erythro-myeloid progenitor-derived fetal monocytes give rise to adult tissue-resident macrophages. Immunity 42:665–678. https://doi.org/10.1016/j.immuni.2015.03.011
Horvath S, Mirnics K (2014) Immune system disturbances in schizophrenia. Biol Psychiatry 75:316–323. https://doi.org/10.1016/j.biopsych.2013.06.010
Hoshiko M, Arnoux I, Avignone E, Yamamoto N, Audinat E (2012) Deficiency of the microglial receptor CX3CR1 impairs postnatal functional development of thalamocortical synapses in the barrel cortex. J Neurosci 32:15106–15111. https://doi.org/10.1523/jneurosci.1167-12.2012

Hristovska I, Pascual O (2015) Deciphering resting microglial morphology and process motility from a synaptic prospect. Front Integr Neurosci 9:73. https://doi.org/10.3389/fnint.2015.00073
Isgren A et al (2017) Markers of neuroinflammation and neuronal injury in bipolar disorder: relation to prospective clinical outcomes. Brain Behav Immun. https://doi.org/10.1016/j.bbi.2017.05.002
Johnson MH (2001) Functional brain development in humans. Nat Rev Neurosci 2:475–483. https://doi.org/10.1038/35081509
Johnson JD, Zimomra ZR, Stewart LT (2013) Beta-adrenergic receptor activation primes microglia cytokine production. J Neuroimmunol 254:161–164. https://doi.org/10.1016/j.jneuroim.2012.08.007
Kenk M et al (2015) Imaging neuroinflammation in gray and white matter in schizophrenia: an in-vivo PET study with [18F]-FEPPA. Schizophr Bull 41:85–93. https://doi.org/10.1093/schbul/sbu157
Kettenmann H, Hanisch UK, Noda M, Verkhratsky A (2011) Physiology of microglia. Physiol Rev 91:461–553. https://doi.org/10.1152/physrev.00011.2010
Khandaker GM, Cousins L, Deakin J, Lennox BR, Yolken R, Jones PB (2015) Inflammation and immunity in schizophrenia: implications for pathophysiology and treatment. Lancet Psychiatry 2:258–270. https://doi.org/10.1016/s2215-0366(14)00122-9
Konrad K, Firk C, Uhlhaas PJ (2013) Brain development during adolescence: neuroscientific insights into this developmental period. Dtsch Arztebl Int 110:425–431. https://doi.org/10.3238/arztebl.2013.0425
Krabbe G, Matyash V, Pannasch U, Mamer L, Boddeke HW, Kettenmann H (2012) Activation of serotonin receptors promotes microglial injury-induced motility but attenuates phagocytic activity. Brain Behav Immun 26:419–428. https://doi.org/10.1016/j.bbi.2011.12.002
Kraepelin E (1890) Über Psychosen nach Influenza. Dtsch Med Wochenschr 16:209–212
Krueger JM (2008) The role of cytokines in sleep regulation. Curr Pharm Des 14:3408–3416
Krueger JM, Majde JA, Rector DM (2011) Cytokines in immune function and sleep regulation. Handb Clin Neurol 98:229–240. https://doi.org/10.1016/b978-0-444-52006-7.00015-0
Kuhn SA et al (2004) Microglia express GABA(B) receptors to modulate interleukin release. Mol Cell Neurosci 25:312–322. https://doi.org/10.1016/j.mcn.2003.10.023
Labrousse VF, Costes L, Aubert A, Darnaudery M, Ferreira G, Amedee T, Laye S (2009) Impaired interleukin-1beta and c-Fos expression in the hippocampus is associated with a spatial memory deficit in P2X(7) receptor-deficient mice. PLoS One 4:e6006. https://doi.org/10.1371/journal.pone.0006006
Laskaris LE et al (2016) Microglial activation and progressive brain changes in schizophrenia. Br J Pharmacol 173:666–680. https://doi.org/10.1111/bph.13364
Lawson LJ, Perry VH, Dri P, Gordon S (1990) Heterogeneity in the distribution and morphology of microglia in the normal adult mouse brain. Neuroscience 39:151–170
Lawson LJ, Perry VH, Gordon S (1992) Turnover of resident microglia in the normal adult mouse brain. Neuroscience 48:405–415
Lee WL, Mason D, Schreiber AD, Grinstein S (2007) Quantitative analysis of membrane remodeling at the phagocytic cup. Mol Biol Cell 18:2883–2892. https://doi.org/10.1091/mbc.E06-05-0450
Lee M, Schwab C, McGeer PL (2011) Astrocytes are GABAergic cells that modulate microglial activity. Glia 59:152–165. https://doi.org/10.1002/glia.21087
Levine J, Barak Y, Chengappa KN, Rapoport A, Rebey M, Barak V (1999) Cerebrospinal cytokine levels in patients with acute depression. Neuropsychobiology 40:171–176. https://doi.org/10.1159/000026615
Li Y, Du XF, Liu CS, Wen ZL, Du JL (2012) Reciprocal regulation between resting microglial dynamics and neuronal activity in vivo. Dev Cell 23:1189–1202. https://doi.org/10.1016/j.devcel.2012.10.027

Liang KJ, Lee JE, Wang YD, Ma W, Fontainhas AM, Fariss RN, Wong WT (2009) Regulation of dynamic behavior of retinal microglia by CX3CR1 signaling. Invest Ophthalmol Vis Sci 50:4444–4451. https://doi.org/10.1167/iovs.08-3357

Lindqvist D et al (2009) Interleukin-6 is elevated in the cerebrospinal fluid of suicide attempters and related to symptom severity. Biol Psychiatry 66:287–292. https://doi.org/10.1016/j.biopsych.2009.01.030

Linnartz B, Kopatz J, Tenner AJ, Neumann H (2012) Sialic acid on the neuronal glycocalyx prevents complement C1 binding and complement receptor-3-mediated removal by microglia. J Neurosci 32:946–952. https://doi.org/10.1523/jneurosci.3830-11.2012

Liu GJ, Nagarajah R, Banati RB, Bennett MR (2009) Glutamate induces directed chemotaxis of microglia. Eur J Neurosci 29:1108–1118. https://doi.org/10.1111/j.1460-9568.2009.06659.x

Liu H, Leak RK, Hu X (2016) Neurotransmitter receptors on microglia. Stroke Vasc Neurol 1:52–58. https://doi.org/10.1136/svn-2016-000012

Luo J et al (2013) Colony-stimulating factor 1 receptor (CSF1R) signaling in injured neurons facilitates protection and survival. J Exp Med 210:157–172. https://doi.org/10.1084/jem.20120412

Maciejewski-Lenoir D, Chen S, Feng L, Maki R, Bacon KB (1999) Characterization of fractalkine in rat brain cells: migratory and activation signals for CX3CR-1-expressing microglia. J Immunol (Baltimore, MD: 1950) 163:1628–1635

Mantovani A, Sica A, Locati M (2005) Macrophage polarization comes of age. Immunity 23:344–346. https://doi.org/10.1016/j.immuni.2005.10.001

Marker DF, Puccini JM, Mockus TE, Barbieri J, Lu SM, Gelbard HA (2012) LRRK2 kinase inhibition prevents pathological microglial phagocytosis in response to HIV-1 Tat protein. J Neuroinflammation 9:261. https://doi.org/10.1186/1742-2094-9-261

Martinez FO, Gordon S (2014) The M1 and M2 paradigm of macrophage activation: time for reassessment. F1000prime Rep 6:13. https://doi.org/10.12703/p6-13

Masgrau R, Guaza C, Ransohoff RM, Galea E (2017) Should we stop saying 'glia' and 'neuroinflammation'? Trends Mol Med 23:486–500. https://doi.org/10.1016/j.molmed.2017.04.005

Matcovitch-Natan O et al (2016) Microglia development follows a stepwise program to regulate brain homeostasis. Science (New York, NY) 353:aad8670. https://doi.org/10.1126/science.aad8670

Mayilyan KR, Weinberger DR, Sim RB (2008) The complement system in schizophrenia. Drug News Perspect 21:200–210. https://doi.org/10.1358/dnp.2008.21.4.1213349

Mazaheri F, Breus O, Durdu S, Haas P, Wittbrodt J, Gilmour D, Peri F (2014) Distinct roles for BAI1 and TIM-4 in the engulfment of dying neurons by microglia. Nat Commun 5:4046. https://doi.org/10.1038/ncomms5046

Meltzer A, Van de Water J (2017) The role of the immune system in autism spectrum disorder. Neuropsychopharmacology 42:284–298. https://doi.org/10.1038/npp.2016.158

Menninger KA (1919) Psychoses associated with influenza, I: general data: statistical analysis. JAMA 72:235–241

Miller AH, Raison CL (2016) The role of inflammation in depression: from evolutionary imperative to modern treatment target. Nat Rev Immunol 16:22–34. https://doi.org/10.1038/nri.2015.5

Mingam R, De Smedt V, Amedee T, Bluthe RM, Kelley KW, Dantzer R, Laye S (2008) In vitro and in vivo evidence for a role of the P2X7 receptor in the release of IL-1 beta in the murine brain. Brain Behav Immun 22:234–244. https://doi.org/10.1016/j.bbi.2007.08.007

Mittelbronn M, Dietz K, Schluesener HJ, Meyermann R (2001) Local distribution of microglia in the normal adult human central nervous system differs by up to one order of magnitude. Acta Neuropathol 101:249–255

Mizuno T et al (2011) Interleukin-34 selectively enhances the neuroprotective effects of microglia to attenuate oligomeric amyloid-beta neurotoxicity. Am J Pathol 179:2016–2027. https://doi.org/10.1016/j.ajpath.2011.06.011

Monji A et al (2013) Neuroinflammation in schizophrenia especially focused on the role of microglia. Prog Neuro-Psychopharmacol Biol Psychiatry 42:115–121. https://doi.org/10.1016/j.pnpbp.2011.12.002

Muller N, Schwarz MJ (2007) The immune-mediated alteration of serotonin and glutamate: towards an integrated view of depression. Mol Psychiatry 12:988–1000. https://doi.org/10.1038/sj.mp.4002006

Muller N, Riedel M, Gruber R, Ackenheil M, Schwarz MJ (2000) The immune system and schizophrenia. An integrative view. Ann N Y Acad Sci 917:456–467

Na KS, Jung HY, Kim YK (2014) The role of pro-inflammatory cytokines in the neuroinflammation and neurogenesis of schizophrenia. Prog Neuro-Psychopharmacol Biol Psychiatry 48:277–286. https://doi.org/10.1016/j.pnpbp.2012.10.022

Najjar S, Pearlman DM (2015) Neuroinflammation and white matter pathology in schizophrenia: systematic review. Schizophr Res 161:102–112. https://doi.org/10.1016/j.schres.2014.04.041

Nakatomi Y et al (2014) Neuroinflammation in patients with chronic fatigue syndrome/myalgic encephalomyelitis: an (1)(1)C-(R)-PK11195 PET study. J Nucl Med 55:945–950. https://doi.org/10.2967/jnumed.113.131045

Neniskyte U, Brown GC (2013) Lactadherin/MFG-E8 is essential for microglia-mediated neuronal loss and phagoptosis induced by amyloid beta. J Neurochem 126:312–317. https://doi.org/10.1111/jnc.12288

Neniskyte U, Gross CT (2017) Errant gardeners: glial-cell-dependent synaptic pruning and neurodevelopmental disorders. Nat Rev Neurosci 18:658–670. https://doi.org/10.1038/nrn.2017.110

Nimmerjahn A, Kirchhoff F, Helmchen F (2005) Resting microglial cells are highly dynamic surveillants of brain parenchyma in vivo. Science (New York, NY) 308:1314–1318. https://doi.org/10.1126/science.1110647

Noda M, Nakanishi H, Nabekura J, Akaike N (2000) AMPA-kainate subtypes of glutamate receptor in rat cerebral microglia. J Neurosci 20:251–258

Noda M et al (2011) Fractalkine attenuates excito-neurotoxicity via microglial clearance of damaged neurons and antioxidant enzyme heme oxygenase-1 expression. J Biol Chem 286:2308–2319. https://doi.org/10.1074/jbc.M110.169839

Noda M et al (2014) FGF-2 released from degenerating neurons exerts microglial-induced neuroprotection via FGFR3-ERK signaling pathway. J Neuroinflammation 11:76. https://doi.org/10.1186/1742-2094-11-76

Orkin SH, Zon LI (2008) Hematopoiesis: an evolving paradigm for stem cell biology. Cell 132:631–644. https://doi.org/10.1016/j.cell.2008.01.025

Orr AG, Orr AL, Li XJ, Gross RE, Traynelis SF (2009) Adenosine A(2A) receptor mediates microglial process retraction. Nat Neurosci 12:872–878. https://doi.org/10.1038/nn.2341

Orsini F, De Blasio D, Zangari R, Zanier ER, De Simoni MG (2014) Versatility of the complement system in neuroinflammation, neurodegeneration and brain homeostasis. Front Cell Neurosci 8:380. https://doi.org/10.3389/fncel.2014.00380

Paolicelli RC, Ferretti MT (2017) Function and dysfunction of microglia during brain development: consequences for synapses and neural circuits. Front Synaptic Neurosci 9:9. https://doi.org/10.3389/fnsyn.2017.00009

Paolicelli RC et al (2011) Synaptic pruning by microglia is necessary for normal brain development. Science (New York, NY) 333:1456–1458. https://doi.org/10.1126/science.1202529

Paolicelli RC, Bisht K, Tremblay ME (2014) Fractalkine regulation of microglial physiology and consequences on the brain and behavior. Front Cell Neurosci 8:129. https://doi.org/10.3389/fncel.2014.00129

Parkhurst CN et al (2013) Microglia promote learning-dependent synapse formation through brain-derived neurotrophic factor. Cell 155:1596–1609. https://doi.org/10.1016/j.cell.2013.11.030

Parkin J, Cohen B (2001) An overview of the immune system. Lancet (London, England) 357:1777–1789. https://doi.org/10.1016/s0140-6736(00)04904-7

Patel M, Pelletier A, Cote JF (2011) Opening up on ELMO regulation: New insights into the control of Rac signaling by the DOCK180/ELMO complex. Small GTPases 2:268–275. https://doi.org/10.4161/sgtp.2.5.17716

Perez-Cerda F, Sanchez-Gomez MV, Matute C (2015) Pio del Rio Hortega and the discovery of the oligodendrocytes. Front Neuroanat 9:92. https://doi.org/10.3389/fnana.2015.00092

Perry VH, Nicoll JA, Holmes C (2010) Microglia in neurodegenerative disease. Nat Rev Neurol 6:193–201. https://doi.org/10.1038/nrneurol.2010.17

Pinto Junior VL, Rebelo MC, Gomes RN, Assis EF, Castro-Faria-Neto HC, Boia MN (2011) IL-6 and IL-8 in cerebrospinal fluid from patients with aseptic meningitis and bacterial meningitis: their potential role as a marker for differential diagnosis. Braz J Infect Dis 15:156–158

Pont-Lezica L, Beumer W, Colasse S, Drexhage H, Versnel M, Bessis A (2014) Microglia shape corpus callosum axon tract fasciculation: functional impact of prenatal inflammation. Eur J Neurosci 39:1551–1557. https://doi.org/10.1111/ejn.12508

Presumey J, Bialas AR, Carroll MC (2017) Complement system in neural synapse elimination in development and disease. Adv Immunol 135:53–79. https://doi.org/10.1016/bs.ai.2017.06.004

Ransohoff RM (2016) A polarizing question: do M1 and M2 microglia exist? Nat Neurosci 19:987–991. https://doi.org/10.1038/nn.4338

Ransohoff RM, Cardona AE (2010) The myeloid cells of the central nervous system parenchyma. Nature 468:253–262. https://doi.org/10.1038/nature09615

Ransohoff RM, Engelhardt B (2012) The anatomical and cellular basis of immune surveillance in the central nervous system. Nat Rev Immunol 12:623–635. https://doi.org/10.1038/nri3265

Ransohoff RM, Perry VH (2009) Microglial physiology: unique stimuli, specialized responses. Annu Rev Immunol 27:119–145. https://doi.org/10.1146/annurev.immunol.021908.132528

Reemst K, Noctor SC, Lucassen PJ, Hol EM (2016) The indispensable roles of microglia and astrocytes during brain development. Front Hum Neurosci 10:566. https://doi.org/10.3389/fnhum.2016.00566

Reu P et al (2017) The lifespan and turnover of microglia in the human brain. Cell Rep 20:779–784. https://doi.org/10.1016/j.celrep.2017.07.004

Rocha e Silva M (1978) A brief survey of the history of inflammation. Agents Actions 8:45–49

Rosales C, Uribe-Querol E (2017) Phagocytosis: a fundamental process in immunity. Biomed Res Int 2017:9042851. https://doi.org/10.1155/2017/9042851

Salter MW, Beggs S (2014) Sublime microglia: expanding roles for the guardians of the CNS. Cell 158:15–24. https://doi.org/10.1016/j.cell.2014.06.008

Salter MW, Stevens B (2017) Microglia emerge as central players in brain disease. Nat Med 23:1018–1027. https://doi.org/10.1038/nm.4397

Santoni G, Cardinali C, Morelli MB, Santoni M, Nabissi M, Amantini C (2015) Danger- and pathogen-associated molecular patterns recognition by pattern-recognition receptors and ion channels of the transient receptor potential family triggers the inflammasome activation in immune cells and sensory neurons. J Neuroinflammation 12:21. https://doi.org/10.1186/s12974-015-0239-2

Schafer DP et al (2012) Microglia sculpt postnatal neural circuits in an activity and complement-dependent manner. Neuron 74:691–705. https://doi.org/10.1016/j.neuron.2012.03.026

Scheffel J et al (2012) Toll-like receptor activation reveals developmental reorganization and unmasks responder subsets of microglia. Glia 60:1930–1943. https://doi.org/10.1002/glia.22409

Schwartz M, Baruch K (2014) The resolution of neuroinflammation in neurodegeneration: leukocyte recruitment via the choroid plexus. EMBO J 33:7–22. https://doi.org/10.1002/embj.201386609

Schwieler L et al (2015) Increased levels of IL-6 in the cerebrospinal fluid of patients with chronic schizophrenia--significance for activation of the kynurenine pathway. J Psychiatry Neurosci 40:126–133

Scott A, Khan KM, Cook JL, Duronio V (2004) What is "inflammation"? Are we ready to move beyond Celsus? Br J Sports Med 38:248–249

Serhan CN, Savill J (2005) Resolution of inflammation: the beginning programs the end. Nat Immunol 6:1191–1197. https://doi.org/10.1038/ni1276
Setiawan E et al (2015) Role of translocator protein density, a marker of neuroinflammation, in the brain during major depressive episodes. JAMA Psychiat 72:268–275. https://doi.org/10.1001/jamapsychiatry.2014.2427
Shankaran M et al (2007) Measurement of brain microglial proliferation rates in vivo in response to neuroinflammatory stimuli: application to drug discovery. J Neurosci Res 85:2374–2384. https://doi.org/10.1002/jnr.21389
Sheng J, Ruedl C, Karjalainen K (2015) Most tissue-resident macrophages except microglia are derived from fetal hematopoietic stem cells. Immunity 43:382–393. https://doi.org/10.1016/j.immuni.2015.07.016
Sheridan GK et al (2014) CX3CL1 is up-regulated in the rat hippocampus during memory-associated synaptic plasticity. Front Cell Neurosci 8:233. https://doi.org/10.3389/fncel.2014.00233
Shigemoto-Mogami Y, Hoshikawa K, Goldman JE, Sekino Y, Sato K (2014) Microglia enhance neurogenesis and oligodendrogenesis in the early postnatal subventricular zone. J Neurosci 34:2231–2243. https://doi.org/10.1523/jneurosci.1619-13.2014
Sierra A et al (2010) Microglia shape adult hippocampal neurogenesis through apoptosis-coupled phagocytosis. Cell Stem Cell 7:483–495. https://doi.org/10.1016/j.stem.2010.08.014
Sierra A, Abiega O, Shahraz A, Neumann H (2013) Janus-faced microglia: beneficial and detrimental consequences of microglial phagocytosis. Front Cell Neurosci 7:6. https://doi.org/10.3389/fncel.2013.00006
Sierra A, Beccari S, Diaz-Aparicio I, Encinas JM, Comeau S, Tremblay ME (2014) Surveillance, phagocytosis, and inflammation: how never-resting microglia influence adult hippocampal neurogenesis. Neural Plast 2014:610343. https://doi.org/10.1155/2014/610343
Soderlund J et al (2009) Activation of brain interleukin-1beta in schizophrenia. Mol Psychiatry 14:1069–1071. https://doi.org/10.1038/mp.2009.52
Sordillo PP, Sordillo LA, Helson L (2016) Bifunctional role of pro-inflammatory cytokines after traumatic brain injury. Brain Inj 30:1043–1053. https://doi.org/10.3109/02699052.2016.1163618
Squarzoni P et al (2014) Microglia modulate wiring of the embryonic forebrain. Cell Rep 8:1271–1279. https://doi.org/10.1016/j.celrep.2014.07.042
Stelmasiak Z, Koziol-Montewka M, Dobosz B, Rejdak K, Bartosik-Psujek H, Mitosek-Szewczyk K, Belniak-Legiec E (2000) Interleukin-6 concentration in serum and cerebrospinal fluid in multiple sclerosis patients. Med Sci Monit 6:1104–1108
Stephan AH, Barres BA, Stevens B (2012) The complement system: an unexpected role in synaptic pruning during development and disease. Annu Rev Neurosci 35:369–389. https://doi.org/10.1146/annurev-neuro-061010-113810
Stevens B et al (2007) The classical complement cascade mediates CNS synapse elimination. Cell 131:1164–1178. https://doi.org/10.1016/j.cell.2007.10.036
Suridjan I et al (2014) Quantitative imaging of neuroinflammation in human white matter: a positron emission tomography study with translocator protein 18 kDa radioligand, [18F]-FEPPA. Synapse (New York, NY) 68:536–547. https://doi.org/10.1002/syn.21765
Svahn AJ, Becker TS, Graeber MB (2014) Emergent properties of microglia. Brain Pathol (Zurich, Switzerland) 24:665–670. https://doi.org/10.1111/bpa.12195
Szalay G et al (2016) Microglia protect against brain injury and their selective elimination dysregulates neuronal network activity after stroke. Nat Commun 7:11499. https://doi.org/10.1038/ncomms11499
Takahashi K, Yamamura F, Naito M (1989) Differentiation, maturation, and proliferation of macrophages in the mouse yolk sac: a light-microscopic, enzyme-cytochemical, immunohistochemical, and ultrastructural study. J Leukoc Biol 45:87–96

Takahashi K, Rochford CD, Neumann H (2005) Clearance of apoptotic neurons without inflammation by microglial triggering receptor expressed on myeloid cells-2. J Exp Med 201:647–657. https://doi.org/10.1084/jem.20041611
Tay TL et al (2017) A new fate mapping system reveals context-dependent random or clonal expansion of microglia. Nat Neurosci 20:793–803. https://doi.org/10.1038/nn.4547
Thion MS, Garel S (2017) On place and time: microglia in embryonic and perinatal brain development. Curr Opin Neurobiol 47:121–130. https://doi.org/10.1016/j.conb.2017.10.004
Tonchev AB, Yamashima T, Zhao L, Okano H (2003) Differential proliferative response in the postischemic hippocampus, temporal cortex, and olfactory bulb of young adult macaque monkeys. Glia 42:209–224. https://doi.org/10.1002/glia.10209
Torres L et al (2016) Dynamic microglial modulation of spatial learning and social behavior. Brain Behav Immun 55:6–16. https://doi.org/10.1016/j.bbi.2015.09.001
Tremblay ME (2011) The role of microglia at synapses in the healthy CNS: novel insights from recent imaging studies. Neuron Glia Biol 7:67–76. https://doi.org/10.1017/s1740925x12000038
Tremblay ME, Lowery RL, Majewska AK (2010) Microglial interactions with synapses are modulated by visual experience. PLoS Biol 8:e1000527. https://doi.org/10.1371/journal.pbio.1000527
Tremblay ME, Stevens B, Sierra A, Wake H, Bessis A, Nimmerjahn A (2011) The role of microglia in the healthy brain. J Neurosci 31:16064–16069. https://doi.org/10.1523/jneurosci.4158-11.2011
Ueno M, Fujita Y, Tanaka T, Nakamura Y, Kikuta J, Ishii M, Yamashita T (2013) Layer V cortical neurons require microglial support for survival during postnatal development. Nat Neurosci 16:543–551. https://doi.org/10.1038/nn.3358
Underhill DM, Goodridge HS (2012) Information processing during phagocytosis. Nat Rev Immunol 12:492–502. https://doi.org/10.1038/nri3244
van Kesteren CF et al (2017) Immune involvement in the pathogenesis of schizophrenia: a meta-analysis on postmortem brain studies. Transl Psychiatry 7:e1075. https://doi.org/10.1038/tp.2017.4
Wakatsuki S, Araki T (2017) Specific phospholipid scramblases are involved in exposure of phosphatidylserine, an "eat-me" signal for phagocytes, on degenerating axons. Commun Integr Biol 10:e1296615. https://doi.org/10.1080/19420889.2017.1296615
Wakselman S, Bechade C, Roumier A, Bernard D, Triller A, Bessis A (2008) Developmental neuronal death in hippocampus requires the microglial CD11b integrin and DAP12 immunoreceptor. J Neurosci 28:8138–8143. https://doi.org/10.1523/jneurosci.1006-08.2008
Wang AK, Miller BJ (2017) Meta-analysis of cerebrospinal fluid cytokine and tryptophan catabolite alterations in psychiatric patients: comparisons between schizophrenia, bipolar disorder, and depression. Schizophr Bull. https://doi.org/10.1093/schbul/sbx035
Wang Y, Neumann H (2010) Alleviation of neurotoxicity by microglial human Siglec-11. J Neurosci 30:3482–3488. https://doi.org/10.1523/jneurosci.3940-09.2010
Watkins CC, Sawa A, Pomper MG (2014) Glia and immune cell signaling in bipolar disorder: insights from neuropharmacology and molecular imaging to clinical application. Transl Psychiatry 4:e350. https://doi.org/10.1038/tp.2013.119
Wolf SA, Boddeke HW, Kettenmann H (2017) Microglia in physiology and disease. Annu Rev Physiol 79:619–643. https://doi.org/10.1146/annurev-physiol-022516-034406
Wu Y, Dissing-Olesen L, MacVicar BA, Stevens B (2015) Microglia: dynamic mediators of synapse development and plasticity. Trends Immunol 36:605–613. https://doi.org/10.1016/j.it.2015.08.008
Xing C, Lo EH (2017) Help-me signaling: non-cell autonomous mechanisms of neuroprotection and neurorecovery. Prog Neurobiol 152:181–199. https://doi.org/10.1016/j.pneurobio.2016.04.004
Yirmiya R, Goshen I (2011) Immune modulation of learning, memory, neural plasticity and neurogenesis. Brain Behav Immun 25:181–213. https://doi.org/10.1016/j.bbi.2010.10.015

Yirmiya R, Winocur G, Goshen I (2002) Brain interleukin-1 is involved in spatial memory and passive avoidance conditioning. Neurobiol Learn Mem 78:379–389
Yolken RH, Torrey EF (2008) Are some cases of psychosis caused by microbial agents? A review of the evidence. Mol Psychiatry 13:470–479. https://doi.org/10.1038/mp.2008.5
Zhan Y et al (2014) Deficient neuron-microglia signaling results in impaired functional brain connectivity and social behavior. Nat Neurosci 17:400–406. https://doi.org/10.1038/nn.3641
Zhang M, Xu G, Liu W, Ni Y, Zhou W (2012) Role of fractalkine/CX3CR1 interaction in light-induced photoreceptor degeneration through regulating retinal microglial activation and migration. PLoS One 7:e35446. https://doi.org/10.1371/journal.pone.0035446
Zhang H, Li F, Yang Y, Chen J, Hu X (2015) SIRP/CD47 signaling in neurological disorders. Brain Res 1623:74–80. https://doi.org/10.1016/j.brainres.2015.03.012

Epidemiological Studies of Prenatal and Childhood Infection and Schizophrenia

Håkan Karlsson and Christina Dalman

Contents

Abstract Certain infectious agents can target the brain and interfere with its growth, development, and/or function. A number of studies indicate that exposure to common infectious agents during fetal and postnatal life may also contribute to the later development of schizophrenia and other non-affective psychoses. Epidemiological studies of maternal infections during pregnancy have provided somewhat contradictory results with regard to infections in general but have reported surprisingly consistent associations with specific maternal exposures such as *Toxoplasma gondii*. Childhood is also beginning to emerge as a sensitive period for the influence of infections including infectious agents not known to target the brain. Recent studies have associated childhood infections not only with a later diagnosis of schizophrenia but also with impaired cognitive function. Importantly, independent studies indicate that the associations between early life infection and the later development of schizophrenia are not explained by factors shared between related individuals or by genetic liability for schizophrenia.

Keywords Childhood · Cognition · Infection · Pregnancy · Schizophrenia

---

H. Karlsson (✉)
Department of Neuroscience, Karolinska Institutet, Stockholm, Sweden
e-mail: hakan.karlsson.2@ki.se

C. Dalman
Department of Public Health Sciences, Karolinska Institutet, Stockholm, Sweden

Curr Topics Behav Neurosci (2020) 44: 35–48
DOI 10.1007/7854_2018_87
Published Online: 10 March 2019

1 Introduction

Some infections of the mother during pregnancy are recognized to target the fetus or newborn to interfere with its growth or development, including that of the brain. The most well-recognized agents with such teratogenic properties are *Toxoplasma gondii*, rubella virus, cytomegalovirus (CMV), and herpes simplex viruses, commonly referred to as the TORCH agents (Bale 2009). Congenital infections by these agents are fortunately rare even among infected mothers without pre-existing immunity. Moreover, the recent Zika virus epidemic in South America illustrates that emerging infections are a severe threat to human health (Alvarado and Schwartz 2017). Maternal infections can also be passed on to the offspring during or after birth (e.g., during breastfeeding) (Bale 2009). After birth, the neonate is increasingly exposed to not only its mother's infections but the infectious agents prevalent in the child's environment, which is likely to vary greatly across time, geographic regions, and socioeconomic groups.

1.1 The Timing of an Infection Affects the Outcome

Among children exposed to infection, the outcome can vary depending on the timing of the exposure in relation to their developmental stage. Whereas infections early in gestation may cause severe CNS malformations, later infections may cause more subtle symptoms (e.g., impairments in hearing or cognition), comprehensively reviewed in Bale (2009). In terms of neurodevelopmental processes, neurogenesis and neuronal migration in the developing cortex are largely complete by birth. The postnatal period is characterized by a very rapid growth, primarily involving glial cell proliferation and differentiation, synapse formation, and myelination, and the human brain reaches 90–95% of its adult volume by age 6. While regional gray matter volumes tend to decrease during adolescence, in part due to synaptic pruning, the formation of myelin sheaths around nerve fibers begins during fetal life and continues through the first few decades of life. Postnatal neurodevelopmental processes occur at different time points and rates in different cortical and subcortical regions, and all contribute to the changes in cognitive abilities and processing of social interactions and emotions that occur during the first decades of life (Semple et al. 2013; Houston et al. 2014).

Not only the brain but also the immune system undergoes development from fetal life through neonatal life, childhood, adolescence to adulthood. For the purpose of infections, the developmental stage of the immune system will also determine the outcome of an exposure. Many infections, primarily of viral origin such as hepatitis B virus (HBV), are largely asymptomatic when occurring during early life in part due to an inability of the neonate to mount an adequate response to eradicate the infections, which can lead to persistence and chronic infection to cause symptoms only later on in life. Older children and adults infected with HBV, on the other hand, often develop acute disease symptoms (e.g., jaundice) due to an efficient destruction of infected (liver) cells and clearance of the infection. Other agents can cause life-threatening conditions in neonates, e.g., herpes simplex virus, but are usually asymptomatic in older children or adults. The developmental processes occurring both in the immune system and in the brain throughout childhood thus may have a substantial influence on the short- and long-term outcomes of an infection (Prendergast et al. 2012).

1.2 Other Factors Affecting the Outcome

In addition to the age of the child at infection, other host characteristics will influence the outcome of an exposure to an infection. These include pre-existing immunity, due to, e.g., vaccination or passive immunization by maternally derived antibodies, and genetic variation. Genetic variation, primarily in the MHC region on chromosome 6 (dense with genes encoding proteins involved in immune recognition and functioning, e.g., HLA molecules), is documented to influence susceptibility, progression, and recovery during infections in adult individuals and will likely also influence the outcome of infections occurring during early life (Dendrou et al. 2018; McLaren et al. 2013). In addition, genetic variation not only in the host but also in the infectious agent itself is a likely determinant of the outcome of an infection. Variation between strains of influenza A virus is well known (Petrova and Russell 2018) and can correlate with differences in cellular tropism and neuro-invasive properties (Ward 1996). In Sweden, the prevalence of different strains of mumps virus has varied with more neurovirulent strains being prevalent during the 1970s and early 1980s as compared to later years (Tecle et al. 1998).

It is established that some known infections during fetal life can interfere with brain growth and development. Whether infections during infancy and early childhood can also affect neurodevelopmental processes or even contribute to risk of schizophrenia or other severe mental illnesses with onsets decades later have been the focus of many investigations (see below).

The fact that still no specific agent or developmental stage has been conclusively linked to schizophrenia has prompted an epidemiological approach to the study of the role of infections in the development of schizophrenia. This approach has during the last four decades provided invaluable information using increasingly sophisticated methods and designs to address issues regarding reverse causality, misclassification, and confounding by socioeconomic conditions, other familial factors, or even genetic variation. Using population-based registers and biobanks with long follow-up times, available in countries like Denmark, Finland, and Sweden, researchers are beginning to dissect the various factors and their potential interactions involved in the etiology of schizophrenia as well as other severe and increasingly prevalent neuropsychiatric illnesses, such as autism spectrum disorder and attention-deficit hyperactivity disorder.

2 Epidemiological Studies of Infections During Pregnancy

In their seminal article published in 1988, Mednick et al. (1988) reported that offspring to pregnant women exposed to the 1957 epidemic of influenza A virus in Helsinki were at increased risk to develop schizophrenia as compared to offspring to women who were pregnant during preceding years when the same flu strain was not present in the population. The study had a purely ecological design and hence no information on if case mothers were actually infected during pregnancy, only that they were pregnant when the virus was prevalent in the population. Subsequent studies with information on maternal exposure at the individual level have, taken together, albeit not convincingly, supported this original observation; see Munk-Jorgensen and Ewald (2001), Selten et al. (2010) and references therein. Brown et al. (2004) reported a weak association between maternal influenza during pregnancy and the later development of schizophrenia in the offspring. This group of researchers used a large US birth cohort (born 1959–1966) to identify cases and controls and accessed stored maternal serum samples collected prospectively during pregnancy. They subsequently investigated these sera for the presence of antibodies to the relevant strains of the virus. A recent meta-analysis including this, and subsequent serological studies (Canetta et al. 2014; Ellman et al. 2009), did not support a significant effect of maternal influenza during pregnancy on schizophrenia risk in the offspring (Selten and Termorshuizen 2017).

Similar population-based approaches with nested case-control comparisons have been used in Denmark and Sweden to identify patients and comparison subjects along with prospectively collected blood samples from the neonatal period. Studies employing prospectively collected maternal serum samples during pregnancy with a nested case-control design based on large prospective cohort studies have, in addition to the United States, also been conducted in Finland. Antibodies of class G (IgG) are actively transported across the placenta during pregnancy to allow passive immunization of the newborn during the first months of life. Detection of IgG directed at specific infectious agents either in neonatal blood or in maternal serum samples allows researchers to determine maternal exposures to various agents at some time point (well) before sampling. These studies have mainly focused on the "usual suspects," i.e., infectious agents with affinity to the nervous system and the ability to establish chronic infections such as herpes viruses and *Toxoplasma gondii*. Using neonatal dried blood spots, schizophrenia or psychosis risk associated with maternal exposure to *T. gondii* has been reported from both Denmark (Mortensen et al. 2007) and Sweden (Blomstrom et al. 2012), whereas risk associated with maternal infections with herpes simplex type 2 virus (HSV-2) and CMV has been reported separately from Denmark (Mortensen et al. 2010) and Sweden (Blomstrom et al. 2012). Some, but not all, studies of maternal sera from the United States reported associations with HSV-2 (Buka et al. 2001, 2008; Brown et al. 2006), whereas risk for psychotic illness in the offspring has been consistently reported also for maternal exposure to *T. gondii* (Brown et al. 2005; Xiao et al. 2009). A more recent study from Finland failed to detect a significant association with HSV-2

(Cheslack-Postava et al. 2015). In summary, the current literature appear consistent with regard to the association between maternal *T. gondii* exposure and risk for psychotic illness in the offspring, but far less so with regard to the risk associated with herpesviruses.

Purely register-based studies examining potential association between maternal infections during pregnancy and the later development of schizophrenia and other non-affective psychoses in large populations have also been conducted. Such studies rely on prospective clinical ascertainments and registration of psychiatric diagnosis in regional or national health-care registers and on the clinical ascertainment/diagnosis of infections in clinical in- and/or outpatient settings during the time of exposure studied.

These kinds of register-based studies of maternal infections during pregnancy have been conducted in cohorts ranging in size from 8,000 to entire national populations with up to two million individuals and have explored maternal infections ranging from genital/reproductive infections (Babulas et al. 2006) and respiratory infections (Brown et al. 2000) to viral, bacterial, or any type of infection (Blomstrom et al. 2015; Nielsen et al. 2013; Sorensen et al. 2009). According to the larger population-based studies, maternal infections during pregnancy recorded in the in- and outpatient care system are rare. In the Swedish patient register, only slightly more than 1% of pregnant women are hospitalized for any type of infection during pregnancy, which likely results in a misclassification of the exposure. While a study in the Danish population reported a slight risk associated with any type of infection (RR 1.2, 95% CI 1.0–1.4), the effect appeared to be larger among mothers with a history of psychiatric disorders suggestive of an interaction between infections and genetic vulnerability for psychiatric diseases (Nielsen et al. 2013). Similar observations were made by Clarke et al. (2009) regarding pyelonephritis during pregnancy and also in another study of a large Swedish population of almost two million individuals (Blomstrom et al. 2015). In our study (Blomstrom et al. 2015), we observed no major risk of infections during pregnancy after taking potential confounding by parental psychiatric history and health-care seeking behaviors into account. Intriguingly, maternal infections during, but not before, pregnancy interacted significantly with maternal, but not paternal, psychiatric history. These observations suggest that additional risk for psychotic illness in the offspring is contributed by the intrauterine environment of women with psychiatric illness and infection. Suvisaari et al. (2013) made a similar observation in their Finnish high-risk cohort where the incidence of maternal infections during pregnancy was similar among mothers with schizophrenia spectrum disorders and comparison mothers but still appeared to significantly contribute to the development of schizophrenia spectrum disorders in the offspring to the affected mothers. In their study of the Mater University Study of Pregnancy pre-birth cohort, Betts et al. (2014) identified those who experienced psychotic symptoms by age 21. They studied the potential association with vaginal infections during pregnancy, based on maternal recall rather than a clinically ascertained diagnosis. They did not find a clear effect of such infections on subsequent psychotic symptoms in the offspring but did observe that maternal vaginal infections during pregnancy conferred increased susceptibility to childhood diseases (including infections) in the offspring, which were, in turn, associated with psychotic symptoms in the offspring. These investigators were not able to consider the influence of a family history of psychiatric illness on these associations. We have briefly examined the association between maternal infection during pregnancy and the occurrence of diagnosed infections during childhood in the offspring (Blomstrom et al. 2015). We indeed observed that maternal infections, both during and before pregnancy, increased the odds of childhood infections in the offspring. Interestingly, we also observed a significant interaction between maternal infections during, but not before, pregnancy and childhood infections in the risk for non-affective psychosis in the offspring.

In conclusion, maternal infections in general during pregnancy appear to be only weakly associated with schizophrenia or other psychotic illnesses in the offspring. Further studies are needed to understand the mechanisms underlying the risks associated with specific agents and the interaction between infections during pregnancy and maternal psychiatric disorders.

3 Epidemiological Studies of Infections During Childhood

As described in the introduction, the postnatal period, from the time of birth to young adulthood, or even up to the time of the appearance of the first psychotic symptoms, entails developmental processes, which, if disturbed or delayed, would arguably be relevant for studies aiming at identifying environmental factors contributing to the pathogenesis of schizophrenia. For example, cannabis use in this period appears to be a true risk factor for chronic psychotic disorders, including schizophrenia, and not only a consequence of premorbid behaviors determined by genetic liability to disease (Marconi et al. 2016).

For natural reasons, many of the published studies of childhood infections have focused on those involving the central nervous system (CNS). Rantakallio et al. (1997) explored the association between registered infections involving the CNS up to age 14 and subsequent risk for a registered diagnosis of schizophrenia up to age 27 in the 1966 Northern Finland birth cohort with >11,000 births. An OR of 4.8 (95% CI 1.6–14.0) of schizophrenia for those exposed to viral CNS infections was reported but based on only four exposed cases illustrating the rarity of such diagnoses in medical registers. A subsequent follow-up study of this cohort with additional later onset cases resulted in a weaker and no longer significant association between infections during childhood and schizophrenia (Koponen et al. 2004). Using data from the UK National Child Development Study, a prospective population-based cohort of >17,000 children born in 1958, Leask et al. (2002) reported a strong association (OR 7.8) between meningitis during childhood and schizophrenia with a very wide CI 1.0–59.0 due to the detection of only a single exposed case. To examine rare exposures properly, we explored the Swedish population born 1973–1985 (>1.1 million individuals) followed until the end of 2002 with regard to the development of psychotic illness (Dalman et al. 2008). Registered diagnoses involving CNS infections before age 12 were rare in this population (<0.8%) with the majority involving viruses. A weak association between viral CNS infections and non-affective psychoses was observed (OR 1.3, 95% CI 0.8–2.0) following adjustment for differences in sex, age, urban living, season of birth, and parental psychotic disorders (Dalman et al. 2008). In their subsequent report, Weiser et al. (2010) did not detect any significant association between meningitis infection up to age 16 and later hospitalization for schizophrenia using the Israeli National Psychiatric Hospitalization Registry. In their case-control design, they however used children with a registered diagnosis of gastroenteritis as a comparison group. Whether this group may also have been at increased risk for developing schizophrenia in comparison to children never hospitalized for infections during childhood was not investigated. A meta-analysis published in 2012 concluded that viral CNS infections during childhood confer increased risk of adult psychotic illness and that mechanisms may include both direct effects of pathogens and the effects of inflammatory response on the developing brain (Khandaker et al. 2012).

To address whether infections in general, and not only CNS infections, during childhood confer schizophrenia or non-affective psychosis risk, Liang and Chikritzhs (2012) investigated the potential association between all registered infections during the first 3 years of life and subsequent schizophrenia among males born 1980–1984 in Western Australia (>51,000 individuals). They reported a significant association with two or more hospitalizations for any infections or with one hospitalization for intestinal or respiratory infections (Liang and Chikritzhs 2012). At the time, two large population-based register studies were conducted in Denmark and Sweden. In our Swedish study, a large number of potential confounders such as male sex, birth in an urban environment, parental migration, parental age at birth of the child, parental psychiatric illness, parental socioeconomic status, and inpatient care during childhood for reasons other than psychiatric or infectious disease were identified and taken into account (Blomstrom et al. 2013). A weak risk associated with any type of infection during the period between birth and age 13 was observed, HR 1.10 (95% CI 1.03–1.18). The fairly narrow confidence interval reflects the fact that 1,114 exposed cases were identified in this the largest population studied to date. A somewhat stronger association between any type of infection and schizophrenia was observed in the Danish population (843,390 individuals born 1981–1996) after adjustments for essentially the same covariates as included in the Swedish study except parental socioeconomic status and hospitalization for "other" reasons, RR 1.41 (95% CI 1.32–1.51) (Nielsen et al. 2014). None of these studies observed an overall significant risk for psychotic illness associated with CNS infections, perhaps due to a lack of power despite the considerable sizes of the two study populations.

Population-based serological studies of specific infections among individuals who will later develop schizophrenia or other non-affective disorders have not been performed due to a general lack of prospectively collected samples during childhood. Khandaker et al. however investigated the potential association between serological evidence of Epstein-Barr virus (EBV) infections (by age 4) and psychotic experiences (by age 13 and reported by 15% of participants) in 400 individuals sampled from the longitudinal ALSPAC birth cohort (Khandaker et al. 2014). EBV is a member of the herpesvirus family that causes usually asymptomatic infection among young individuals after which the virus establishes a latent state and lifelong persistence. These investigators reported that those infected with EBV were more likely to experience psychotic symptoms than the unexposed comparison group (Khandaker et al. 2014).

3.1 Childhood Infections and Development of Cognitive Abilities in Psychotic Illness

Low premorbid cognitive ability is a well-established risk factor for schizophrenia and other non-affective psychoses with an estimated 3.7% risk increase in schizophrenia risk for every point decrease in IQ (Khandaker et al. 2011). With regard to cognitive abilities, hospitalizations for infections have been associated with slightly poorer performance on scales measuring childhood emotional and cognitive development (Kariuki et al. 2016) as well as on cognitive tests at age 18 (Benros et al. 2015) in large population-based studies. In light of recent reports indicating that the low premorbid cognitive function observed in schizophrenia does not appear to be fully explained by shared familial factors (Kendler et al. 2015, 2016), we recently investigated the potential association between registered hospitalizations for infections during childhood, IQ at age 18, and the later development of schizophrenia and other non-affective psychoses in approximately 650,000 Swedish males (Khandaker et al. 2018). We observed that infections before age 5, but not later, were associated with slight but significant reductions in IQ at age 18, see Fig. 1, and with increased risk for the later development of non-affective psychoses. These relations were similar between individuals in the general population and in a comparison between full siblings suggesting that the associations were not fully explained by shared familial factors. IQ appeared to both mediate and moderate the effects of early childhood infections (Khandaker et al. 2018). This study thus suggests that early childhood infection may increase the risk of non-affective psychosis, not only by interfering with cognitive development but also by exaggerating the effects of cognitive vulnerability to psychosis. Both CNS infections and non-CNS infections were associated with cognitive deficits at age 18, in agreement with observations in the previous Danish study (Benros et al. 2015). This study included only males conscripted by the Swedish military and can thus not be generalized to females and individuals not eligible for mandatory screening by the Swedish military. The potential genetic confounding of the association between prior infections and schizophrenia was recently directly addressed in a case-control study nested in the population born since 1981 in Denmark (Benros et al. 2016). Benros et al. reported that polygenetic risk for schizophrenia and infections both conferred risk for schizophrenia, independent of each other suggesting that common genetic variation associated with schizophrenia risk is not explaining the association between infections and schizophrenia (Benros et al. 2016).

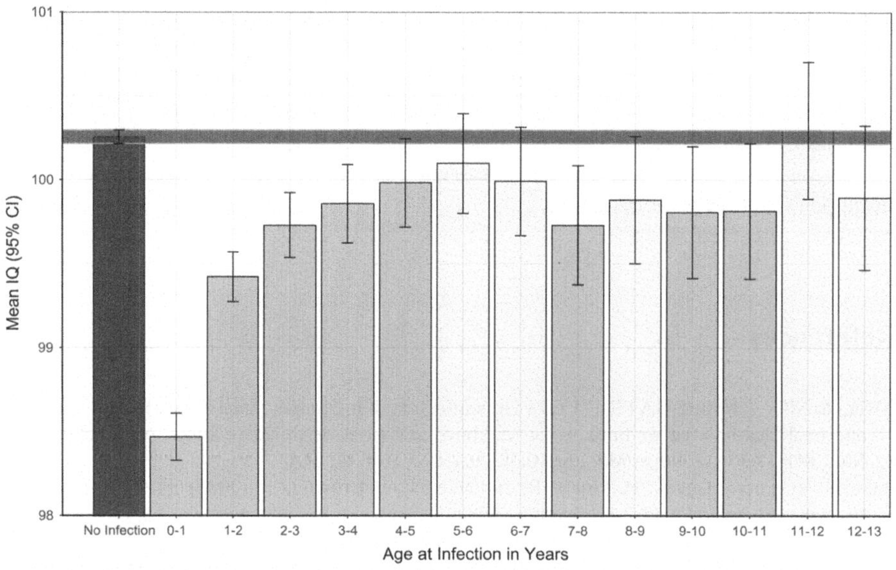

Fig. 1 Mean IQ (95% CI) at conscription for participants exposed to infection in childhood grouped by age at infection. The black bar indicates mean IQ for the unexposed group (i.e., no infection at any age). The gray bars indicate mean IQ for participants exposed to infection grouped by age at infection. Dark gray bars indicate a statistically significant difference in mean IQ for exposure to infection in that particular age compared with unexposed group. Reprinted from Khandaker et al. (2018)

In conclusion, studies of hospitalization for a wide range of infections after birth appear to be consistently associated with a later diagnosis of schizophrenia. Contrary to earlier reports, more recent studies indicate that risk does not appear to be limited to infections targeting the central nervous system. Recent studies are also beginning to address the important issue of familial or genetic confounding of these associations and thus far indicate that the risk conferred by infections is independent of both familial and genetic risk for schizophrenia.

4 Final Remarks

While family and twin studies clearly indicate a high heritability of schizophrenia, they also support important roles for shared and non-shared environmental factors (Lichtenstein et al. 2009; Sullivan et al. 2003). To move forward with regard to the risk for schizophrenia associated with maternal infections during pregnancy, we need to conduct large studies with objective measures of genetic risk among parents and comprehensive assessments of acute infectious exposures occurring during pregnancy. Moreover, we need to further understand the potential role played by chronic maternal infections, particularly those infections established before pregnancy that can be reactivated during pregnancy. Recent studies also indicate that the role of infections occurring after birth needs to be further investigated. The studies suggesting that the reported associations between postnatal infections and schizophrenia are not fully explained by familial or genetic confounding tentatively suggest that infections can in fact be involved in the etiology of schizophrenia and other non-affective psychoses. We, however, need a far better understanding of "when, what, and who" before these observations will be useful to devise preventive strategies for psychotic disorders.

References

Alvarado MG, Schwartz DA (2017) Zika virus infection in pregnancy, microcephaly, and maternal and fetal health: what we think, what we know, and what we think we know. Arch Pathol Lab Med 141(1):26–32. https://doi.org/10.5858/arpa.2016-0382-RA

Babulas V, Factor-Litvak P, Goetz R, Schaefer CA, Brown AS (2006) Prenatal exposure to maternal genital and reproductive infections and adult schizophrenia. Am J Psychiatry 163(5):927–929. https://doi.org/10.1176/appi.ajp.163.5.927

Bale JF Jr (2009) Fetal infections and brain development. Clin Perinatol 36(3):639–653. https://doi.org/10.1016/j.clp.2009.06.005

Benros ME, Sorensen HJ, Nielsen PR, Nordentoft M, Mortensen PB, Petersen L (2015) The association between infections and general cognitive ability in young men – a nationwide study. PLoS One 10(5):e0124005. https://doi.org/10.1371/journal.pone.0124005

Benros ME, Trabjerg BB, Meier S, Mattheisen M, Mortensen PB, Mors O et al (2016) Influence of polygenic risk scores on the association between infections and schizophrenia. Biol Psychiatry 80(8):609–616. https://doi.org/10.1016/j.biopsych.2016.04.008

Betts KS, Williams GM, Najman JM, Scott J, Alati R (2014) Maternal prenatal infection, early susceptibility to illness and adult psychotic experiences: a birth cohort study. Schizophr Res 156(2–3):161–167. https://doi.org/10.1016/j.schres.2014.04.013

Blomstrom A, Karlsson H, Wicks S, Yang S, Yolken RH, Dalman C (2012) Maternal antibodies to infectious agents and risk for non-affective psychoses in the offspring – a matched case-control study. Schizophr Res 140(1–3):25–30. https://doi.org/10.1016/j.schres.2012.06.035S0920-9964(12)00352-0

Blomstrom A, Karlsson H, Svensson A, Frisell T, Lee BK, Dal H et al (2013) Hospital admission with infection during childhood and risk for psychotic illness – a population-based cohort study. Schizophr Bull 40:1518–1525. https://doi.org/10.1093/schbul/sbt195

Blomstrom A, Karlsson H, Gardner R, Jorgensen L, Magnusson C, Dalman C (2015) Associations between maternal infection during pregnancy, childhood infections and the risk of subsequent psychotic disorder – a Swedish cohort study of nearly 2 million individuals. Schizophr Bull: sbv112. https://doi.org/10.1093/schbul/sbv112

Brown AS, Schaefer CA, Wyatt RJ, Goetz R, Begg MD, Gorman JM et al (2000) Maternal exposure to respiratory infections and adult schizophrenia spectrum disorders: a prospective birth cohort study. Schizophr Bull 26(2):287–295. http://www.ncbi.nlm.nih.gov/pubmed/10885631

Brown AS, Begg MD, Gravenstein S, Schaefer CA, Wyatt RJ, Bresnahan M et al (2004) Serologic evidence of prenatal influenza in the etiology of schizophrenia. Arch Gen Psychiatry 61 (8):774–780. http://www.ncbi.nlm.nih.gov/entrez/query.fcgi?cmd=Retrieve&db=PubMed&dopt=Citation&list_uids=15289276

Brown AS, Schaefer CA, Quesenberry CP Jr, Liu L, Babulas VP, Susser ES (2005) Maternal exposure to toxoplasmosis and risk of schizophrenia in adult offspring. Am J Psychiatry 162 (4):767–773. http://www.ncbi.nlm.nih.gov/entrez/query.fcgi?cmd=Retrieve&db=PubMed&dopt=Citation&list_uids=15800151

Brown AS, Schaefer CA, Quesenberry CP Jr, Shen L, Susser ES (2006) No evidence of relation between maternal exposure to herpes simplex virus type 2 and risk of schizophrenia? Am J Psychiatry 163(12):2178–2180. http://www.ncbi.nlm.nih.gov/entrez/query.fcgi?cmd=Retrieve&db=PubMed&dopt=Citation&list_uids=17151171

Buka SL, Tsuang MT, Torrey EF, Klebanoff MA, Bernstein D, Yolken RH (2001) Maternal infections and subsequent psychosis among offspring. Arch Gen Psychiatry 58(11):1032–1037. http://www.ncbi.nlm.nih.gov/entrez/query.fcgi?cmd=Retrieve&db=PubMed&dopt=Citation&list_uids=11695949

Buka SL, Cannon TD, Torrey EF, Yolken RH (2008) Maternal exposure to herpes simplex virus and risk of psychosis among adult offspring. Biol Psychiatry 63(8):809–815. http://www.ncbi.nlm.nih.gov/entrez/query.fcgi?cmd=Retrieve&db=PubMed&dopt=Citation&list_uids=17981263

Canetta SE, Bao Y, Co MD, Ennis FA, Cruz J, Terajima M et al (2014) Serological documentation of maternal influenza exposure and bipolar disorder in adult offspring. Am J Psychiatry 171:557. https://doi.org/10.1176/appi.ajp.2013.130709431827767

Cheslack-Postava K, Brown AS, Chudal R, Suominen A, Huttunen J, Surcel HM et al (2015) Maternal exposure to sexually transmitted infections and schizophrenia among offspring. Schizophr Res 166(1–3):255–260. https://doi.org/10.1016/j.schres.2015.05.012

Clarke MC, Tanskanen A, Huttunen M, Whittaker JC, Cannon M (2009) Evidence for an interaction between familial liability and prenatal exposure to infection in the causation of schizophrenia. Am J Psychiatry 166(9):1025–1030. https://doi.org/10.1176/appi.ajp.2009.08010031

Dalman C, Allebeck P, Gunnell D, Harrison G, Kristensson K, Lewis G et al (2008) Infections in the CNS during childhood and the risk of subsequent psychotic illness: a cohort study of more than one million Swedish subjects. Am J Psychiatry 165(1):59–65. http://www.ncbi.nlm.nih.gov/entrez/query.fcgi?cmd=Retrieve&db=PubMed&dopt=Citation&list_uids=18056223

Dendrou CA, Petersen J, Rossjohn J, Fugger L (2018) HLA variation and disease. Nat Rev Immunol 18:325. https://doi.org/10.1038/nri.2017.143

Ellman LM, Yolken RH, Buka SL, Torrey EF, Cannon TD (2009) Cognitive functioning prior to the onset of psychosis: the role of fetal exposure to serologically determined influenza infection. Biol Psychiatry 65:1040. http://www.ncbi.nlm.nih.gov/entrez/query.fcgi?cmd=Retrieve&db=PubMed&dopt=Citation&list_uids=19195645

Houston SM, Herting MM, Sowell ER (2014) The neurobiology of childhood structural brain development: conception through adulthood. Curr Top Behav Neurosci 16:3–17. https://doi.org/10.1007/7854_2013_265

Kariuki M, Raudino A, Green MJ, Laurens KR, Dean K, Brinkman SA et al (2016) Hospital admission for infection during early childhood influences developmental vulnerabilities at age 5 years. J Paediatr Child Health 52(9):882–888. https://doi.org/10.1111/jpc.13239

Kendler KS, Ohlsson H, Sundquist J, Sundquist K (2015) IQ and schizophrenia in a Swedish national sample: their causal relationship and the interaction of IQ with genetic risk. Am J Psychiatry 172(3):259–265. https://doi.org/10.1176/appi.ajp.2014.14040516

Kendler KS, Ohlsson H, Mezuk B, Sundquist JO, Sundquist K (2016) Observed cognitive performance and deviation from familial cognitive aptitude at age 16 years and ages 18 to 20 years and risk for schizophrenia and bipolar illness in a Swedish national sample. JAMA Psychiat 73(5):465–471. https://doi.org/10.1001/jamapsychiatry.2016.0053

Khandaker GM, Barnett JH, White IR, Jones PB (2011) A quantitative meta-analysis of population-based studies of premorbid intelligence and schizophrenia. Schizophr Res 132(2–3):220–227. https://doi.org/10.1016/j.schres.2011.06.017

Khandaker GM, Zimbron J, Dalman C, Lewis G, Jones PB (2012) Childhood infection and adult schizophrenia: a meta-analysis of population-based studies. Schizophr Res 139(1–3):161–168. https://doi.org/10.1016/j.schres.2012.05.023

Khandaker GM, Stochl J, Zammit S, Lewis G, Jones PB (2014) Childhood Epstein-Barr Virus infection and subsequent risk of psychotic experiences in adolescence: a population-based prospective serological study. Schizophr Res 158(1–3):19–24. https://doi.org/10.1016/j.schres.2014.05.019

Khandaker GM, Dalman C, Kappelmann N, Stochl J, Dal H, Kosidou K et al (2018) Association of childhood infection with IQ and adult nonaffective psychosis in Swedish men: a population-based longitudinal cohort and co-relative study. JAMA Psychiatry 75:356. https://doi.org/10.1001/jamapsychiatry.2017.4491

Koponen H, Rantakallio P, Veijola J, Jones P, Jokelainen J, Isohanni M (2004) Childhood central nervous system infections and risk for schizophrenia. Eur Arch Psychiatry Clin Neurosci 254(1):9–13. https://doi.org/10.1007/s00406-004-0485-2

Leask SJ, Done DJ, Crow TJ (2002) Adult psychosis, common childhood infections and neurological soft signs in a national birth cohort. Br J Psychiatry 181:387–392. http://www.ncbi.nlm.nih.gov/pubmed/12411263

Liang W, Chikritzhs T (2012) Early childhood infections and risk of schizophrenia. Psychiatry Res 200(2–3):214–217. https://doi.org/10.1016/j.psychres.2012.06.007

Lichtenstein P, Yip BH, Bjork C, Pawitan Y, Cannon TD, Sullivan PF et al (2009) Common genetic determinants of schizophrenia and bipolar disorder in Swedish families: a population-based study. Lancet 373(9659):234–239. http://www.ncbi.nlm.nih.gov/entrez/query.fcgi?cmd=Retrieve&db=PubMed&dopt=Citation&list_uids=19150704

Marconi A, Di Forti M, Lewis CM, Murray RM, Vassos E (2016) Meta-analysis of the association between the level of Cannabis use and risk of psychosis. Schizophr Bull 42(5):1262–1269. https://doi.org/10.1093/schbul/sbw003

McLaren PJ, Fellay J, Telenti A (2013) European genetic diversity and susceptibility to pathogens. Hum Hered 76(3–4):187–193. https://doi.org/10.1159/000357758

Mednick SA, Machon RA, Huttunen MO, Bonett D (1988) Adult schizophrenia following prenatal exposure to an influenza epidemic. Arch Gen Psychiatry 45(2):189–192

Mortensen PB, Norgaard-Pedersen B, Waltoft BL, Sorensen TL, Hougaard D, Torrey EF et al (2007) Toxoplasma gondii as a risk factor for early-onset schizophrenia: analysis of filter paper blood samples obtained at birth. Biol Psychiatry 61(5):688–693. http://www.ncbi.nlm.nih.gov/entrez/query.fcgi?cmd=Retrieve&db=PubMed&dopt=Citation&list_uids=16920078

Mortensen PB, Pedersen CB, Hougaard DM, Norgaard-Petersen B, Mors O, Borglum AD et al (2010) A Danish National Birth Cohort study of maternal HSV-2 antibodies as a risk factor for schizophrenia in their offspring. Schizophr Res 122(1–3):257–263. https://doi.org/10.1016/j.schres.2010.06.010

Munk-Jorgensen P, Ewald H (2001) Epidemiology in neurobiological research: exemplified by the influenza-schizophrenia theory. Br J Psychiatry 178(Suppl 40):S30–S32. http://www.ncbi.nlm.nih.gov/cgi-bin/Entrez/referer?http://bjp.rcpsych.org/cgi/content/abstract/178/40/s30

Nielsen PR, Laursen TM, Mortensen PB (2013) Association between parental hospital-treated infection and the risk of schizophrenia in adolescence and early adulthood. Schizophr Bull 39(1):230–237. https://doi.org/10.1093/schbul/sbr149sbr149

Nielsen PR, Benros ME, Mortensen PB (2014) Hospital contacts with infection and risk of schizophrenia: a population-based cohort study with linkage of Danish national registers. Schizophr Bull 40(6):1526–1532. https://doi.org/10.1093/schbul/sbt200

Petrova VN, Russell CA (2018) The evolution of seasonal influenza viruses. Nat Rev Microbiol 16(1):47–60. https://doi.org/10.1038/nrmicro.2017.118

Prendergast AJ, Klenerman P, Goulder PJ (2012) The impact of differential antiviral immunity in children and adults. Nat Rev Immunol 12(9):636–648. https://doi.org/10.1038/nri3277

Rantakallio P, Jones P, Moring J, Von Wendt L (1997) Association between central nervous system infections during childhood and adult onset schizophrenia and other psychoses: a 28-year follow-up. Int J Epidemiol 26(4):837–843

Selten JP, Termorshuizen F (2017) The serological evidence for maternal influenza as risk factor for psychosis in offspring is insufficient: critical review and meta-analysis. Schizophr Res 183:2–9. https://doi.org/10.1016/j.schres.2016.11.006

Selten JP, Frissen A, Lensvelt-Mulders G, Morgan VA (2010) Schizophrenia and 1957 pandemic of influenza: meta-analysis. Schizophr Bull 36(2):219–228. https://doi.org/10.1093/schbul/sbp147

Semple BD, Blomgren K, Gimlin K, Ferriero DM, Noble-Haeusslein LJ (2013) Brain development in rodents and humans: identifying benchmarks of maturation and vulnerability to injury across species. Prog Neurobiol 106-107:1–16. https://doi.org/10.1016/j.pneurobio.2013.04.001

Sorensen HJ, Mortensen EL, Reinisch JM, Mednick SA (2009) Association between prenatal exposure to bacterial infection and risk of schizophrenia. Schizophr Bull 35(3):631–637. https://doi.org/10.1093/schbul/sbn121

Sullivan PF, Kendler KS, Neale MC (2003) Schizophrenia as a complex trait: evidence from a meta-analysis of twin studies. Arch Gen Psychiatry 60(12):1187–1192. https://doi.org/10.1001/archpsyc.60.12.1187

Suvisaari JM, Taxell-Lassas V, Pankakoski M, Haukka JK, Lonnqvist JK, Hakkinen LT (2013) Obstetric complications as risk factors for schizophrenia spectrum psychoses in offspring of mothers with psychotic disorder. Schizophr Bull 39(5):1056–1066. https://doi.org/10.1093/schbul/sbs109

Tecle T, Johansson B, Jejcic A, Forsgren M, Orvell C (1998) Characterization of three co-circulating genotypes of the small hydrophobic protein gene of mumps virus. J Gen Virol 79(Pt 12):2929–2937. https://doi.org/10.1099/0022-1317-79-12-2929

Ward AC (1996) Neurovirulence of influenza A virus. J Neurovirol 2(3):139–151. http://www.ncbi.nlm.nih.gov/pubmed/8799206

Weiser M, Werbeloff N, Levine A, Livni G, Schreiber S, Halperin D et al (2010) CNS infection in childhood does not confer risk for later schizophrenia: a case-control study. Schizophr Res 124(1–3):231–235. https://doi.org/10.1016/j.schres.2010.08.025

Xiao J, Buka SL, Cannon TD, Suzuki Y, Viscidi RP, Torrey EF et al (2009) Serological pattern consistent with infection with type I Toxoplasma gondii in mothers and risk of psychosis among adult offspring. Microbes Infect 11(13):1011–1018. http://www.ncbi.nlm.nih.gov/entrez/query.fcgi?cmd=Retrieve&db=PubMed&dopt=Citation&list_uids=19638313

Cytokines, Oxidative Stress and Cellular Markers of Inflammation in Schizophrenia

Rachel Upthegrove and Golam M. Khandaker

Contents

Abstract In this article, we review current evidence linking immune dysfunction in schizophrenia and related psychotic disorders focusing particularly on circulating cytokines, oxidative stress and cellular markers of inflammation in various stages on illness from drug-naïve first episode psychosis to chronic schizophrenia. Acute psychotic episode is associated with low-grade systemic inflammation in some patients, as reflected by increased concentrations of cytokines and other inflammatory markers in peripheral blood. Evidence from general population-based

R. Upthegrove (✉)
Institute for Mental Health, University of Birmingham, Birmingham, UK

Birmingham Early Intervention Service, Birmingham Women's and Children's NHS Trust, Birmingham, UK
e-mail: r.upthegrove@bham.ac.uk

G. M. Khandaker
Department of Psychiatry, University of Cambridge, Cambridge, UK

Cambridgeshire and Peterborough NHS Foundation Trust, Cambridge, UK

Curr Topics Behav Neurosci (2020) 44: 49–66
DOI 10.1007/7854_2018_88
Published Online: 22 May 2019

longitudinal cohort studies reporting an association between elevated inflammatory markers in childhood/adolescence and risk of schizophrenia and related psychosis subsequently in adulthood suggest that inflammation could be a causal risk factor for psychosis rather than simply be a consequence of illness. Mendelian randomization studies also suggest that associations between IL-6, CRP and schizophrenia are likely to be causal. In addition, we discuss evidence for disruptions in oxidative stress markers and CSF cytokine levels in schizophrenia, and potential reasons for reported trans-diagnostic associations for inflammatory cytokines including role of early-life adversity/maltreatment. We argue that low-grade inflammation is a clinically useful feature, because it is associated with poor response to antipsychotic medication in first episode psychosis. We discuss clinical implications for immunological understanding of schizophrenia including scope for clinical trials of anti-inflammatory agents and notable gaps in current knowledge, and offer suggestions for future research.

Keywords Cytokine · Inflammation · Innate immunity · Schizophrenia · Psychotic disorder

Abbreviations

ALSPAC	Avon Longitudinal Study of Parents and Children
BBB	Blood-brain barrier
CNS	Central nervous system
CRP	C-reactive protein
CSF	Cerebrospinal fluid
FEP	First episode psychosis
Glx	Glutamate/glutamine ratio
GSH	Glutathione
HPA	Hypothalamic-pituitary-adrenal
IL-6	Interleukin-6
IL-8	Interleukin-8
MHC	Major histocompatibility complex
ROS	Reactive oxygen series
TNF-α	Tumour necrosis factor alpha

1 Introduction

Schizophrenia can be understood as a neurodevelopmental disorder (Murray and Lewis 1987; Weinberger 1987), with onset usually in early adulthood (van Os et al. 2009). Biological research into the pathogenesis of schizophrenia has focused on brain structure, function and neurotransmitter abnormalities (Jones et al. 2005) and genetic risk (Consortium 2011). However, there is also an accepted environmental impact, with a gene and environmental (GxE) combined effect related to increased risk, precipitation of illness and/or poorer outcomes. Notable environmental factors include childhood trauma, social and economic deprivation, minority status and stressful life events (van Os et al. 2010). Increasing evidence suggests a role for the involvement of immunological processes in mediating the genetic and environmental risk for schizophrenia. Indeed, schizophrenia has been associated with an abnormal activation of the immune system for many years (Dameshek 1930; Müller et al. 2012; Ganguli et al. 1994). Previous reviews have summarized evidence linking schizophrenia with abnormalities in various components of the immune system; see Khandaker et al. (2015) and Khandaker and Dantzer (2016), but here, we focus primarily on the evidence on circulating inflammatory cytokines, oxidative stress and cellular markers of inflammation.

2 Cytokines as Key Mediators of Immune Response

Cytokines are the key signalling molecules that coordinate both innate and adaptive arms of the immune system and exert effects in the periphery and the brain. The immune response is a highly coordinated process involving an array of cell types that protect the body from harm while maintaining tolerance to self-antigens and beneficial organisms. The first arm is our "innate" defence mechanisms, older in evolutionary terms, and considered to be a first-line defence. Its cellular components include neutrophils, basophils and eosinophils, monocytes and macrophages, dendritic cells and natural killer (NK) cells, which recognize and promote defence against pathogens but lack the sophistication to adapt compared to other more recent additions to the immune system (Upthegrove and Barnes 2014). The innate humoral component is made up of acute phase proteins such as C-reactive protein (CRP), cytokines such as interleukin-6 (IL-6) and the compliment cascade, which allow phagocytic cells to clear pathogens (see below).

The second arm of our immune system is the "adaptive" system, which acts on re-exposure to a known pathogen. The prime cellular components of the adaptive system include T and B lymphocytes. Antibodies produced by B lymphocytes comprise the main humoral part of adaptive immunity. There is considerable "crosstalk" between the two major arms of the immune system. T cells comprise key components of the T helper 1 (Th1) system and the T helper 2 (Th2) system. The Th1 system is polarized towards the production of pro-inflammatory (activating) cytokines such as interleukin-2 (IL-2), interferon-γ (INF-γ) and tumour necrosis factor (TNFα). The Th2 system promotes the generation and maintenance of antibody-mediated immune responses as well as production of anti-inflammatory cytokines such as interleukin-4 (IL-4), interleukin-10 (IL-10) and interleukin-13 (IL-13). However, cytokines often have pleotropic effect, as, for example, IL-6 has both pro- and anti-inflammatory properties. More recently, a key role for the Th17 system has been discovered, in regulation of immune response, so this system is important for pathogenesis of a number of immune-related disorders

(Janeway et al. 2001). It is now understood that both innate and adaptive systems are able to form and retain memory, which influences immune response on re-exposure to a stimulus.

Changes in cytokines and their receptor levels have been reported in the blood and cerebrospinal fluid (CSF) of patients with schizophrenia (Upthegrove et al. 2014) – see below. Previously, the brain was thought to be protected from peripheral inflammatory responses due to the blood-brain barrier (BBB). However, it is now clear that cytokines and other circulating inflammatory mediators can reach and influence the brain in a number of ways. Peripheral immune-to-brain communication pathways include direct entry through leaky circumventricular areas in the BBB, the lymphatic system, infiltration of immune cells to the brain and retrograde axonal transport of immune signal via cranial nerves, for example, the vagus nerve (for a review, see Khandaker and Dantzer 2016). Peripheral or systemic inflammation is therefore relevant for neuropsychiatric disorders such as schizophrenia.

3 Evidence for Cytokine Alteration in Different Stages of Schizophrenia

Here, we review the evidence for aberrant cytokines in peripheral blood and in CSF in patients with schizophrenia and related psychotic disorders. We also discuss evidence from epidemiological prospective cohort studies linking elevated inflammatory marker levels in childhood/adolescence with risk of psychosis subsequently in adulthood. There is notable heterogeneity in existing studies of inflammatory cytokines in schizophrenia as, for example, studies have used patients in different stages of illness. Levels of inflammatory markers could be influenced by neuroleptic and other drugs, alcohol and illicit drug use, sex, smoking, body mass index (BMI) and comorbid physical illness (Upthegrove and Barnes 2014). Therefore, we have put particular emphasis on stage of illness and effects of potential confounders such as those listed above.

3.1 Drug Naïve Psychosis

Studies of medication naive patients are particularly useful to gain a better understanding of inflammatory cytokine alteration in schizophrenia. It is well known that antipsychotic medication can influence the immune system. Drzyga and colleagues carried out an in vitro study showing that antipsychotic drugs affect immune cell function, which often occurs very shortly after initial exposure to drug (Drzyzga et al. 2006). However, there are mixed results and differing effects, including either stimulatory or inhibitory actions. Relatedly, other in vitro studies suggest that suppression of cytokine mediated microglial activity may partly underpin the efficacy of some antipsychotic drugs (Bian et al. 2008). For example, aripiprazole suppresses apoptosis of rodent oligodendrocytes by IFN-γ-activated microglia and inhibition of TNF-α secretion from IFN-γ-activated microglia (Seki et al. 2013). Clozapine, the most effective antipsychotic medication, influences the immune system. Its effects on white blood cell (WBC) count are well known. The drug may have immediate (Røge et al. 2012) and longer-term effects on IL-6, CRP (Kluge et al. 2009) and high-sensitivity CRP (hs-CRP) levels (Löffler et al. 2010) in schizophrenia patients.

In a systematic review and meta-analysis published in 2014, we included 14 studies that together assessed levels of 20 different cytokines and cytokine receptors in 570 neuroleptic naive patients. The majority of these patients had a diagnosis of schizophrenia or schizophreniform disorder (81%). Highly significant effect sizes were seen for IL-1β, IL-6, sIL2r and TNFα, suggesting that an increase in these cytokines in first episode psychosis (FEP) patients, compared with controls, is unrelated to antipsychotic drugs (Fig. 1). These cytokines play key roles in orchestrating innate immune response; IL-1β and TNF-α are responsible for stimulating IL-6 production, while IL-6 signals hepatocytes to produce acute phase proteins such as CRP. Some increase in levels of IL-2, IL-4 and IFN-γ were also seen, but differences in these cytokine levels were not statistically significant. These cytokines were measured in studies with small samples and in fewer studies altogether leading to low statistical power.

More recently, a study by Noto et al. reported that comorbid depression might influence cytokine levels in FEP patients. While increased levels of IL-6, IL-10 and TNFα were found in 55 FEP patients overall, compared with controls, patients with depression showed higher IL-4 and TNFα levels compared with those without depression (Noto et al. 2015).

3.2 Acute Psychosis

In an extensive meta-analysis published in 2011, Miller et al. explored cytokine function by phase of illness in schizophrenia. Assessing 40 studies, they found IL-1β, IL-6 and TGF-β were raised in the acute phase of illness (both in relapse patients and in first episode psychosis) and reduced with successful treatment (Miller et al. 2011). IL-6 correlated to total level of psychopathology in two out of five studies (Miller et al. 2011). TNF-α and IL-6 levels were analysed in most studies (97 and 156 total studies, respectively). It was proposed that these cytokines could be state-dependent markers of inflammation, resolving with symptom reduction. However, elevated levels of IL-6 in childhood measured years before onset of psychosis is associated with psychotic symptoms in early-adulthood (Khandaker et al. 2014) (see below). Levels of IL-6 and TNF-α are also associated with childhood maltreatment (Baumeister et al. 2016), so these cytokines could also be trait markers for psychosis.

In first episode psychosis, Mondelli et al. measured BDNF, IL-6 and TNF-α in 46 patients. Compared to healthy controls, patients had reduced BDNF gene expression and increased IL-6 and TNF-α. History of childhood trauma was

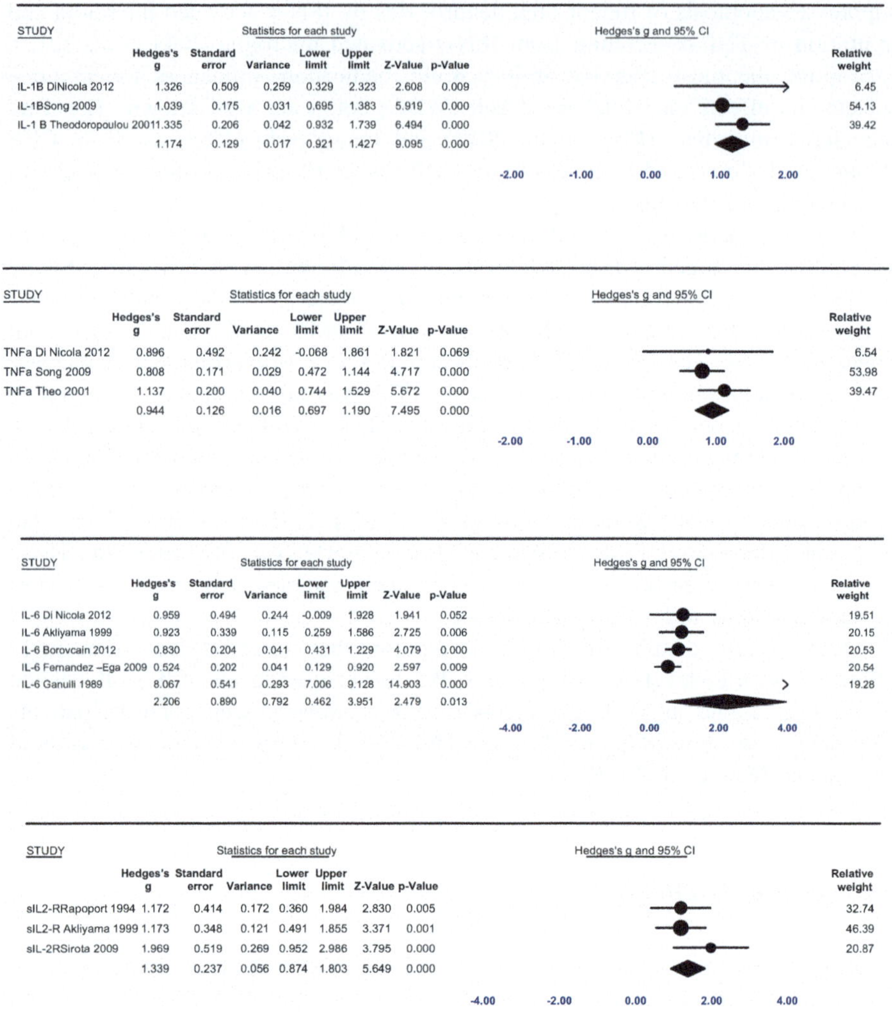

STUDY	Hedges's g	Standard error	Variance	Lower limit	Upper limit	Z-Value	p-Value	Relative weight
IL-1B DiNicola 2012	1.326	0.509	0.259	0.329	2.323	2.608	0.009	6.45
IL-1BSong 2009	1.039	0.175	0.031	0.695	1.383	5.919	0.000	54.13
IL-1 B Theodoropoulou 2001	1.335	0.206	0.042	0.932	1.739	6.494	0.000	39.42
	1.174	0.129	0.017	0.921	1.427	9.095	0.000	

STUDY	Hedges's g	Standard error	Variance	Lower limit	Upper limit	Z-Value	p-Value	Relative weight
TNFa Di Nicola 2012	0.896	0.492	0.242	-0.068	1.861	1.821	0.069	6.54
TNFa Song 2009	0.808	0.171	0.029	0.472	1.144	4.717	0.000	53.98
TNFa Theo 2001	1.137	0.200	0.040	0.744	1.529	5.672	0.000	39.47
	0.944	0.126	0.016	0.697	1.190	7.495	0.000	

STUDY	Hedges's g	Standard error	Variance	Lower limit	Upper limit	Z-Value	p-Value	Relative weight
IL-6 Di Nicola 2012	0.959	0.494	0.244	-0.009	1.928	1.941	0.052	19.51
IL-6 Akliyama 1999	0.923	0.339	0.115	0.259	1.586	2.725	0.006	20.15
IL-6 Borovcain 2012	0.830	0.204	0.041	0.431	1.229	4.079	0.000	20.53
IL-6 Fernandez –Ega 2009	0.524	0.202	0.041	0.129	0.920	2.597	0.009	20.54
IL-6 Ganuili 1989	8.067	0.541	0.293	7.006	9.128	14.903	0.000	19.28
	2.206	0.890	0.792	0.462	3.951	2.479	0.013	

STUDY	Hedges's g	Standard error	Variance	Lower limit	Upper limit	Z-Value	p-Value	Relative weight
sIL2-RRapoport 1994	1.172	0.414	0.172	0.360	1.984	2.830	0.005	32.74
sIL2-R Akliyama 1999	1.173	0.348	0.121	0.491	1.855	3.371	0.001	46.39
sIL-2RSirota 2009	1.969	0.519	0.269	0.952	2.986	3.795	0.000	20.87
	1.339	0.237	0.056	0.874	1.803	5.649	0.000	

Fig. 1 Cytokine profile in medication-naive first episode psychosis: patients have higher IL-1β, TNF-α, IL-6 and sIL-2r levels compared with controls. Modified from Upthegrove et al. (2014)

associated with lower BDNF mediated through IL-6 (Mondelli et al. 2015). In a more recent review, Goldsmith et al. investigated acute and chronic cytokine changes in schizophrenia, bipolar disorder and depression, which included 40 studies on acute schizophrenia (Goldsmith et al. 2016). In meta-analysis, IL-6, sIL2r, IL-1RA and TNFα were all significantly raised in acute schizophrenia, bipolar disorder and depression. There was more heterogeneity in FEP samples than acute relapse of established schizophrenia. No publication bias was reported for IL-6 (Goldsmith et al. 2016). These findings suggest that the association of increased inflammatory cytokines transcend traditional diagnostic boundaries; for a discussion on the trans-diagnostic effect of inflammation please, see below.

Another meta-analysis reported elevated serum levels of CRP in FEP and chronic schizophrenia irrespective of medication status (Fernandes et al. 2016). With regard to association with specific symptoms, an association between CRP levels and positive symptoms, but not negative symptoms of psychosis, was found. However, Johnsen et al. investigated CRP in acute psychosis, reporting a particular association with cognitive dysfunctions rather than positive symptoms (Johnsen et al. 2016). Studies in healthy volunteers and non-human primates have reported association of inflammatory markers with anhedonia-like behaviour and reward alterations (Harrison et al. 2016; Capuron et al. 2012). So overall, patient and animal studies indicate an association of inflammatory markers with positive, negative and cognitive symptoms.

3.3 Chronic Schizophrenia

TNF-α, IL-12, INF-γ and sIL2r have been reported to be elevated in both acute illness and stable "outpatients" with chronic schizophrenia. Goldsmith et al. found that, compared with controls, the levels of IL-6 were significantly increased in chronic schizophrenia, euthymic (but not depressed) bipolar disorder and major depressive disorder (Goldsmith et al. 2016). IL-1β and sIL2R were significantly increased in chronic schizophrenia and euthymic bipolar disorder. TNF-α has been suggested as a trait marker of neuroinflammation (Goldsmith et al. 2016). Goldsmith et al. review confirmed that TNF-α was raised in acute schizophrenia compared to controls and remained so after treatment.

In meta-analysis of five studies, Miller and Culpeper found that 28% of patients with chronic schizophrenia have an elevated CRP (Miller et al. 2013). In a recent study of 295 patients with schizophrenia and 192 with bipolar disorder, CRP was elevated in the schizophrenia even after adjusting for age, gender, race, maternal education, smoking status and BMI, but this was not found in bipolar disorder (Dickerson et al. 2013). The association between CRP levels and cognitive functioning in patients with predominantly chronic schizophrenia has been reported in one cross-sectional study (Dickerson et al. 2007). Together these studies suggest poorer cognitive function may be associated with enduring neuroinflammation.

4 Population-Based Longitudinal and Mendelian Randomization Studies Examining Causality

Psychological stress can activate an innate immune response (Maes et al. 1998), so cytokine elevation during acute psychosis could be a consequence of illness rather than be its cause (i.e., reverse causality). Therefore, longitudinal studies need to establish, or refute, a potentially causal role of inflammation in the pathogenesis of psychosis. Evidence from population-based longitudinal cohort studies from the UK, Finland and Sweden has linked higher levels of IL-6, CRP and erythrocyte sedimentation rate (ESR) in childhood/adolescence with risk of psychotic symptoms or diagnosis of schizophrenia subsequently in adulthood (Khandaker et al. 2014; Kappelmann et al. 2018a; Metcalf et al. 2017). Using data from the Avon Longitudinal Study of Parents and Children (ALSPAC), a general population-representative birth cohort, Khandaker et al. have reported that higher levels of IL-6 in childhood at age 9 years are associated with increased risk of psychotic symptoms at early adulthood at age 18 years in a linear dose-response fashion (Khandaker et al. 2014) (Fig. 2). Evidence for this association remained after controlling for sex, BMI, social class, ethnicity and childhood psychological and behavioural problems preceding the measurement of IL-6. Although, in ALSPAC, CRP was not associated with psychosis risk, Khandaker and colleagues found that in the Northern Finland Birth Cohort (NFBC) 1986, higher levels of CRP in adolescence were associated with increased risk of hospitalization with a diagnosis of schizophrenia subsequently in adulthood. Furthermore, there was evidence that higher CRP levels in adolescence were associated with earlier age at illness onset (Metcalf et al. 2017). These findings are consistent with a Danish study reporting that higher CRP at baseline is associated with increased risk of late- and very-late-onset schizophrenia subsequently at follow-up (Wium-Andersen et al. 2014). More recently, we have conducted a longitudinal study based on Swedish male conscripts which found that higher ESR, a marker of systemic inflammation, in early adulthood is associated with increased risk of schizophrenia subsequently in adulthood (Kappelmann et al. 2018a).

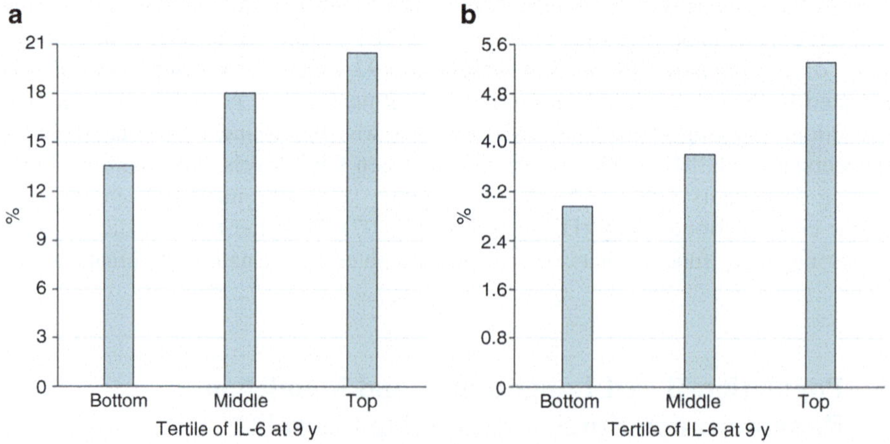

Fig. 2 Depression and psychotic experiences at age 18 years in the Avon Longitudinal Study of Parents and Children. Reproduced with permission from: Khandaker et al. (2014). Note: Samples of depression (**a**) and psychotic experiences (**b**) were divided by tertiles of interleukin-6 (IL-6) in participants at age 9 years. Cut-off values for the top and bottom thirds of the distribution of IL-6 values in the total sample (cases and noncases combined) were 1.08 and 0.57 pg/mL, respectively

Although previous longitudinal studies have controlled for key confounders such as sex, BMI, and social class, residual confounding still might explain the association between inflammatory markers and psychosis. Recently, Khandaker and colleagues have used genetic association analysis informed by Mendelian randomization (MR) which indicates that residual confounding is unlikely to explain the association between IL-6 and psychosis fully. MR is based on the idea that if a biomarker is causally related to an illness, genetic variant(s) regulating levels/activity of that biomarker should also be associated with the illness (Davey Smith and Ebrahim 2003). Using data from the ALSPAC birth cohort, we have shown that a genetic variant in the IL-6 receptor gene (*IL-6R* Asp358Ala; rs2228145) that is known to dampen down inflammation by impairing the activity of IL-6 is protective for severe depression and/or psychosis (Khandaker et al. 2018a). The genetic variant is strongly associated with serum IL-6 and CRP levels, but not with any common confounders of the inflammation-psychosis relationship such as sex, social class, ethnicity and BMI. Genetic variants segregate at random during meiosis and are unrelated to sociodemographic and other confounders. Therefore, an association between psychosis and a genetic variant that regulates IL-6 activity strongly indicates that the IL-6/IL-6R pathways are causally linked to psychosis. Similarly, using MR analysis of psychiatric genomic consortium (PGC) data, Hartwig et al. has reported that IL-6 and CRP are likely to be causally linked with schizophrenia (Hartwig et al. 2017).

Together these studies suggest that reverse causality and residual confounding are unlikely explanations for previously observed association between inflammatory markers, particularly IL-6, CRP and psychosis. These associations are likely to be causal, so these biomarkers could be novel targets for intervention and prevention of psychotic disorders.

5 CSF Cytokine Alteration in Schizophrenia

Meta-analysis of studies of CSF from patients with schizophrenia suggests that levels of inflammatory cytokines are increased in patients with psychosis. A meta-analysis of 16 studies published in 2018 by Wang et al. reported that CSF levels of IL-1β, IL-6 and IL-8 were significantly elevated in schizophrenia patients compared with controls (Wang and Miller 2018). Similar findings were also observed for depression. Whether levels of inflammatory markers in peripheral blood correspond with neuroinflammaton is an important question. This meta-analysis also reported that many CSF alterations are also concordant with those in the peripheral blood, particularly for schizophrenia (Wang and Miller 2018). This provides some validity to the use of peripheral markers of inflammation in schizophrenia research. This finding is consistent with a study by Coughlin et al. that measured IL-6 in the CSF of patients with recent-onset schizophrenia also undergoing a positron emission tomography (PET) study of translocator protein 18 kDa (TSPO), a marker of microglial activation using [^{11}C] DPA-713 (Coughlin et al. 2016). While non-significant results were seen in the TSPO analysis, IL-6 levels were significantly raised in patients compared to controls. Furthermore, the study found IL-6 levels in the CSF correlated significantly with circulating IL-6.

6 Oxidative Stress Markers in Acute and Chronic Schizophrenia

6.1 First Episode Psychosis

Decreased levels of the antioxidant, glutathione (GSH), have been reported in patients with acute psychosis compared with controls (Wood et al. 2009; Raffa et al. 2011). GSH is an antioxidant which provides within cell protection and when depleted contributes to oxidative stress. Flatlow, Buckly and Miller reported a decrease in GSH in acutely relapsed patients with schizophrenia together with other antioxidants such as superoxide dismutase (SOD) and catalase (CAT); however, there was significant heterogeneity in reported studies such that SOD and CAT but not GSH were reduced in FEP (Flatow et al. 2013). This may suggest a more persevered oxidative defence earlier in the course of illness.

As a substantial minority of patients with FEP do not go on to develop an enduring mental illness, it may be that those with a protective oxidative defence to cellular stress have better outcomes (Lally et al. 2017). This hypothesis remains to be tested. However Wang et al. investigated cysteine, a semi-essential amino acid and a precursor of GSH glutathione, as a potential indicator of preserved cognitive function in FEP with some positive findings (Wang et al. 2018). The functional consequences of increased oxidative stress or reduced defence against this oxidative stress in the brain are still to be fully understood. Changes in neuronal membrane permeability eventually leading to cell death could contribute to grey matter volume (GMV) loss seen in schizophrenia (Mahadik and Mukherjee 1996). In respect to known neurochemical changes in psychosis, it has been proposed that oxidative stress results in over activation of NMDA and altered dopamine receptor function. While schizophrenia is no longer understood within a simplistic conclusion of hyperdopaminergia, rather consisting of regionally specific prefrontal hypodopaminergia and subcortical hyperdopaminergia, the cumulative effect of overactive or more readily available dopamine may account for positive symptoms (Howes and Kapur 2009). NMDA glutamate receptors are downregulated by oxidation leading to disinhibition of pyramidal cells and unregulated glutamatergic excess (Traynelis et al. 2010).

6.2 Chronic Schizophrenia

There is substantial evidence of impaired oxidative defence in chronic schizophrenia, as reviewed by Flatlow et al. (2013). Negative symptoms have been associated with low levels of GSH, and positive symptoms have been positively correlated with SOD activity. A study by Fraguas et al. assessed the relationship between GMV and GSH in the brains of patients with schizophrenia, with a progressive decline in GMV correlated to declining circulating GSH (Fraguas et al. 2012). Thus, the deficit state of some patients with schizophrenia may be related to a specific lack of defence against oxidative stress.

There is a great deal of clinical and phenomenological commonality between schizophrenia and depression (Upthegrove et al. 2017). In schizophrenia, there is substantial glutamate dysfunction including clear evidence for Glx abnormality in brain areas such as the posterior medial prefrontal cortex (pmPFC). Increased Glx is seen in younger patients or more acute phases of illness, whereas reduced Glx has been reported in patients who are older and have residual negative symptoms (Marsman et al. 2013). Just as in schizophrenia, there is substantial evidence for increase in peripheral markers of inflammation in MDD. However, comorbid depression doesn't solely account for increased inflammation in schizophrenia, as evidence for association persists after excluding or controlling for depression (Khandaker et al. 2017). Nevertheless, it is possible that the presence of both psychosis and depression has a greater impact on immune dysfunction than the sum of each individually. Poorer clinical outcomes including increased risk of relapse, low quality of life and poor functional outcomes for patients with comorbid depression and schizophrenia support this idea (McGinty et al. 2018).

Early changes in neurochemicals such as Glx may result in more significant impact for those individuals with deficient defence against this inflammatory challenge (Bian et al. 2008). The effect of inflammation includes the generation of reactive oxygen species such as superoxide, hydroxyl and peroxyl. The substantial evidence of impaired oxidative defence in early psychosis with diminished levels of GSH (Wood et al. 2009; Raffa et al. 2011; Jiménez-Fernández et al. 2015) show that the response of brain glutamate to inflammation and associated oxidative stress will also depend on the strength of defences against it. Thus, clinically poorer outcomes, such as seen with some subjects with depression and schizophrenia, may be related not just to the effect of chronic inflammation but also impaired defence against this.

6.3 Trans-diagnostic Effect of Inflammation and Potential Role of Early-Life Adversity

Studies of inflammatory markers in peripheral blood and CSF suggest that inflammation is associated with a number of psychiatric disorders including schizophrenia and related psychoses (Khandaker et al. 2015; Upthegrove et al. 2014; Wang and Miller 2018; Miller et al. 2011), depression (Dantzer et al. 2008; Miller et al. 2009), anxiety (Wohleb et al. 2014), post-traumatic stress disorder (PTSD) (Eraly et al. 2014), autism (Brown et al. 2014), Alzheimer's disease and other dementias (Schmidt et al. 2002). However, possible reasons for this apparent trans-diagnostic association of inflammation are unknown. We have recently reported that the apparent trans-diagnostic effect may arise from association of inflammation with symptoms that are commonly shared between disorders (Khandaker et al. 2018b). Using a symptom-level data on ten positive and ten negative symptoms of psychosis assessed in adolescent participants from the ALSPAC birth cohort, we have shown at the group level positive and negative symptom dimension scores were associated with serum CRP levels in a similar fashion. At individual symptom level, CRP was associated with particularly auditory hallucinations and anhedonia. Auditory hallucinations can occur in psychosis, depression and anxiety disorders (APA 2013). Anhedonia is both an important negative symptom for psychosis and a core feature of depression (APA 2013).

Association between inflammation and anhedonia is supported by experimental studies. In non-human primates, chronic, low-dose peripheral interferon administration reduces striatal dopamine release in association with anhedonia-like behaviour (Felger et al. 2013). In healthy volunteers, inflammation induces hedonic alterations (decreased preference for reward and increased avoidance of punishment) (Harrison et al. 2016), which resemble anhedonia. Other reasons for this apparent trans-diagnostic effect could be shared genes that contribute to inflammation and risk of depression and schizophrenia. Genetic overlap between schizophrenia, bipolar disorder and depression is well established.

Shared risk factors, particularly early-life adversity, could be another explanation for the apparent trans-diagnostic effect of inflammation. Childhood abuse/maltreatment may programme the immune system leading with increased concentrations inflammatory markers in adulthood (Baumeister et al. 2016), which, in turn, may increase psychiatric risk. In the ALSPAC birth cohort, maternal/parental depression is associated with higher levels of IL-6 and CRP in childhood and with higher risk of depression and psychosis in early adulthood in offspring (Khandaker et al. 2018c). Furthermore, childhood IL-6 levels mediate the association between prenatal depression and offspring psychosis risk. These findings are consistent with the developmental programming hypothesis by David Barker, which posits that exposure to stress during critical period of development may program certain physiological system(s) leading to increased risk of chronic illnesses of adult life (Barker 1993). Early-life adversity is associated with coronary heart disease and type 2 diabetes, which are common comorbidities for schizophrenia and depression. Young adults with psychotic symptoms display evidence of dysglycaemia, which is linked with levels of IL-6 in childhood (Perry et al. 2018). Therefore, whether programming of innate immune response by early-life adversity may explain the comorbidity between schizophrenia, depression, coronary heart disease and type 2 diabetes is an interesting hypothesis that needs investigating.

6.4 *Therapeutic Implications for Low-Grade Inflammation in Schizophrenia*

Mondelli et al. have reported an association between innate immune activation and poorer treatment response in 57 patients with FEP: nonresponders (as defined by an absence of clinically significant symptom response in keep with remission criteria at 12 weeks) had a significantly higher IL-6 and INF-γ at baseline. They also reported an aberrant cortisol wakening response and suggest this combination of markers may be an early signal of poor outcome (Mondelli et al. 2015). Indeed, inflammatory mechanisms, as outlined above, have been cited as one of the potential mechanisms of effect of clozapine in treatment-resistant schizophrenia.

As well as poor treatment response, makers of inflammation may indicate poorer physical health outcomes. Russell et al. investigated 53 FEP patients and showed that FEP patients with raised CRP were at more risk of developing short-term metabolic abnormalities including dyslipidaemia, independent of weight gain (Russell et al. 2015). As mentioned earlier, in the ALSPAC birth cohort, young adults with psychotic symptoms displayed evidence of dysglycaemia, which was associated with childhood IL-6 levels (Perry et al. 2018). Thus, the potential for stratifying treatment approach; early targeting of potential treatment resistance or heightened monitoring from adverse effects of antipsychotics shows some promise in a personalized approach to FEP and schizophrenia.

Because inflammation is associated with BMI, smoking, alcohol use, physical comorbidity, antipsychotic treatment and treatment-induced weight gain, further work is needed to understand whether and how measuring inflammation in clinical setting could be useful for predicting response to antidepressant/antipsychotic treatment and for identifying patients who are likely to benefit from immunomodulatory treatments.

Inflammation is unlikely to be relevant for illness pathogenesis in all patients with psychosis. For depression, clinical trials indicate that anti-inflammatory drugs may be helpful for patients who show evidence of inflammation (Raison et al. 2013; Kappelmann et al. 2018b). Existing RCTs of anti-inflammatory drugs for schizophrenia have yielded mixed results (Deakin et al. 2018; Girgis et al. 2018; Miller et al. 2016), possibly due to imprecise targeting of patients. Patients with psychosis who do show evidence of inflammation may be more suitable candidate for RCTs of anti-inflammatory drugs in future.

7 Conclusions and Future Directions

For a number of years now, there have been considerable efforts to have a better understanding of the immunological and inflammatory aspects of schizophrenia, in the hope that this might lead to novel approaches to diagnosis and treatment. See Boxes 1 and 2 for key clinical findings and questions for future research. Some aspects are becoming clearer. For example, accumulating evidence now confirms that inflammation could play a causal role in psychosis rather than being an epiphenomenon or result of treatment, other confounders or illness itself. An immunological understanding of schizophrenia could be clinically useful. Inflammation is associated with poor response to antipsychotics; comorbid physical illness, such as type 2 diabetes mellitus (Pradhan et al. 2001); and increased all-cause mortality (Zacho et al. 2010). Therefore, measuring inflammation levels (e.g. CRP test) as part of routine clinical assessment of psychosis could identify treatable causes of inflammation and potentially guide antipsychotic treatment decision. However, the "one-size-fits-all" approach to drug therapy is unlikely to be effective for immune therapies, so more personalized approach is needed.

Box 1 Key Clinical Findings

- Patients with schizophrenia show evidence of low-grade inflammation detectable in peripheral blood.
- Inflammation appears to predate the onset of illness and be independent of medication treatment.
- Inflammation is associated with poor response to antipsychotic medication.

Box 2 Key Clinical Questions

- What are the relationships between peripheral markers of inflammation and structural or functional brain changes?
- Which patients could benefit from anti-inflammatory therapies?
- Could measuring inflammation levels in clinical practice help better monitoring of psychiatric and physical health?

A key challenge for future is to determine precisely which patients are likely to benefit from anti-inflammatory therapies. This would require a concerted approach including immune target identification using genomic and other methods, deep immuno-phenotyping of psychosis patients to identify cellular source of inflammation and clinical studies to identify effect of inflammation on symptom dimensions, followed by experimental medicine and animal studies to examine the effects of novel immune-modulating agents on the brain and behaviour. With regard to certain pathways where there is sufficient evidence for a causal association with schizophrenia, e.g. IL-6/IL-6R pathway, the field now needs experimental medicine studies based on selected patient groups to test whether targeting these pathways with immuno-modulating drugs improves clinical/clinically relevant outcome measures.

Another important avenue for future research would be to understand how inflammation influences developmental trajectories of neuropsychiatric symptoms, cognition or functional outcome over the life course. Population-based prospective studies and animal experiments would be useful for this purpose.

In summary, many of the major advances in our understanding of the links between immune system and schizophrenia suggest that immuno-psychiatry is a promising field, which could transform our understanding of illness pathogenesis and approaches to treatment and prevention for schizophrenia. To be successful, the field requires collaborative working among many experts including those from psychiatry, neuroscience, immunology, neurobiology, genomics, data science, epidemiology and clinical trial.

Acknowledgement Dr. Khandaker acknowledges funding support from the Wellcome Trust (201486/Z/16/Z), MQ: Transforming Mental Health (MQDS17/40) and MRC, UK (MC_PC_17213).

Declaration of Interest The authors have no competing financial interests in relation to the work described.

References

APA (2013) Diagnostic and statistical manual of mental disorders, 5th edn. American Psychiatric Publishing, Arlington

Barker DJP (1993) Fetal and infant origins of adult disease. British Medical Journal, London

Baumeister D et al (2016) Childhood trauma and adulthood inflammation: a meta-analysis of peripheral C-reactive protein, interleukin-6 and tumour necrosis factor-alpha. Mol Psychiatry 21(5):642–649

Bian Q et al (2008) The effect of atypical antipsychotics, perospirone, ziprasidone and quetiapine on microglial activation induced by interferon-γ. Prog Neuro-Psychopharmacol Biol Psychiatry 32(1):42–48

Brown AS et al (2014) Elevated maternal C-reactive protein and autism in a national birth cohort. Mol Psychiatry 19(2):259–264

Capuron L et al (2012) Dopaminergic mechanisms of reduced basal ganglia responses to hedonic reward during interferon alfa administration. Arch Gen Psychiatry 69(10):1044–1053

Consortium SPG-WAS (2011) Genome-wide association study identifies five new schizophrenia loci. Nat Genet 43(10):969–976

Coughlin J et al (2016) In vivo markers of inflammatory response in recent-onset schizophrenia: a combined study using [11 C] DPA-713 PET and analysis of CSF and plasma. Transl Psychiatry 6(4):e777

Dameshek W (1930) White blood cells in dementia praecox and dementia paralytica. Arch Neurol Psychiatr 24:855

Dantzer R et al (2008) From inflammation to sickness and depression: when the immune system subjugates the brain. Nat Rev Neurosci 9(1):46–56

Davey Smith G, Ebrahim S (2003) 'Mendelian randomization': can genetic epidemiology contribute to understanding environmental determinants of disease? Int J Epidemiol 32(1):1–22

Deakin B et al (2018) The benefit of minocycline on negative symptoms of schizophrenia in patients with recent-onset psychosis (BeneMin): a randomised, double-blind, placebo-controlled trial. Lancet Psychiatry 5(11):885–894

Dickerson F et al (2007) C-reactive protein is associated with the severity of cognitive impairment but not of psychiatric symptoms in individuals with schizophrenia. Schizophr Res 93(1):261–265

Dickerson F et al (2013) C-reactive protein is elevated in schizophrenia. Schizophr Res 143(1):198–202

Drzyzga Ł et al (2006) Cytokines in schizophrenia and the effects of antipsychotic drugs. Brain Behav Immun 20(6):532–545

Eraly SA et al (2014) Assessment of plasma C-reactive protein as a biomarker of posttraumatic stress disorder risk. JAMA Psychiat 71(4):423–431

Felger JC et al (2013) Chronic interferon-alpha decreases dopamine 2 receptor binding and striatal dopamine release in association with anhedonia-like behavior in nonhuman primates. Neuropsychopharmacology 38(11):2179–2187

Fernandes B et al (2016) C-reactive protein is increased in schizophrenia but is not altered by antipsychotics: meta-analysis and implications. Mol Psychiatry 21(4):554–564

Flatow J, Buckley P, Miller BJ (2013) Meta-analysis of oxidative stress in schizophrenia. Biol Psychiatry 74(6):400–409

Fraguas D et al (2012) Decreased glutathione levels predict loss of brain volume in children and adolescents with first-episode psychosis in a two-year longitudinal study. Schizophr Res 137(1):58–65

Ganguli R et al (1994) Serum interleukin-6 concentration in schizophrenia: elevation associated with duration of illness. Psychiatry Res 51(1):1–10

Girgis RR et al (2018) A randomized, double-blind, placebo-controlled clinical trial of tocilizumab, an interleukin-6 receptor antibody, for residual symptoms in schizophrenia. Neuropsychopharmacology 43(6):1317–1323

Goldsmith DR, Rapaport MH, Miller BJ (2016) A meta-analysis of blood cytokine network alterations in psychiatric patients: comparisons between schizophrenia, bipolar disorder and depression. Mol Psychiatry 21(12):1696–1709

Harrison NA et al (2016) A Neurocomputational account of how inflammation enhances sensitivity to punishments versus rewards. Biol Psychiatry 80(1):73–81

Hartwig FP et al (2017) Inflammatory biomarkers and risk of schizophrenia: a 2-sample Mendelian randomization study. JAMA Psychiat 74(12):1226–1233

Howes OD, Kapur S (2009) The dopamine hypothesis of schizophrenia: version III – the final common pathway. Schizophr Bull 35(3):549–562

Janeway CA et al (2001) Immunobiology: the immune system in health and disease, 5th edn. Garland Science, New York

Jiménez-Fernández S et al (2015) Oxidative stress and antioxidant parameters in patients with major depressive disorder compared to healthy controls before and after antidepressant treatment: results from a meta-analysis. J Clin Psychiatry 76(12):1658–1667

Johnsen E et al (2016) The serum level of C-reactive protein (CRP) is associated with cognitive performance in acute phase psychosis. BMC Psychiatry 16(1):60

Jones AL et al (2005) Immune dysregulation and self-reactivity in schizophrenia: do some cases of schizophrenia have an autoimmune basis? Immunol Cell Biol 83(1):9–17

Kappelmann N et al (2018a) Systemic inflammation and intelligence in early adulthood and subsequent risk of schizophrenia and other non-affective psychoses: a longitudinal cohort and co-relative study. Psychol Med 49:1–8

Kappelmann N et al (2018b) Antidepressant activity of anti-cytokine treatment: a systematic review and meta-analysis of clinical trials of chronic inflammatory conditions. Mol Psychiatry 23(2):335–343

Khandaker GM, Dantzer R (2016) Is there a role for immune-to-brain communication in schizophrenia? Psychopharmacology 233(9):1559–1573

Khandaker GM et al (2014) Association of serum interleukin 6 and C-reactive protein in childhood with depression and psychosis in young adult life: a population-based longitudinal study. JAMA Psychiat 71(10):1121–1128

Khandaker GM et al (2015) Inflammation and immunity in schizophrenia: implications for pathophysiology and treatment. Lancet Psychiatry 2(3):258–270

Khandaker GM, Dantzer R, Jones PB (2017) Immunopsychiatry: important facts. Psychol Med 47(13):2229–2237

Khandaker GM et al (2018a) Association between a functional interleukin 6 receptor genetic variant and risk of depression and psychosis in a population-based birth cohort. Brain Behav Immun 69:264–272

Khandaker GM et al (2018b) Positive and negative symptoms of psychosis associated with circulating C-reactive protein: unravelling trans-diagnostic effect of inflammation (under review)

Khandaker GM et al (2018c) Influence of prenatal maternal depression on circulating inflammatory markers and risks of depression and psychosis in offspring: a prospective birth cohort study. Psychoneuroendocrinology (in press)

Kluge M et al (2009) Effects of clozapine and olanzapine on cytokine systems are closely linked to weight gain and drug-induced fever. Psychoneuroendocrinology 34(1):118–128

Lally J et al (2017) Remission and recovery from first-episode psychosis in adults: systematic review and meta-analysis of long-term outcome studies. Br J Psychiatry 211:350–358

Löffler S et al (2010) Clozapine therapy raises serum concentrations of high sensitive C-reactive protein in schizophrenic patients. Int Clin Psychopharmacol 25(2):101–106

Maes M et al (1998) The effects of psychological stress on humans: increased production of pro-inflammatory cytokines and a Th1-like response in stress-induced anxiety. Cytokine 10(4):313–318

Mahadik SP, Mukherjee S (1996) Free radical pathology and antioxidant defense in schizophrenia: a review. Schizophr Res 19(1):1–17

Marsman A et al (2013) Glutamate in schizophrenia: a focused review and meta-analysis of 1H-MRS studies. Schizophr Bull 39(1):120–129

McGinty J, Haque MS, Upthegrove R (2018) Depression during first episode psychosis and subsequent suicide risk: a systematic review and meta-analysis of longitudinal studies. Schizophr Res 195:58–66

Metcalf SA et al (2017) Serum C-reactive protein in adolescence and risk of schizophrenia in adulthood: a prospective birth cohort study. Brain Behav Immun 59:253–259

Miller AH, Maletic V, Raison CL (2009) Inflammation and its discontents: the role of cytokines in the pathophysiology of major depression. Biol Psychiatry 65(9):732–741

Miller BJ et al (2011) Meta-analysis of cytokine alterations in schizophrenia: clinical status and antipsychotic effects. Biol Psychiatry 70(7):663–671

Miller BJ, Culpepper N, Rapaport MH (2013) C-reactive protein levels in schizophrenia: a review and meta-analysis. Clin Schizophr Relat Psychoses 7(4):223–230

Miller BJ et al (2016) An open-label, pilot trial of adjunctive tocilizumab in schizophrenia. J Clin Psychiatry 77(2):275–276

Mondelli V et al (2015) Cortisol and inflammatory biomarkers predict poor treatment response in first episode psychosis. Schizophr Bull 41(5):1162–1170

Müller N et al (2012) Impaired monocyte activation in schizophrenia. Psychiatry Res 198:341–346

Murray RM, Lewis SW (1987) Is schizophrenia a neurodevelopmental disorder? BMJ (Clin Res Ed) 295(6600):681–682

Noto C et al (2015) Effects of depression on the cytokine profile in drug naive first-episode psychosis. Schizophr Res 164(1):53–58

Perry BI et al (2018) Dysglycaemia, inflammation and psychosis: findings from the UK ALSPAC birth cohort. Schizophr Bull. https://doi.org/10.1093/schbul/sby040

Pradhan AD et al (2001) C-reactive protein, interleukin 6, and risk of developing type 2 diabetes mellitus. JAMA 286(3):327–334

Raffa M et al (2011) Decreased glutathione levels and impaired antioxidant enzyme activities in drug-naive first-episode schizophrenic patients. BMC Psychiatry 11(1):1

Raison CL et al (2013) A randomized controlled trial of the tumor necrosis factor antagonist infliximab for treatment-resistant depression: the role of baseline inflammatory biomarkers. JAMA Psychiat 70(1):31–41

Røge R et al (2012) Immunomodulatory effects of clozapine and their clinical implications: what have we learned so far? Schizophr Res 140(1):204–213

Russell A et al (2015) Inflammation and metabolic changes in first episode psychosis: preliminary results from a longitudinal study. Brain Behav Immun 49:25–29

Schmidt R et al (2002) Early inflammation and dementia: a 25-year follow-up of the Honolulu-Asia aging study. Ann Neurol 52(2):168–174

Seki Y et al (2013) Pretreatment of aripiprazole and minocycline, but not haloperidol, suppresses oligodendrocyte damage from interferon-gamma-stimulated microglia in co-culture model. Schizophr Res 151(1–3):20–28

Traynelis SF et al (2010) Glutamate receptor ion channels: structure, regulation, and function. Pharmacol Rev 62(3):405–496

Upthegrove R, Barnes NM (2014) The immune system and schizophrenia: an update for clinicians. Adv Psychiatr Treat 20(2):83–91

Upthegrove R, Manzanares-Teson N, Barnes NM (2014) Cytokine function in medication-naive first episode psychosis: a systematic review and meta-analysis. Schizophr Res 155(1–3):101–108

Upthegrove R, Marwaha S, Birchwood M (2017) Depression and schizophrenia: cause, consequence or trans-diagnostic issue? Schizophr Bull 43:240–244

van Os J et al (2009) A systematic review and meta-analysis of the psychosis continuum: evidence for a psychosis proneness-persistence-impairment model of psychotic disorder. Psychol Med 39(2):179

van Os J, Kenis G, Rutten BP (2010) The environment and schizophrenia. Nature 468(7321):203–212

Wang AK, Miller BJ (2018) Meta-analysis of cerebrospinal fluid cytokine and tryptophan catabolite alterations in psychiatric patients: comparisons between schizophrenia, bipolar disorder, and depression. Schizophr Bull 44(1):75–83

Wang L-J et al (2018) Increased serum levels of cysteine in patients with schizophrenia: a potential marker of cognitive function preservation. Schizophr Res 192:391–397

Weinberger DR (1987) Implications of normal brain development for the pathogenesis of schizophrenia. Arch Gen Psychiatry 44(7):660–669

Wium-Andersen MK, Orsted DD, Nordestgaard BG (2014) Elevated C-reactive protein associated with late- and very-late-onset schizophrenia in the general population: a prospective study. Schizophr Bull 40(5):1117–1127

Wohleb ES et al (2014) Re-establishment of anxiety in stress-sensitized mice is caused by monocyte trafficking from the spleen to the brain. Biol Psychiatry 75(12):970–981

Wood SJ et al (2009) Neurobiology of schizophrenia spectrum disorders: the role of oxidative stress. Ann Acad Med Singap 38(5):396–401

Zacho J, Tybjaerg-Hansen A, Nordestgaard BG (2010) C-reactive protein and all-cause mortality – the Copenhagen city heart study. Eur Heart J 31(13):1624–1632

From Infection to the Microbiome: An Evolving Role of Microbes in Schizophrenia

Emily G. Severance and Robert H. Yolken

Contents

Abstract The study of microorganisms such as bacteria, viruses, archaea, fungi, and protozoa in the context of psychiatric disorders may be surprising to some. This intersection of disciplines, however, has a rich history and is currently revitalized by newfound functions of the microbiome and the gut-brain axis in human diseases. Schizophrenia, in particular, fits this model as a disorder with gene and environmental roots that may be anchored in the immune system. In this context, the combination of a precisely timed pathogen exposure in a person with genetically encoded altered immunity may have especially destructive consequences for the central nervous system (CNS). Furthermore, significant components of immunity, such as the development of the immune response and the concept of immune tolerance, are largely dictated by the commensal residents of the microbiome. When this community of microbes is imbalanced, perhaps as the result of a pathogen invasion, stress, or immune gene deficiency, a pathological cycle of localized inflammation, endothelial barrier compromise, translocation of gut-derived products, and systemic inflammation may ensue. If these pathologies enable access of gut and microbial metabolites and immune molecules to the CNS across the blood-brain

E. G. Severance (✉)
Johns Hopkins University School of Medicine, Baltimore, MD, USA

Stanley Division of Developmental Neurovirology, Department of Pediatrics, Johns Hopkins University School of Medicine, Baltimore, MD, USA
e-mail: eseverance@jhmi.edu

R. H. Yolken
Johns Hopkins University School of Medicine, Baltimore, MD, USA

Curr Topics Behav Neurosci (2020) 44: 67–84
DOI 10.1007/7854_2018_84
Published Online: 8 March 2019

barrier (BBB), and studies of the gut-brain axis support this hypothesis, a worsening of cognitive deficits and psychiatric symptoms is predicted to occur in susceptible individuals with schizophrenia. In this chapter, we review the role of microbes in various stages of this model and how these organisms may contribute to documented phenotypes of schizophrenia. An increased understanding of the role of pathogens and the microbiome in psychiatric disorders will better guide the development of microbial and immune-based therapeutics for disease prevention and treatment.

Keywords Gastrointestinal · Host-pathogen interactions · Microbiota · Neuroimmune · Psychiatry

1 Introduction

Schizophrenia is a chronic, debilitating, and etiologically complex psychiatric disorder that is likely the product of various combinations of interacting genetic and environmental influences (Demjaha et al. 2012; European Network of National Networks studying Gene-Environment Interactions in Schizophrenia (EU-GEI) et al. 2014; Kavanagh et al. 2015; Modinos et al. 2013; Nimgaonkar et al. 2017; Schizophrenia Working Group of the Psychiatric Genomics Consortium 2014; Tsuang 2000). Currently favored hypotheses regarding its causes converge on a key role of the immune system, the dysfunction of which is reflected in the body's peripheral organs and in the central nervous system (CNS). While schizophrenia is not a classic example of a disorder associated with immunity, evidence points toward an immune gene susceptibility that may be further compounded by environmental factors challenging the immune system. For example, genetic studies have implicated a series of immune genes in the chromosome 6 region that contains the MHC/HLA and complement C4 genes as important susceptibility loci associated with schizophrenia (Mayilyan et al. 2008; Sekar et al. 2016; Shi et al. 2009; Stefansson et al. 2009). A role for a microbial component in the etiology, pathogenesis, and pathophysiology of schizophrenia has been examined in various forms and would also be consistent with immune-related hypotheses for this disorder (Crow 1983; Dickerson et al. 2017a; Torrey and Peterson 1973, 1976; Yolken and Torrey 2008). Defining the nature of this microbial contribution to a host phenotype as heterogeneous as schizophrenia has been challenging and has changed in focus over the years without a successful consensus regarding cause and effect. In this review, we cover the evolving microbial landscape pertinent to this disorder and in particular highlight the shift from a search for an incontrovertible pathogen to understanding microbiome-mediated modulations of the gut-brain axis. An overview of some putative mechanisms by which pathogens and commensal microbes might contribute to schizophrenia pathophysiology is diagrammed in Fig. 1.

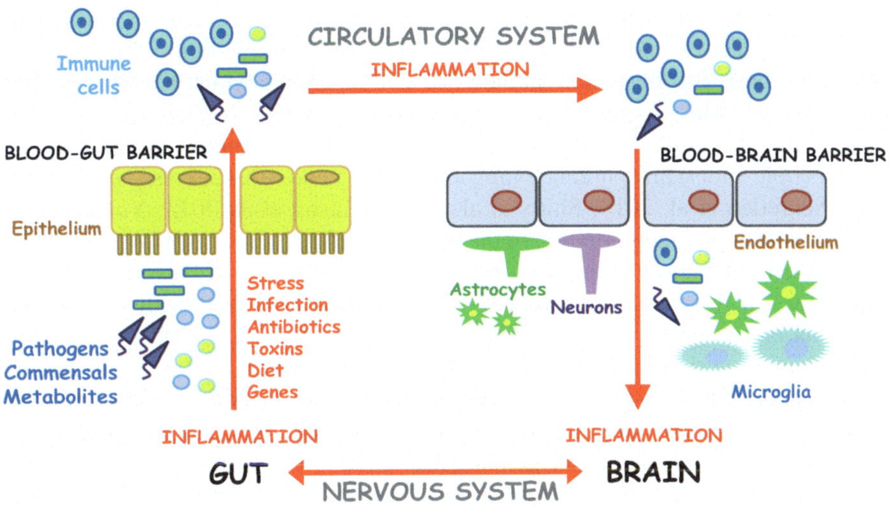

Fig. 1 Pathogens, commensal microbes, and the gut-brain axis in schizophrenia. The proposed model illustrates an overview of how neurotropic pathogens and microbial dysbioses can create an inflammatory environment in the GI tract, a process which leads to systemic inflammation and loss of integrity of the blood-gut and blood-brain barriers. Permeabilized barriers lead to the translocation of resident microbes, metabolites, and toxic products, activation of the immune response, and access to the brain for these gut-derived and immune molecules. The brain's own immune machinery becomes activated as glial cells respond to the intruders. This immune activation both peripherally and centrally includes the complement pathway, components of which can function to modify synaptic connections. The gut-brain axis is bi-directional, and through the vagus nerve, a direct neural conduit joins the enteric and central nervous systems

2 Search for a Pathogen

Jean-Étienne Esquirol (1845) was among the first who suggested that an infectious component might be relevant to psychoses, as based on his observations, psychotic episodes seemed to progress over time in a manner similar to an epidemic-like process (Esquirol 1845). Toward the end of the nineteenth century, Emile Kraepelin hypothesized that dementia praecox, the term for schizophrenia before it was so named by Eugen Bleuler in 1911, was the product of autointoxication. This autotoxication was characterized by the presence of an infectious reservoir that caused accumulation of toxins systemically, which could ultimately detrimentally affect the brain (Noll 2004; Yolken and Torrey 2008). In later years, epidemics of psychosis were reported following the 1918 influenza outbreak with additional observations that psychoses often were comorbid to typhoid, tuberculosis, diphtheria, syphilis, and other encephalitis-type states, thus supporting the earlier observations (Kirch 1993; Menninger 1919, 1926; Torrey and Peterson 1973, 1976; Yolken and Torrey 2008). The possible role of a specific pathogenic organism that might cause a brain disorder such as schizophrenia is exemplified by investigations of neurotropic viruses, such as the herpes simplex viruses (HSV), cytomegalovirus

(CMV), Epstein-Barr virus, measles, and rubella (Alam et al. 2017; Crow 1978; Meyer 2014; Torrey and Peterson 1973; Yolken and Torrey 2008). These studies formed the basis for the viral hypothesis of schizophrenia, which is still prevalent and supported today. Exposure to these neurotropic viruses, furthermore, was also found in multiple studies to be associated with deficits in cognition and gray matter loss in people with schizophrenia (Nimgaonkar and Yolken 2012; Prasad et al. 2011, 2012; Schretlen et al. 2010; Shirts et al. 2008; Watson et al. 2013; Yolken et al. 2011). Interestingly, incorporation of the human leukocyte antigen (HLA) typing has recently identified, through imputations of genome-wide association study (GWAS) data, those HLA types that were most significantly associated with neurotropic infections including the viruses, CMV and HSV1 (Parks et al. 2018). These HLA associations were significant predominantly in healthy controls and were not present in the schizophrenia group, indicating a possible disease-specific alteration of these HLA pathways. Also intriguing is the concept of retrovirus and retrotransposon integration into places within the genome that are involved with regulating cerebral growth and other functions (Crow 1984). Indeed, other viruses demonstrating significant associations with schizophrenia and psychoses over the years have included human endogenous retrovirus, as well as polio, influenza, coronaviruses, and Borna disease virus (Arias et al. 2012; Azami et al. 2018; Dickerson et al. 2010a; Karlsson et al. 2001, 2004; Khandaker et al. 2014; Leweke et al. 2004; Mednick et al. 1988; Perron et al. 2012; Prasad et al. 2007; Severance et al. 2011; Suvisaari et al. 1999). Direct sequencing of brain tissue with the aim to detect viral sequences in post-mortem brains has generally been less successful than antibody-based efforts to document seroprevalence rates in people with schizophrenia. For example, in a recent metagenomics screening of post-mortem prefrontal cortex, 156 unique viral RNA fragments were detected, but there were no differences in viral sequences between cases and controls (Tomasik et al. 2018). Difficulties finding viral nucleic acids in the brain and thus establishing potentially significant differences between diagnostic groups could be due to any number of variables including the time since infection, sensitivity of the assays, and a highly localized, and therefore well-hidden, infectious agent.

This search for a pathogen which would demonstrate a concrete connection between infectious disease agents and schizophrenia etiology or pathophysiology has not been limited to viruses. A possible bacterial basis for schizophrenia was put forth mid-twentieth century based on observed cutaneous reactions to the Rosenow antibody-antigen skin reaction. These findings suggested that schizophrenia may result following exposure to alpha-hemolytic streptococci, although not all individuals with the disorder were affected (Gurassa and Fleischhacker 1958; Rosenow 1948). In hindsight, these mixed results at the time and those garnered well into the future in similar studies, likely merely reflected the heterogeneity of the disorder and collectively suggested that there were subsets of individuals with schizophrenia who were affected in this manner. Later in this chapter, we will focus further on the bacterial contribution to schizophrenia, in particular with respect to the body's microbiome and gut-brain axis. Another microbe, the neurotropic parasite, *Toxoplasma gondii*, has been repeatedly implicated in schizophrenia etiopathogenesis, and this relationship is reviewed in numerous analyses and meta-analyses (Arias et al. 2012; Monroe et al. 2015; Severance et al. 2016b; Torrey et al. 2007, 2012). Exposure to this parasite has been associated with important clinical effects such as decreased cognition, suicidal behavior, and severity of the psychotic symptoms (Dickerson et al. 2017b; Esshili et al. 2016; Hamdani et al. 2017; Kannan et al. 2017; Lindgren et al. 2018). Studies have also uncovered evidence for heightened exposures to fungal species such as the yeasts, *Candida albicans* and *Saccharomyces cerevisiae*, in individuals with schizophrenia (Severance et al. 2012, 2016a). For *C. albicans*, cognitive deficits and worse psychiatric symptoms have been reported in those who were seropositive (Severance et al. 2016a).

3 Is It the Infectious Disease Process?

To date, an undisputed, causative pathogen has not been singularly identified, in spite of intensive effort and technical advances in deep sequencing of the genome and transcriptome. The rationale for studying pathogenic microorganisms in schizophrenia has been based on the hypothesis that a given microbe or its products are neurotropic and thus potentially directly pathogenic to brain neurons and tissue. In the absence of a conclusive etiological pathogenic species, immune activation as a process is a logical next appropriate focus of these investigations. Epidemiological studies have surveyed for the presence of infections, irrespective of a specific infectious agent, as a risk factor for the development of schizophrenia. For example, in a large cohort study of the Swedish National Register, viral but not bacterial CNS infections during childhood were found to result in the later development of schizophrenia and nonaffective psychoses (Dalman et al. 2008). A similar study of the Danish National Hospital Register indicated an increased risk of schizophrenia in individuals who had hospital contact due to an infection, with specifically bacterial infection showing the highest risk (Nielsen et al. 2014). In other study populations, urinary tract infections, likely of bacterial origin, were found at higher rates in people with schizophrenia or acute psychosis, and these infections were associated with acute relapse of psychosis (Carson et al. 2017; Graham et al. 2014; Miller et al. 2013). A variation of this type of investigation comes from epidemiological studies examining the use of anti-infective agents in schizophrenia. In one such study, the use of antibiotics, but not antivirals, antimycotics, or anti-parasitic agents, was associated with an increased risk for schizophrenia. Furthermore, in this study, if the infection required hospitalization, the risk for developing the disorder was even greater (Kohler et al. 2017). Although the results of these studies are varied in terms of the relative contribution of the type of pathogen (bacterial, viral, fungal), collectively, all point toward microbial infection as an informative comorbidity for at least a portion of those with schizophrenia.

It has long been hypothesized that schizophrenia is a neurodevelopmental disorder (Murray and Lewis 1987; Weinberger 1987); therefore, it seems likely that it is the process of immune system activation or its dysregulation at sensitive pre-, peri-, and postnatal time-points which may dictate the degree of pathogenicity that will result in the subsequent development of schizophrenia. Numerous mouse models have been developed to illustrate altered behavior or brain biochemistry in offspring of mothers whose immune system has been challenged experimentally during pregnancy (Brown and Derkits 2010; Estes and McAllister 2016; Labouesse et al. 2015; Meyer 2014). In humans, studies of specific pathogens in this context are made possible by the availability of maternal sera drawn during pregnancy or neonatal blood spots obtained at birth. Findings from these studies do, in fact, reveal that a variety of pathogen exposures are associated with the future development of schizophrenia or psychosis in offspring (Blomstrom et al. 2016; Brown et al. 2004; Brown and Derkits 2010; Buka et al. 2008; Ellman et al. 2009; Estes and McAllister 2016; Khandaker et al. 2013; Mortensen et al. 2010; Xiao et al. 2009). Furthermore, prenatal exposure to maternal sinusitis, tonsillitis, pneumonia, as well as genital and other reproductive infections was also associated with the subsequent development of schizophrenia (Babulas et al. 2006; Sorensen et al. 2009). In another study of the Danish National Register, it was found that prenatal infection and peri-pubertal psychological trauma not only each increased the risk of schizophrenia with some sex-specific differences, but the combination of these two factors acted in synergy to compound that risk for disease (Debost et al. 2017). It has also been shown that activation of innate immunity including cytokines and components of the complement pathway were elevated in mothers whose adult offspring developed schizophrenia or psychoses as adults (Allswede et al. 2016; Severance et al. 2014).

4 Gut Inflammation

Immune activation due to infection or another source, or immune dysregulation in general, appears to be as relevant to schizophrenia as exposure to a specific pathogen. Studies of innate immunity in schizophrenia support a low-grade inflammatory component peripherally and in the CNS, which is prevalent in the disorder (Bechter 2013; Catts et al. 2014; Dickerson et al. 2016; Fillman et al. 2013, 2014, 2016; Kirkpatrick and Miller 2013; Miller et al. 2011; Muller 2016; Severance et al. 2012, 2013). The source of this inflammation, however, remains unknown, as does whether this inflammation reflects a pathophysiology of the disease state or a comorbidity resulting from lifestyle choices or medication.

Interestingly, even older than the hypothesis that infection is at the crux of schizophrenia is the hypothesis that all diseases begin in the gut (Hippocrates). The tenets of Hippocratic medicine premised that health was based on four balanced humors, black bile, yellow bile, phlegm, and blood. One of these humors, black bile, referred to the temperament of melancholy, or what we now know as depression (Jackson 2001). In the mid-nineteenth century, purgatives and emetics were suggested treatments for psychiatric symptoms (Prichard 1837). Other historical accounts support a pervasive gastrointestinal (GI) inflammatory state present in individuals with psychoses and schizophrenia, even well before these psychiatric disorders were described by the earliest versions of our current psychiatric classification systems, the *Diagnostic and Statistical Manual of Mental Disorders* (DSM; 1952) and the International Classification of Diseases (ICD; 1949) (APA 1952; WHO 1949; Alander et al. 2005; Buscaino 1953; Hemmings 2004; Reiter 1926; Schneck 1946). Reports of this GI inflammation likewise preceded the 1950's discovery of modern antipsychotics that are often indicated as the cause of GI comorbidities due to strong anticholinergic effects contributing to decreased bowel motility and constipation (Dean 2010; Dome et al. 2007; McNamara et al. 2011; Watanabe et al. 2010). Indeed, over the years, a number of enteropathic disorders have been studied for association with schizophrenia including celiac disease, gluten intolerance, ulcerative colitis, Crohn's disease, and irritable bowel syndrome (Dohan 1970; Eaton et al. 2004; Gupta et al. 1997; Makikyro et al. 1998; Severance et al. 2015b, 2016c). Serological measures of antibodies directed against *Saccharomyces cerevisiae* (ASCA), which are used clinically to diagnose inflammatory bowel diseases, are also elevated in schizophrenia and especially so in those early-stage patients who were medication-naïve (Ashorn et al. 2009; Desplat-Jego et al. 2007; Kotze et al. 2010; Mallant-Hent et al. 2006; Oshitani et al. 2000; Severance et al. 2012). Likewise, antibodies directed against other antigens that contribute to GI inflammation, such as antigenic foods and gut pathogens, are elevated, in schizophrenia (Dickerson et al. 2010b; Kelly et al. 2018; Severance et al. 2010, 2012, 2016c). The well-studied parasite, *T. gondii*, to which seroprevalence is increased in schizophrenia, is, in fact, a routinely used laboratory tool to model inflammatory bowel diseases in experimental rodents (Craven et al. 2012; Grainger et al. 2013; Hand et al. 2012; Heimesaat et al. 2006).

GI inflammation leads to permeability of the endothelial blood-gut barrier and the potential crossing of microbes, microbial-generated toxins or metabolites, and food-related peptides and antigens into the circulation (Brenchley et al. 2006; Lambert 2009; Sandler and Douek 2012). The translocation of GI-related products has been the focus of studies of depression (Maes et al. 2008, 2012a, b) and to a more limited extent in schizophrenia (Caso et al. 2016; Karakula-Juchnowicz et al. 2016; Severance et al. 2013; Weber et al. 2018). Two surrogate biomarkers of the bacterial translocation process, soluble CD14 (sCD14) and lipopolysaccharide (LPS) binding protein (LBP), were found to be intercorrelated with each other, with a general marker of inflammation, C-reactive protein, and with antibodies to food antigens in individuals with schizophrenia (Severance et al. 2013). In this study and in a follow-up investigation, levels of sCD14 were significantly upregulated not only in individuals with established schizophrenia but also in individuals with pre-onset schizophrenia, as identified based on blood samples and medical records from a US military cohort. In both studies, LBP levels did not match the elevated sCD14 suggesting that additional pathogenic mechanisms related to bacterial translocation and dysregulated monocyte activation may be operative in schizophrenia (Severance et al. 2013; Weber et al. 2018).

5 Microbial Dysbiosis and the Brain

Microbial translocation reflects a gut commensal community that is imbalanced or dysbiotic and that fosters a cycle of inflammation, barrier compromise, and bowel dysfunction. A healthy gut is required for digestion, nutrient absorption, metabolism, maintenance of gut-blood barrier integrity, and development of host immunity (Ismail and Hooper 2005; Round et al. 2010; Smith and Garrett 2011; Sommer and Backhed 2013). Gut function is coordinated by a diverse community of bacteria, viruses, fungi, and archaea, which are at equilibrium with host cell activities (Dinan and Cryan 2015; Sandhya et al. 2016). This equilibrium can be disrupted by stress, diet, antibiotics, toxins, infectious agents, and products generated by host genetics (Sandhya et al. 2016). Thus, for schizophrenia, dysbiosis of the gut microbiome is important to document because it provides a mechanism of GI-localized inflammation that has systemic consequences that are relevant to neuroinflammation and the brain. Importantly, translocated GI products act as triggers of the body's systemic immune machinery, such as the complement pathway, put in motion to clear antigens perceived as foreign from the bloodstream (Brenchley et al. 2006; Lambert 2009; Sandler and Douek 2012). Complement also has important functions in the brain which include the removal of inappropriate synapses, and the genetic and functional associations of this pathway with schizophrenia have been reported and reviewed elsewhere (Nimgaonkar et al. 2017; Presumey et al. 2017; Sekar et al. 2016). Physical access to the brain is a converging and critical consideration in this context, both with respect to translocated gut products and immune molecules. Endothelial barrier defects at both the blood-gut and blood-brain barriers present pathologies that are consistent with a compromised gut-brain pathway operative in schizophrenia (Kannan et al. 2017). Findings from studies employing various approaches suggest an altered function of endothelial cells and BBB permeability associated with schizophrenia (Greene et al. 2017; Khandaker and Dantzer 2016; Severance et al. 2015a). For example, markers of endothelial cell activation including the selectin family of adhesion molecules have been found to be elevated in schizophrenia (Iwata et al. 2007; Khandaker and Dantzer 2016). This endothelial cell activation in the BBB has been shown to follow systemic inflammation and is associated with the translocation of inflammatory cells into the brain (D'Mello and Swain 2014; Khandaker and Dantzer 2016). Accompanying this activation are increased monocyte levels and monocyte infiltration of the BBB which are consistent with the elevations of sCD14 reported in the previous section.

The ability to interrogate rodent models in a germ-free setting has provided much insight regarding the possible mechanisms by which gut microbes are actively engaged in biological pathways that regulate the gut-brain axis. Importantly, these studies allow associations to be made and solidified without a plethora of confounding variables that often accompany and cloud results from clinical studies. Summarily, in the absence of a gut microbiome, the brain fails to develop normally (Sampson and Mazmanian 2015). Altered brain biochemistry, cognition, and behaviors are repeatedly demonstrated following manipulations of gut microbiota in germ-free and/or pathogen-specific animals (Collins et al. 2012; Diaz Heijtz et al. 2011; Erny et al. 2015; Foster and McVey Neufeld 2013; Hsiao et al. 2013; Luczynski et al. 2016; Stilling et al. 2014). In the germ-free setting, such abnormalities included alterations of myelination, microglial regulation, neurogenesis, and neurotransmitter abundances such as serotonin and precursor tryptophan and trophic factors. These deficits were recovered with further manipulations or corrections of bacterial compositions, vagotomy, and administration of probiotics and/or antibiotics. As relevant to schizophrenia, a revealing set of experiments were those that showed how directly the gut microbiota can impact BBB permeability (Braniste et al. 2014). The absence of a microbiome increased BBB permeability, and this defect was restored following transplantation of germ-free animals with a normal microbiota. Thus, garnered from these studies is evidence of some of the most promising pathways in support of a gut-brain axis including the following: (1) the parasympathetic nervous system and related enteric innervation including the vagus nerve, (2) the neuroendocrine system including stress hormones and the HPA axis, (3) metabolic pathways including microbially generated short-chain fatty acids that bind to G protein-coupled receptors and that are epigenetic modulators, (4) the circulatory system which enables the delivery of gut-generated neuroactive metabolites and neurotransmitters to the vicinity of the brain, and (5) the immune system which is extensively referenced throughout this chapter (Alam et al. 2017; Berger et al. 2009; Dinan et al. 2018; El Aidy et al. 2014).

Of interest are metagenomic and 16S rRNA gene sequencing studies of the oropharyngeal and fecal microbiomes in people with schizophrenia and psychoses compared to controls (Castro-Nallar et al. 2015; Schwarz et al. 2018; Shen et al. 2018; Yolken et al. 2015). In the oropharyngeal microbiome, the genera lactobacilli and bifidobacteria were more abundant in schizophrenia compared to controls, and intriguingly, these are the genera that help to modulate inflammation (Castro-Nallar et al. 2015). Similarly, the oropharyngeal microbiome in schizophrenia contained altered levels of the phage, *Lactobacillus* phiadh, which infects *Lactobacillus gasseri*, a bacteria that functions in part to maintain epithelial cell integrity and to modulate the immune system (Yolken et al. 2015). Differences in fecal lactobacilli were also observed in patients with first-episode psychosis compared to controls, and numbers of these taxa were particularly elevated in those who were most treatment resistant (Schwarz et al. 2018). In another study of the fecal microbiome, case-control differences in numerous taxa were observed including an elevation of the phylum, *Proteobacteria*, and those taxa that functioned in metabolic pathways (Shen et al. 2018).

Clinical trials of probiotics in schizophrenia can be similarly informative regarding potentially correcting a microbe- or gut-based pathology. In a randomized, placebo-controlled trial of adjunctive probiotics in schizophrenia, improved GI function was reported, but there was no change in the severity of psychiatric symptoms associated with probiotic treatment (Dickerson et al. 2014). Serologically, there were significant alterations in an array of immune proteins that pathway analyses indicated were suggestive of improved GI epithelial and immune pathologies associated with probiotic treatment (Tomasik et al. 2015). Of interest also is how other non-bacterial components of the microbiome might influence these clinical trial findings. For example, in healthy people, commensal yeast species cohabitate with resident bacteria in a homeostatic balance. If this balance is shifted perhaps by diet or antibiotics, bacterial dysbioses, species depletion, and yeast overgrowth can result (Kim and Sudbery 2011). In the probiotic trial cited above, we found evidence for improvement in psychiatric symptoms associated with probiotics, but only in those who were not positive for these invasive yeast infections (Severance et al. 2017). *C. albicans* was, in fact, particularly overrepresented in individuals with schizophrenia compared to controls, and these yeast-positive individuals had correspondingly more cognitive impairments and severe psychiatric symptoms (Severance et al. 2016a, 2017).

6 Conclusions

As such, we are only just beginning to unravel the extent to which microbes regulate human health and disease. Disciplines as dissimilar as gastroenterology, oncology, dermatology, endocrinology, hepatology, neuroscience, and psychiatry are all actively engaged in researching the microbiome. As summarized in this chapter, microbes are associated with schizophrenia etiology, pathogenesis, and pathophysiology in a diversity of ways, ranging from infection-based pathologies to alterations of the gut-brain axis. Infection, inflammation, and gut dysbioses are all treatable conditions, but to develop an effective therapeutic applicable to schizophrenia, it is critical to identify the source of the pathology and to identify those individuals who are impacted. The surge of interest and effort directed toward understanding the microbiome will hopefully accelerate the improvement of methods for manipulating microbiota and lead to novel agents to prevent and treat a wide range of human disorders.

Acknowledgments This work was supported by a NIMH P50 Silvio O. Conte Center at Johns Hopkins (grant# MH-94268) and by the Stanley Medical Research Institute.

References

Alam R, Abdolmaleky HM, Zhou JR (2017) Microbiome, inflammation, epigenetic alterations, and mental diseases. Am J Med Genet B Neuropsychiatr Genet 174:651–660

Alander T, Svardsudd K, Johansson SE, Agreus L (2005) Psychological illness is commonly associated with functional gastrointestinal disorders and is important to consider during patient consultation: a population-based study. BMC Med 3:8

Allswede DM, Buka SL, Yolken RH, Torrey EF, Cannon TD (2016) Elevated maternal cytokine levels at birth and risk for psychosis in adult offspring. Schizophr Res 172:41–45

APA (1952) Diagnostic and statistical manual of mental disorders (DSM-I), 1st edn. American Psychiatric Association, Washington

Arias I, Sorlozano A, Villegas E, de Dios Luna J, McKenney K, Cervilla J et al (2012) Infectious agents associated with schizophrenia: a meta-analysis. Schizophr Res 136:128–136

Ashorn S, Valineva T, Kaukinen K, Ashorn M, Braun J, Raukola H et al (2009) Serological responses to microbial antigens in celiac disease patients during a gluten-free diet. J Clin Immunol 29:190–195

Azami M, Jalilian FA, Khorshidi A, Mohammadi Y, Tardeh Z (2018) The association between Borna disease virus and schizophrenia: a systematic review and meta-analysis. Asian J Psychiatr 34:67–73

Babulas V, Factor-Litvak P, Goetz R, Schaefer CA, Brown AS (2006) Prenatal exposure to maternal genital and reproductive infections and adult schizophrenia. Am J Psychiatry 163:927–929

Bechter K (2013) Updating the mild encephalitis hypothesis of schizophrenia. Prog Neuro-Psychopharmacol Biol Psychiatry 42:71–91

Berger M, Gray JA, Roth BL (2009) The expanded biology of serotonin. Annu Rev Med 60:355–366

Blomstrom A, Karlsson H, Gardner R, Jorgensen L, Magnusson C, Dalman C (2016) Associations between maternal infection during pregnancy, childhood infections, and the risk of subsequent psychotic disorder – a Swedish cohort study of nearly two million individuals. Schizophr Bull 42:125–133

Braniste V, Al-Asmakh M, Kowal C, Anuar F, Abbaspour A, Toth M et al (2014) The gut microbiota influences blood-brain barrier permeability in mice. Sci Transl Med 6:263ra158

Brenchley JM, Price DA, Schacker TW, Asher TE, Silvestri G, Rao S et al (2006) Microbial translocation is a cause of systemic immune activation in chronic HIV infection. Nat Med 12:1365–1371

Brown AS, Derkits EJ (2010) Prenatal infection and schizophrenia: a review of epidemiologic and translational studies. Am J Psychiatry 167:261–280

Brown AS, Begg MD, Gravenstein S, Schaefer CA, Wyatt RJ, Bresnahan M et al (2004) Serologic evidence of prenatal influenza in the etiology of schizophrenia. Arch Gen Psychiatry 61:774–780

Buka SL, Cannon TD, Torrey EF, Yolken RH, Collaborative Study Group on the Perinatal Origins of Severe Psychiatric Disorders (2008) Maternal exposure to herpes simplex virus and risk of psychosis among adult offspring. Biol Psychiatry 63:809–815

Buscaino V (1953) Patologia extraneurale della schizofrenia. Fegato, tubo digerente, sistema reticolo-endoteliale. Acta Neurol 8:1–60

Carson CM, Phillip N, Miller BJ (2017) Urinary tract infections in children and adolescents with acute psychosis. Schizophr Res 183:36–40

Caso JR, Balanza-Martinez V, Palomo T, Garcia-Bueno B (2016) The microbiota and gut-brain axis: contributions to the immunopathogenesis of schizophrenia. Curr Pharm Des 22:6122–6133

Castro-Nallar E, Bendall ML, Perez-Losada M, Sabuncyan S, Severance EG, Dickerson FB et al (2015) Composition, taxonomy and functional diversity of the oropharynx microbiome in individuals with schizophrenia and controls. PeerJ 3:e1140

Catts VS, Wong J, Fillman SG, Fung SJ, Shannon Weickert C (2014) Increased expression of astrocyte markers in schizophrenia: association with neuroinflammation. Aust N Z J Psychiatry 48:722–734

Collins SM, Surette M, Bercik P (2012) The interplay between the intestinal microbiota and the brain. Nat Rev Microbiol 10:735–742

Craven M, Egan CE, Dowd SE, McDonough SP, Dogan B, Denkers EY et al (2012) Inflammation drives dysbiosis and bacterial invasion in murine models of ileal Crohn's disease. PLoS One 7: e41594

Crow TJ (1978) Viral causes of psychiatric disease. Postgrad Med J 54:763–767

Crow TJ (1983) Is schizophrenia an infectious disease? Lancet 1:173–175

Crow TJ (1984) A re-evaluation of the viral hypothesis: is psychosis the result of retroviral integration at a site close to the cerebral dominance gene? Br J Psychiatry 145:243–253

Dalman C, Allebeck P, Gunnell D, Harrison G, Kristensson K, Lewis G et al (2008) Infections in the CNS during childhood and the risk of subsequent psychotic illness: a cohort study of more than one million Swedish subjects. Am J Psychiatry 165:59–65

Dean B (2010) Understanding the role of inflammatory-related pathways in the pathophysiology and treatment of psychiatric disorders: evidence from human peripheral studies and CNS studies. Int J Neuropsychopharmacol 14:997–1012

Debost JP, Larsen JT, Munk-Olsen T, Mortensen PB, Meyer U, Petersen L (2017) Joint effects of exposure to prenatal infection and peripubertal psychological trauma in schizophrenia. Schizophr Bull 43:171–179

Demjaha A, MacCabe JH, Murray RM (2012) How genes and environmental factors determine the different neurodevelopmental trajectories of schizophrenia and bipolar disorder. Schizophr Bull 38:209–214

Desplat-Jego S, Johanet C, Escande A, Goetz J, Fabien N, Olsson N et al (2007) Update on anti-Saccharomyces cerevisiae antibodies, anti-nuclear associated anti-neutrophil antibodies and antibodies to exocrine pancreas detected by indirect immunofluorescence as biomarkers in chronic inflammatory bowel diseases: results of a multicenter study. World J Gastroenterol: WJG 13:2312–2318

Diaz Heijtz R, Wang S, Anuar F, Qian Y, Bjorkholm B, Samuelsson A et al (2011) Normal gut microbiota modulates brain development and behavior. Proc Natl Acad Sci U S A 108:3047–3052

Dickerson F, Stallings C, Origoni A, Copp C, Khushalani S, Yolken R (2010a) Antibodies to measles in individuals with recent onset psychosis. Schizophr Res 119:89–94

Dickerson F, Stallings C, Origoni A, Vaughan C, Khushalani S, Leister F et al (2010b) Markers of gluten sensitivity and celiac disease in recent-onset psychosis and multi-episode schizophrenia. Biol Psychiatry 68:100–104

Dickerson FB, Stallings C, Origoni A, Katsafanas E, Savage CL, Schweinfurth LA et al (2014) Effect of probiotic supplementation on schizophrenia symptoms and association with gastrointestinal functioning: a randomized, placebo-controlled trial. Prim Care Companion CNS Disord 16. https://doi.org/10.4088/PCC.13m01579

Dickerson F, Stallings C, Origoni A, Schroeder J, Katsafanas E, Schweinfurth L et al (2016) Inflammatory markers in recent onset psychosis and chronic schizophrenia. Schizophr Bull 42:134–141

Dickerson F, Severance E, Yolken R (2017a) The microbiome, immunity, and schizophrenia and bipolar disorder. Brain Behav Immun 62:46–52

Dickerson F, Wilcox HC, Adamos M, Katsafanas E, Khushalani S, Origoni A et al (2017b) Suicide attempts and markers of immune response in individuals with serious mental illness. J Psychiatr Res 87:37–43

Dinan TG, Cryan JF (2015) The impact of gut microbiota on brain and behaviour: implications for psychiatry. Curr Opin Clin Nutr Metab Care 18:552–558

Dinan TG, Cryan JF, Stanton C (2018) Gut microbes and brain development have black box connectivity. Biol Psychiatry 83:97–99

D'Mello C, Swain MG (2014) Liver-brain interactions in inflammatory liver diseases: implications for fatigue and mood disorders. Brain Behav Immun 35:9–20

Dohan FC (1970) Coeliac disease and schizophrenia. Lancet 1:897–898

Dome P, Teleki Z, Kotanyi R (2007) Paralytic ileus associated with combined atypical antipsychotic therapy. Prog Neuro-Psychopharmacol Biol Psychiatry 31:557–560

Eaton W, Mortensen PB, Agerbo E, Byrne M, Mors O, Ewald H (2004) Coeliac disease and schizophrenia: population based case control study with linkage of Danish national registers. BMJ 328:438–439

El Aidy S, Dinan TG, Cryan JF (2014) Immune modulation of the brain-gut-microbe axis. Front Microbiol 5:146

Ellman LM, Yolken RH, Buka SL, Torrey EF, Cannon TD (2009) Cognitive functioning prior to the onset of psychosis: the role of fetal exposure to serologically determined influenza infection. Biol Psychiatry 65:1040–1047

Erny D, Hrabe de Angelis AL, Jaitin D, Wieghofer P, Staszewski O, David E et al (2015) Host microbiota constantly control maturation and function of microglia in the CNS. Nat Neurosci 18:965–977

Esquirol JE (1845) Mental maladies, a treatise on insanity. Lea and Blanchard, Philadelphia

Esshili A, Thabet S, Jemli A, Trifa F, Mechri A, Zaafrane F et al (2016) Toxoplasma gondii infection in schizophrenia and associated clinical features. Psychiatry Res 245:327–332

Estes ML, McAllister AK (2016) Maternal immune activation: implications for neuropsychiatric disorders. Science 353:772–777

European Network of National Networks studying Gene-Environment Interactions in Schizophrenia (EU-GEI), van Os J, Rutten BP, Myin-Germeys I, Delespaul P, Viechtbauer W et al (2014) Identifying gene-environment interactions in schizophrenia: contemporary challenges for integrated, large-scale investigations. Schizophr Bull 40:729–736

Fillman SG, Cloonan N, Catts VS, Miller LC, Wong J, McCrossin T et al (2013) Increased inflammatory markers identified in the dorsolateral prefrontal cortex of individuals with schizophrenia. Mol Psychiatry 18:206–214

Fillman SG, Sinclair D, Fung SJ, Webster MJ, Shannon Weickert C (2014) Markers of inflammation and stress distinguish subsets of individuals with schizophrenia and bipolar disorder. Transl Psychiatry 4:e365

Fillman SG, Weickert TW, Lenroot RK, Catts SV, Bruggemann JM, Catts VS et al (2016) Elevated peripheral cytokines characterize a subgroup of people with schizophrenia displaying poor verbal fluency and reduced Broca's area volume. Mol Psychiatry 21:1090–1098

Foster JA, McVey Neufeld KA (2013) Gut-brain axis: how the microbiome influences anxiety and depression. Trends Neurosci 36:305–312

Graham KL, Carson CM, Ezeoke A, Buckley PF, Miller BJ (2014) Urinary tract infections in acute psychosis. J Clin Psychiatry 75:379–385

Grainger JR, Wohlfert EA, Fuss IJ, Bouladoux N, Askenase MH, Legrand F et al (2013) Inflammatory monocytes regulate pathologic responses to commensals during acute gastrointestinal infection. Nat Med 19:713–721

Greene C, Kealy J, Humphries MM, Gong Y, Hou J, Hudson N et al (2017) Dose-dependent expression of claudin-5 is a modifying factor in schizophrenia. Mol Psychiatry 23:2156–2166

Gupta S, Masand PS, Kaplan D, Bhandary A, Hendricks S (1997) The relationship between schizophrenia and irritable bowel syndrome (IBS). Schizophr Res 23:265–268

Gurassa WP, Fleischhacker HH (1958) An investigation of the Rosenow antibody antigen skin reaction in schizophrenia. J Neurol Neurosurg Psychiatry 21:141–145

Hamdani N, Daban-Huard C, Godin O, Laouamri H, Jamain S, Attiba D et al (2017) Effects of cumulative herpesviridae and Toxoplasma gondii infections on cognitive function in healthy, bipolar, and schizophrenia subjects. J Clin Psychiatry 78:e18–e27

Hand TW, Dos Santos LM, Bouladoux N, Molloy MJ, Pagan AJ, Pepper M et al (2012) Acute gastrointestinal infection induces long-lived microbiota-specific t cell responses. Science 337:1553–1556

Heimesaat MM, Bereswill S, Fischer A, Fuchs D, Struck D, Niebergall J et al (2006) Gram-negative bacteria aggravate murine small intestinal th1-type immunopathology following oral infection with Toxoplasma gondii. J Immunol 177:8785–8795

Hemmings G (2004) Schizophrenia. Lancet 364:1312–1313

Hsiao EY, McBride SW, Hsien S, Sharon G, Hyde ER, McCue T et al (2013) Microbiota modulate behavioral and physiological abnormalities associated with neurodevelopmental disorders. Cell 155:1451–1463

Ismail AS, Hooper LV (2005) Epithelial cells and their neighbors. IV. Bacterial contributions to intestinal epithelial barrier integrity. Am J Physiol Gastrointest Liver Physiol 289:G779–G784

Iwata Y, Suzuki K, Nakamura K, Matsuzaki H, Sekine Y, Tsuchiya KJ et al (2007) Increased levels of serum soluble l-selectin in unmedicated patients with schizophrenia. Schizophr Res 89:154–160

Jackson WA (2001) A short guide to humoral medicine. Trends Pharmacol Sci 22:487–489

Kannan G, Gressitt KL, Yang S, Stallings CR, Katsafanas E, Schweinfurth LA et al (2017) Pathogen-mediated NMDA receptor autoimmunity and cellular barrier dysfunction in schizophrenia. Transl Psychiatry 7:e1186

Karakula-Juchnowicz H, Dzikowski M, Pelczarska A, Dzikowska I, Juchnowicz D (2016) The brain-gut axis dysfunctions and hypersensitivity to food antigens in the etiopathogenesis of schizophrenia. Psychiatr Pol 50:747–760

Karlsson H, Bachmann S, Schroder J, McArthur J, Torrey EF, Yolken RH (2001) Retroviral RNA identified in the cerebrospinal fluids and brains of individuals with schizophrenia. Proc Natl Acad Sci U S A 98:4634–4639

Karlsson H, Schroder J, Bachmann S, Bottmer C, Yolken RH (2004) HERV-W-related RNA detected in plasma from individuals with recent-onset schizophrenia or schizoaffective disorder. Mol Psychiatry 9:12–13

Kavanagh DH, Tansey KE, O'Donovan MC, Owen MJ (2015) Schizophrenia genetics: emerging themes for a complex disorder. Mol Psychiatry 20:72–76

Kelly DL, Demyanovich HK, Eaton WW, Cascella N, Jackson J, Fasano A et al (2018) Anti gliadin antibodies (AGA IgG) related to peripheral inflammation in schizophrenia. Brain Behav Immun 69:57–59

Khandaker GM, Dantzer R (2016) Is there a role for immune-to-brain communication in schizophrenia? Psychopharmacology 233:1559–1573

Khandaker GM, Zimbron J, Lewis G, Jones PB (2013) Prenatal maternal infection, neurodevelopment and adult schizophrenia: a systematic review of population-based studies. Psychol Med 43:239–257

Khandaker GM, Stochl J, Zammit S, Lewis G, Jones PB (2014) Childhood Epstein-Barr virus infection and subsequent risk of psychotic experiences in adolescence: a population-based prospective serological study. Schizophr Res 158:19–24

Kim J, Sudbery P (2011) Candida albicans, a major human fungal pathogen. J Microbiol 49:171–177

Kirch DG (1993) Infection and autoimmunity as etiologic factors in schizophrenia: a review and reappraisal. Schizophr Bull 19:355–370

Kirkpatrick B, Miller BJ (2013) Inflammation and schizophrenia. Schizophr Bull 39:1174–1179

Kohler O, Petersen L, Mors O, Mortensen PB, Yolken RH, Gasse C et al (2017) Infections and exposure to anti-infective agents and the risk of severe mental disorders: a nationwide study. Acta Psychiatr Scand 135:97–105

Kotze LM, Nisihara RM, Utiyama SR, Kotze PG, Theiss PM, Olandoski M (2010) Antibodies anti-Saccharomyces cerevisiae (ASCA) do not differentiate Crohn's disease from celiac disease. Arq Gastroenterol 47:242–245

Labouesse MA, Langhans W, Meyer U (2015) Long-term pathological consequences of prenatal infection: beyond brain disorders. Am J Physiol Regul Integr Comp Physiol 309:R1–R12

Lambert GP (2009) Stress-induced gastrointestinal barrier dysfunction and its inflammatory effects. J Anim Sci 87:E101–E108

Leweke FM, Gerth CW, Koethe D, Klosterkotter J, Ruslanova I, Krivogorsky B et al (2004) Antibodies to infectious agents in individuals with recent onset schizophrenia. Eur Arch Psychiatry Clin Neurosci 254:4–8

Lindgren M, Torniainen-Holm M, Harkanen T, Dickerson F, Yolken RH, Suvisaari J (2018) The association between Toxoplasma and the psychosis continuum in a general population setting. Schizophr Res 193:329–335

Luczynski P, Whelan SO, O'Sullivan C, Clarke G, Shanahan F, Dinan TG et al (2016) Adult microbiota-deficient mice have distinct dendritic morphological changes: differential effects in the amygdala and hippocampus. Eur J Neurosci 44:2654–2666

Maes M, Kubera M, Leunis JC (2008) The gut-brain barrier in major depression: intestinal mucosal dysfunction with an increased translocation of LPS from gram negative enterobacteria (leaky gut) plays a role in the inflammatory pathophysiology of depression. Neuro Endocrinol Lett 29:117–124

Maes M, Kubera M, Leunis JC, Berk M (2012a) Increased IgA and IgM responses against gut commensals in chronic depression: further evidence for increased bacterial translocation or leaky gut. J Affect Disord 141:55–62

Maes M, Kubera M, Leunis JC, Berk M, Geffard M, Bosmans E (2012b) In depression, bacterial translocation may drive inflammatory responses, oxidative and nitrosative stress (O&NS), and autoimmune responses directed against O&NS-damaged neoepitopes. Acta Psychiatr Scand 127:344–354

Makikyro T, Karvonen JT, Hakko H, Nieminen P, Joukamaa M, Isohanni M et al (1998) Comorbidity of hospital-treated psychiatric and physical disorders with special reference to schizophrenia: a 28 year follow-up of the 1966 northern Finland general population birth cohort. Public Health 112:221–228

Mallant-Hent R, Mary B, von Blomberg E, Yuksel Z, Wahab PJ, Gundy C et al (2006) Disappearance of anti-Saccharomyces cerevisiae antibodies in coeliac disease during a gluten-free diet. Eur J Gastroenterol Hepatol 18:75–78

Mayilyan KR, Dodds AW, Boyajyan AS, Soghoyan AF, Sim RB (2008) Complement c4b protein in schizophrenia. World J Biol Psychiatry 9:225–230

McNamara RK, Jandacek R, Rider T, Tso P (2011) Chronic risperidone normalizes elevated pro-inflammatory cytokine and C-reactive protein production in omega-3 fatty acid deficient rats. Eur J Pharmacol 652:152–156

Mednick SA, Machon RA, Huttunen MO, Bonett D (1988) Adult schizophrenia following prenatal exposure to an influenza epidemic. Arch Gen Psychiatry 45:189–192

Menninger KA (1919) Psychoses associated with influenza. J Am Med Assoc 72:235–241

Menninger KA (1926) Influenza and schizophrenia: an analysis of post-influenza "dementia praecox" as of 1918 and five years later. Am J Psychiatr 5:469–529

Meyer U (2014) Prenatal poly(i:C) exposure and other developmental immune activation models in rodent systems. Biol Psychiatry 75:307–315

Miller BJ, Buckley P, Seabolt W, Mellor A, Kirkpatrick B (2011) Meta-analysis of cytokine alterations in schizophrenia: clinical status and antipsychotic effects. Biol Psychiatry 70:663–671

Miller BJ, Graham KL, Bodenheimer CM, Culpepper NH, Waller JL, Buckley PF (2013) A prevalence study of urinary tract infections in acute relapse of schizophrenia. J Clin Psychiatry 74:271–277

Modinos G, Iyegbe C, Prata D, Rivera M, Kempton MJ, Valmaggia LR et al (2013) Molecular genetic gene-environment studies using candidate genes in schizophrenia: a systematic review. Schizophr Res 150:356–365

Monroe JM, Buckley PF, Miller BJ (2015) Meta-analysis of anti-Toxoplasma gondii IgM antibodies in acute psychosis. Schizophr Bull 41:989–998

Mortensen PB, Pedersen CB, Hougaard DM, Norgaard-Petersen B, Mors O, Borglum AD et al (2010) A Danish national birth cohort study of maternal HSV-2 antibodies as a risk factor for schizophrenia in their offspring. Schizophr Res 122:257–263

Muller N (2016) What role does inflammation play in schizophrenia? Expert Rev Neurother 16:1337–1340

Murray RM, Lewis SW (1987) Is schizophrenia a neurodevelopmental disorder? Br Med J (Clin Res Ed) 295:681–682

Nielsen PR, Benros ME, Mortensen PB (2014) Hospital contacts with infection and risk of schizophrenia: a population-based cohort study with linkage of Danish national registers. Schizophr Bull 40:1526–1532

Nimgaonkar VL, Yolken RH (2012) Neurotropic infectious agents and cognitive impairment in schizophrenia. Schizophr Bull 38:1135–1136

Nimgaonkar VL, Prasad KM, Chowdari KV, Severance EG, Yolken RH (2017) The complement system: a gateway to gene-environment interactions in schizophrenia pathogenesis. Mol Psychiatry 22:1554–1561
Noll R (2004) Historical review: autointoxication and focal infection theories of dementia praecox. World J Biol Psychiatry 5:66–72
Oshitani N, Hato F, Matsumoto T, Jinno Y, Sawa Y, Hara J et al (2000) Decreased anti-Saccharomyces cerevisiae antibody titer by mesalazine in patients with Crohn's disease. J Gastroenterol Hepatol 15:1400–1403
Parks S, Avramopoulos D, Mulle J, McGrath J, Wang R, Goes FS et al (2018) HLA typing using genome wide data reveals susceptibility types for infections in a psychiatric disease enriched sample. Brain Behav Immun 70:203–213
Perron H, Hamdani N, Faucard R, Lajnef M, Jamain S, Daban-Huard C et al (2012) Molecular characteristics of human endogenous retrovirus type-w in schizophrenia and bipolar disorder. Transl Psychiatry 2:e201
Prasad KM, Shirts BH, Yolken RH, Keshavan MS, Nimgaonkar VL (2007) Brain morphological changes associated with exposure to HSV1 in first-episode schizophrenia. Mol Psychiatry 12:105–113, 101
Prasad KM, Eack SM, Goradia D, Pancholi KM, Keshavan MS, Yolken RH et al (2011) Progressive gray matter loss and changes in cognitive functioning associated with exposure to herpes simplex virus 1 in schizophrenia: a longitudinal study. Am J Psychiatry 168:822–830
Prasad KM, Watson AM, Dickerson FB, Yolken RH, Nimgaonkar VL (2012) Exposure to herpes simplex virus type 1 and cognitive impairments in individuals with schizophrenia. Schizophr Bull 38:1137–1148
Presumey J, Bialas AR, Carroll MC (2017) Complement system in neural synapse elimination in development and disease. Adv Immunol 135:53–79
Prichard JC (1837) A treatise on insanity and other disorders affecting the mind. E.L. Carey & A. Hart, Philadelphia
Reiter P (1926) Extrapyramidal motor disturbances in dementia praecox. Acta Psychiatr Neurol 1:287–304
Rosenow EC (1948) Bacteriologic, etiologic, and serologic studies in epilepsy and schizophrenia; cutaneous reactions to intradermal injection of streptococcal antibody and antigen. Postgrad Med 3:367–376
Round JL, O'Connell RM, Mazmanian SK (2010) Coordination of tolerogenic immune responses by the commensal microbiota. J Autoimmun 34:J220–J225
Sampson TR, Mazmanian SK (2015) Control of brain development, function, and behavior by the microbiome. Cell Host Microbe 17:565–576
Sandhya P, Danda D, Sharma D, Scaria V (2016) Does the buck stop with the bugs?: an overview of microbial dysbiosis in rheumatoid arthritis. Int J Rheum Dis 19:8–20
Sandler NG, Douek DC (2012) Microbial translocation in HIV infection: causes, consequences and treatment opportunities. Nat Rev Microbiol 10:655–666
Schizophrenia Working Group of the Psychiatric Genomics Consortium (2014) Biological insights from 108 schizophrenia-associated genetic loci. Nature 511:421–427
Schneck JM (1946) Gastro-intestinal symptomatology in schizophrenia. Am J Dig Dis 13:257–260
Schretlen DJ, Vannorsdall TD, Winicki JM, Mushtaq Y, Hikida T, Sawa A et al (2010) Neuroanatomic and cognitive abnormalities related to herpes simplex virus type 1 in schizophrenia. Schizophr Res 118:224–231
Schwarz E, Maukonen J, Hyytiainen T, Kieseppa T, Oresic M, Sabunciyan S et al (2018) Analysis of microbiota in first episode psychosis identifies preliminary associations with symptom severity and treatment response. Schizophr Res 192:398–403
Sekar A, Bialas AR, de Rivera H, Davis A, Hammond TR, Kamitaki N et al (2016) Schizophrenia risk from complex variation of complement component 4. Nature 530:177–183

Severance EG, Dickerson FB, Halling M, Krivogorsky B, Haile L, Yang S et al (2010) Subunit and whole molecule specificity of the anti-bovine casein immune response in recent onset psychosis and schizophrenia. Schizophr Res 118:240–247
Severance EG, Dickerson FB, Viscidi RP, Bossis I, Stallings CR, Origoni AE et al (2011) Coronavirus immunoreactivity in individuals with a recent onset of psychotic symptoms. Schizophr Bull 37:101–107
Severance EG, Alaedini A, Yang S, Halling M, Gressitt KL, Stallings CR et al (2012) Gastrointestinal inflammation and associated immune activation in schizophrenia. Schizophr Res 138:48–53
Severance EG, Gressitt KL, Stallings CR, Origoni AE, Khushalani S, Leweke FM et al (2013) Discordant patterns of bacterial translocation markers and implications for innate immune imbalances in schizophrenia. Schizophr Res 148:130–137
Severance EG, Gressitt KL, Buka SL, Cannon TD, Yolken RH (2014) Maternal complement c1q and increased odds for psychosis in adult offspring. Schizophr Res 159:14–19
Severance EG, Gressitt KL, Alaedini A, Rohleder C, Enning F, Bumb JM et al (2015a) Igg dynamics of dietary antigens point to cerebrospinal fluid barrier or flow dysfunction in first-episode schizophrenia. Brain Behav Immun 44:148–158
Severance EG, Prandovszky E, Castiglione J, Yolken RH (2015b) Gastroenterology issues in schizophrenia: why the gut matters. Curr Psychiatry Rep 17:27
Severance EG, Gressitt KL, Stallings CR, Katsafanas E, Schweinfurth LA, Savage CL et al (2016a) Candida albicans exposures, sex specificity and cognitive deficits in schizophrenia and bipolar disorder. NPJ Schizophr 2:16018
Severance EG, Xiao J, Jones-Brando L, Sabunciyan S, Li Y, Pletnikov M et al (2016b) Toxoplasma gondii – a gastrointestinal pathogen associated with human brain diseases. Int Rev Neurobiol 131:143–163
Severance EG, Yolken RH, Eaton WW (2016c) Autoimmune diseases, gastrointestinal disorders and the microbiome in schizophrenia: more than a gut feeling. Schizophr Res 176:23–35
Severance EG, Gressitt KL, Stallings CR, Katsafanas E, Schweinfurth LA, Savage CLG et al (2017) Probiotic normalization of Candida albicans in schizophrenia: a randomized, placebo-controlled, longitudinal pilot study. Brain Behav Immun 62:41–45
Shen Y, Xu J, Li Z, Huang Y, Yuan Y, Wang J et al (2018) Analysis of gut microbiota diversity and auxiliary diagnosis as a biomarker in patients with schizophrenia: a cross-sectional study. Schizophr Res. https://doi.org/10.1016/j.schres.2018.01.002
Shi J, Levinson DF, Duan J, Sanders AR, Zheng Y, Pe'er I et al (2009) Common variants on chromosome 6p22.1 are associated with schizophrenia. Nature 460:753–757
Shirts BH, Prasad KM, Pogue-Geile MF, Dickerson F, Yolken RH, Nimgaonkar VL (2008) Antibodies to cytomegalovirus and herpes simplex virus 1 associated with cognitive function in schizophrenia. Schizophr Res 106:268–274
Smith PM, Garrett WS (2011) The gut microbiota and mucosal t cells. Front Microbiol 2:111
Sommer F, Backhed F (2013) The gut microbiota – masters of host development and physiology. Nat Rev Microbiol 11:227–238
Sorensen HJ, Mortensen EL, Reinisch JM, Mednick SA (2009) Association between prenatal exposure to bacterial infection and risk of schizophrenia. Schizophr Bull 35:631–637
Stefansson H, Ophoff RA, Steinberg S, Andreassen OA, Cichon S, Rujescu D et al (2009) Common variants conferring risk of schizophrenia. Nature 460:744–747
Stilling RM, Dinan TG, Cryan JF (2014) Microbial genes, brain & behaviour – epigenetic regulation of the gut-brain axis. Genes Brain Behav 13:69–86
Suvisaari J, Haukka J, Tanskanen A, Hovi T, Lonnqvist J (1999) Association between prenatal exposure to poliovirus infection and adult schizophrenia. Am J Psychiatry 156:1100–1102
Tomasik J, Yolken RH, Bahn S, Dickerson FB (2015) Immunomodulatory effects of probiotic supplementation in schizophrenia patients: a randomized, placebo-controlled trial. Biomark Insights 10:47–54

Tomasik J, Smits SL, Leweke FM, Eljasz P, Pas S, Kahn RS et al (2018) Virus discovery analyses on post-mortem brain tissue and cerebrospinal fluid of schizophrenia patients. Schizophr Res. https://doi.org/10.1016/j.schres.2018.02.012

Torrey EF, Peterson MR (1973) Slow and latent viruses in schizophrenia. Lancet 2:22–24

Torrey EF, Peterson MR (1976) The viral hypothesis of schizophrenia. Schizophr Bull 2:136–146

Torrey EF, Bartko JJ, Lun ZR, Yolken RH (2007) Antibodies to Toxoplasma gondii in patients with schizophrenia: a meta-analysis. Schizophr Bull 33:729–736

Torrey EF, Bartko JJ, Yolken RH (2012) Toxoplasma gondii and other risk factors for schizophrenia: an update. Schizophr Bull 38:642–647

Tsuang M (2000) Schizophrenia: genes and environment. Biol Psychiatry 47:210–220

Watanabe Y, Someya T, Nawa H (2010) Cytokine hypothesis of schizophrenia pathogenesis: evidence from human studies and animal models. Psychiatry Clin Neurosci 64:217–230

Watson AM, Prasad KM, Klei L, Wood JA, Yolken RH, Gur RC et al (2013) Persistent infection with neurotropic herpes viruses and cognitive impairment. Psychol Med 43:1023–1031

Weber NS, Gressitt KL, Cowan DN, Niebuhr DW, Yolken RH, Severance EG (2018) Monocyte activation detected prior to a diagnosis of schizophrenia in the US Military New Onset Psychosis Project (MNOPP). Schizophr Res. https://doi.org/10.1016/j.schres.2017.12.016

Weinberger DR (1987) Implications of normal brain development for the pathogenesis of schizophrenia. Arch Gen Psychiatry 44:660–669

WHO (1949) Manual of the international statistical classification of diseases, injuries and causes of death. World Health Organization, Geneva

Xiao J, Buka SL, Cannon TD, Suzuki Y, Viscidi RP, Torrey EF et al (2009) Serological pattern consistent with infection with type i Toxoplasma gondii in mothers and risk of psychosis among adult offspring. Microbes Infect 11:1011–1018

Yolken RH, Torrey EF (2008) Are some cases of psychosis caused by microbial agents? A review of the evidence. Mol Psychiatry 13:470–479

Yolken RH, Torrey EF, Lieberman JA, Yang S, Dickerson FB (2011) Serological evidence of exposure to Herpes Simplex Virus type 1 is associated with cognitive deficits in the CATIE schizophrenia sample. Schizophr Res 128:61–65

Yolken RH, Severance EG, Sabunciyan S, Gressitt KL, Chen O, Stallings C et al (2015) Metagenomic sequencing indicates that the oropharyngeal phageome of individuals with schizophrenia differs from that of controls. Schizophr Bull 41:1153–1161

Autoantibodies and Psychosis

Eric Kelleher, Helen Barry, David R. Cotter, Aiden Corvin, and Kieran C. Murphy

Contents

E. Kelleher (✉)
Department of Psychiatry and Neurobehavioral Science, University College Cork, Cork, Ireland
e-mail: eric.kelleher@ucc.ie

H. Barry, D. R. Cotter, and K. C. Murphy
Department of Psychiatry, Royal College of Surgeons of Ireland, Dublin, Ireland

A. Corvin
Department of Psychiatry, Trinity College Dublin, Dublin, Ireland

Curr Topics Behav Neurosci (2020) 44: 85–124
DOI 10.1007/7854_2019_90
Published Online: 11 July 2019

Abstract Research into antibody-mediated disease, in response to immune dysfunction or to tumour development, has rapidly expanded in recent years. Antibodies binding to neuroreceptors can cause psychiatric features, including psychosis, in a minority of patients as well as neurological features. The responsiveness of some of these cases to immunotherapy supports the hypothesis that antibody-associated mechanisms play a role in the pathogenesis of psychotic diseases. The purpose of this chapter is to review autoantibodies that are most likely to be relevant for patients with psychotic symptoms. Herein, we describe receptor structure and mechanism of action, clinical and psychiatric features for the growing number of neuronal surface antibodies, including those to the N-methyl-D-aspartate (NMDA) receptor. The identification of a subgroup of patients with psychiatric features having antibody-mediated disease highlights the importance of considering the diagnosis, particularly in those patients presenting with a first episode of psychosis.

Keywords Autoimmune encephalitis · First episode psychosis · NMDA encephalitis · Organic psychosis · Psychoneuroimmunology

1 Background

Inflammation is increasingly recognised as a risk factor for psychosis (Cannon et al. 2014; Khandaker et al. 2015; Kelleher and Corvin 2013). The influence of inflammation was recognised as early as schizophrenia was conceptualised, with Kraepelin suggesting that dementia praecox was a disease of the brain caused by 'an autointoxication' from the body (Kraepelin 1899; Noll 2004). The study of immunity and psychosis incorporates many landmark studies of psychiatric research. Examples include Wagner-Jauregg's treatment of 'general paralysis of the insane' (in fact neurosyphilis) with induced *Plasmodium vivax* malaria (Wagner-Jauregg and Bruetsch 1946) and the systematic reviews examining the association between winter or spring birth and schizophrenia (Davies et al. 2003; Mcgrath and Welham 1999), maternal infection during pregnancy and schizophrenia (Brown et al. 2004; Mortensen et al. 2007; O'Callaghan et al. 1991), childhood infection and schizophrenia (Khandaker et al. 2018, 2012; Dalman et al. 2008; Benros et al. 2011), childhood/adolescent inflammatory markers and risk of schizophrenia and related psychoses (Khandaker et al. 2014; Metcalf et al. 2017; Kappelmann et al. 2019) and maternal infections and inflammation in animal models and their effects on neurodevelopment (Cotter et al. 1995; Farrelly et al. 2015). From a genetic perspective, genome-wide association studies (GWAS) have identified that several common variants of the locus for the major histocompatibility complex (MHC), a group of cell surface proteins that regulate the adaptive immune system, are strongly associated with schizophrenia (Purcell et al. 2009; Shi et al. 2009; Stefansson et al. 2009). Most recently, emerging evidence suggest that the effects of antibodies binding to neuroreceptors, generated in response to immune dysfunction or in response to tumour development, can lead to psychiatric features, providing additional evidence of this interplay. While research has linked schizophrenia with dysfunction in various aspects of immune system, the purpose of this chapter is to review autoantibodies that are most likely relevant for pathogenesis of psychoses.

2 History of the Influence of Antibodies on Psychosis

Early work by Oppenheim (Schulz and Pruss 2015), Lehmann-Facius (Deakin et al. 2014) and Deny-Brown (Denny-Brown 1948) examined links between the immune system, malignancy, circulating antibodies and subsequent neuropsychiatric features. In 1960, Brierley and colleagues reported three patients with 'subacute encephalitis of later adult life, mainly affecting the limbic areas' (Brierley et al. 1960). Corsellis et al. (1968) later described 'limbic encephalitis' in a case series of patients with either short-term memory loss or dementia in association with bronchial carcinoma. All those affected had degenerative changes concentrated in the temporal parts of the limbic grey matter. In the 1980s, paraneoplastic limbic encephalitis associated with antibodies targeting neuronal epitopes (the part of the antigen the antibody binds to) was identified in patients with central nervous system (CNS) and peripheral nervous system (PNS) syndromes who had an underlying cancer (Graus et al. 1985). Subsequently the concept of an immune-mediated pathogenesis gained relevance after anti-Hu (Dalmau et al. 1992) and other onconeuronal antibodies against intracellular antigens were identified (Dalmau and Bataller 2006), some of them with more syndrome specificity for limbic dysfunction than the anti-Hu immune response (Table 1).

Later work identified that, from a psychiatric perspective, the most relevant antibodies are antibodies to the N-methyl-D-aspartate receptor antibody (NMDAR-Ab) and to a lesser extent leucine-rich glioma-inactivated protein (LGI1-Ab) and contactin-associated protein-2 (CASPR2-Ab). These can present with prominent neuropsychiatric features particularly in the early stages of illness (Dalmau et al. 2007, 2008; Irani et al. 2010a, b; Lai et al. 2010; Vitaliani et al. 2005; Zandi et al. 2011). While most are rare, other autoantibodies that have been linked to encephalitis with psychiatric features include dopamine 2 receptor (D2R) (Dale et al. 2012), gamma-aminobutyric acid (GABA) (Lancaster et al. 2010) and α-amino-3-hydroxy-5-methyl-4-isoxazolepropionic receptor (AMPAR) (Lai et al. 2009)

Table 1 Paraneoplastic antibodies that may associate with limbic encephalitis (Dalmau and Bataller 2006)

Antibody	Syndrome	Cancer
Hu	Limbic encephalitis, encephalomyelitis	SCLC, other
Ma	Limbic, hypothalamic and brainstem encephalitis	Testis, lung, other
CV2/ CRMP5	Limbic, striatal encephalitis (chorea), cerebellar ataxia, peripheral neuropathy, uveitis	SCLC, thymoma
Amphiphysin	Limbic encephalitis, stiff-person syndrome	Breast, SCLC

SCLC small cell lung cancer

amongst others. A summary of relevant paraclinical data linking these autoantibodies with psychiatric features is outlined in Table 2.

3 N-Methyl-D-Aspartate Receptor Antibody (NMDAR-Ab)

Since at least 1995, there have been case reports of women with ovarian teratomas presenting with reduced consciousness, psychiatric symptoms and neurological disease whose symptoms improved following tumour removal (Nokura et al. 1997; Okamura et al. 1997). Initially this was believed to be a paraneoplastic process due to an antibody to an unknown antigen expressed in the hippocampus (Vitaliani et al. 2005). In 2007, antibodies were identified as binding to the N-methyl-D-aspartate receptor (NMDAR) in a case series of four females who presented with initial psychotic/altered behaviour and then developed progressive neurological features. This condition was termed anti-NMDA or NMDAR-Ab encephalitis (Dalmau et al. 2007; Irani et al. 2010b), an immune-mediated disorder that occurs when IgG antibodies bind to the GluN1 subunit of the NMDA receptor causing it to be internalised and destroyed (Hughes et al. 2010).

Due to prominent psychiatric symptoms early on in the course of this illness, psychiatry is the specialty that is often in contact with these patients in early stages of their presentation, in approximately 75% of cases according to one report (Dalmau et al. 2011). As highlighted in the first report in the psychiatric literature by Barry et al. (2011), psychosis features prominently. NMDAR-Ab encephalitis supports the influence of glutamatergic dysregulation and has revitalised efforts into whether specific autoantibody syndromes might explain a subset of patients with schizophrenia or psychosis.

Table 2 Clinical features and associations of antibodies relevant to patients with psychiatric features

Antibodies	Psychiatric clinical features	Associated features/symptoms	Gender/age associations	Paraclinical information	Malignancy associations	Clinical course
NMDAR (to NR1 subunit)	Behavioural change (psychosis and mood changes) follows a viral 'prodromal period'. Catatonia and movement disorders, e.g. dyskinesia may then follow the 'psychiatric' phase, usually within 1 month (Titulaer et al. 2013). Autonomic features	'Viral symptoms' prodromal period (1–2 weeks) Altered conscious level Memory deficits New seizures Autonomic instability Dyskinesias Speech disturbance (including mutism)	Young (median age 21) female (Dalmau et al. 2008; Titulaer et al. 2013; Irani et al. 2010b)	Delta brush sign on EEG in 30% of cases (Schmitt et al. 2012) Abnormal EEG in 90% of patients (Dalmau et al. 2008; Irani et al. 2010b; Titulaer et al. 2013) CSF lymphocytosis in 80% of cases (Irani et al. 2010b; Dalmau et al. 2008) CSF oligoclonal bands in 60% cases (Dalmau et al. 2011) Non-specific MRI brain changes in 30% cases (Titulaer et al. 2013; Irani et al. 2010b, Dalmau et al. 2008)	Ovarian teratoma in 30–50% (Irani et al. 2010b; Titulaer et al. 2013)	10–15% patients relapse (Titulaer et al. 2013) and relapses likely monosymptomatic (Kayser et al. 2013) Cognitive deficits (impaired processing speed, episodic memory, executive functioning) reported several years later (Finke et al. 2012; Mckeon et al. 2017) Estimated mortality rate of 4–6% (Dalmau et al. 2011; Titulaer et al. 2013) up to 12% if untreated (Titulaer et al. 2013)
LGI1	Broad range of psychiatric features including mood and anxiety symptoms (depression, apathy, disinhibition and compulsive behaviour) Psychotic features (hallucinations) also	Seizures (initially subtle focal seizures or faciobrachial dystonic seizures later tonic-clonic seizures) followed by memory deficits Confusion Insomnia Autonomic	Median age 60 years, males more likely affected (Van Sonderen et al. 2016c)	Hyponatraemia in ~60% (Irani et al. 2010a, 2013) Abnormal MRI brain (hippocampal T2 hyperintensity) in~56–75% (Irani et al. 2010a, 2013; Lai et al. 2010; Celicanin et al. 2017)	Usually thymoma in ~10%	35% of patients experience a relapse 2-year case fatality rate of 19% (Van Sonderen et al. 2016c)

(continued)

Table 2 (continued)

Antibodies	Psychiatric clinical features	Associated features/ symptoms	Gender/age associations	Paraclinical information	Malignancy associations	Clinical course
	described Spatial disorientation	dysfunction Morvan's syndrome Isolated epilepsy		EEG abnormal 56–89% (Celicanin et al. 2017, Irani et al. 2011; Van Sonderen et al. 2016c) Normal CSF in 75% of cases (Irani et al. 2013, Van Sonderen et al. 2016c)		
CASPR2	Behavioural disturbance Hallucinations Psychosis	Limbic encephalitis Morvan syndrome (peripheral nerve hyperexcitability) Neuromyotonia Muscle spasms/fasciculations Cognitive impairment Seizures Insomnia Autonomic disturbance	Middle-aged (median age 57)/ elderly males (Irani et al. 2010a, 2012; Lancaster et al. 2011a)	Hyponatremia in 10% MRI brain: 25% cases had T2 hyperintensities of medial temporal lobes (Irani et al. 2010a, 2012; Lancaster et al. 2011a) CSF abnormal in 35%–50% of cases (raised WCC, raised protein and/or unmatched oligoclonal bands) (Van Sonderen et al. 2016c; Irani et al. 2012) EEG abnormal in ~60% cases (epileptic	0–32% have malignancy (usually thymoma) (Irani et al. 2010a; Lancaster et al. 2011a; Klein et al. 2013)	Full recovery in ~40% of cases and partial recovery ~12% cases Approximately 25% of cases with >1-year follow-up relapse (median 19 months) (Van Sonderen et al. 2016c)

				or slow waves) (Lancaster et al. 2011a)		
AMPA receptor (GluA1 or GluA2 subunit)	Behavioural change Psychosis, hallucinations/delusions	Short-term memory loss, confusion and abnormal behaviour Seizures, ataxia, abnormal movements	Middle-aged (median age 60, range 38–87) females (Lai et al. 2009)	Rarely hyponatremia (Hoftberger et al. 2015) MRI brain may show increased signal in bilateral or unilateral mesiotemporal lobes (~75% of cases) EEG: 75% of cases abnormal (diffuse slow activity, theta activity or short waves) (Lai et al. 2009) CSF pleocytosis in 50–90% of cases (Lai et al. 2009; Hoftberger et al. 2015)	Thymoma, breast, small cell lung carcinoma in ~70% of cases (Lai et al. 2009)	Immunotherapy and oncological treatment lead to improvement in 70–90% of cases Relapses occur in approximately 16% of cases (Lai et al. 2009; Hoftberger et al. 2015)
$GABA_a$ receptor	Affective problems (mood and anxiety) Behavioural changes, psychotic features (hallucinations)	Limbic encephalitis, status epilepticus, refractory seizures	Median age 22 years, males more likely affected (Petit-Pedrol et al. 2014)	Extensive cortical-subcortical MRI abnormalities in antibodies in CSF/high serum titres EEG may be abnormal in up to 100% of cases CSF abnormal in 84% of cases (Petit-Pedrol et al. 2014)	Hodgkin's lymphoma	Immunotherapy improves outcomes in 50% of cases (Petit-Pedrol et al. 2014)

(continued)

Table 2 (continued)

Antibodies	Psychiatric clinical features	Associated features/symptoms	Gender/age associations	Paraclinical information	Malignancy associations	Clinical course
$GABA_B$ receptor	Behavioural changes, hallucinations	Limbic encephalitis (memory loss, confusion, seizures) Severe seizures/status epilepticus Rarely opsoclonus-myoclonus or cerebellar ataxia prior to limbic encephalitis (Kim et al. 2014)	Typically middle aged (median age 62) (M/F, 1:1) (Lancaster et al. 2010; Kim et al. 2014)	MRI brain abnormal in ~70% of cases (usually unilateral/bilateral increases in medial temporal lobe FLAIR/T2 signal) CSF demonstrates lymphocytic pleocytosis in 80% of cases EEG invariably abnormal ~90% of cases (Lancaster et al. 2010)	~50% of cases have small cell lung carcinoma	75% have partial or complete response to immunotherapy and oncological treatment where indicated (Hoftberger et al. 2013) 30% mortality rate (due to malignancy or chemotherapy treatment) (Lancaster et al. 2010)
D2R	Psychosis, depression, agitation (Dale et al. 2012; Pathmanandavel et al. 2015)	Basal ganglia encephalitis (parkinsonism, chorea, dystonia)	Children, mean (range) age at onset 6.7 (0.4–15) years	MRI basal ganglia changes in 50% of cases CSF abnormal in 75% of cases EEG either normal or non-specific slowing (Dale et al. 2012)	Unclear, not all patients tested	Limited data Variable recovery with and without treatment
DPPX	Depression/apathy initially Psychosis	Cognitive impairment Diarrhoea/weight loss Hyperexcitability Progressive encephalomyelitis with	Median age 53, males more affected (Boronat et al. 2013; Hara et al. 2017; Tobin et al. 2014)	CSF: Abnormal in 55–100% cases EEG: Slowing or epileptiform activity in 85–100% cases MRI brain: May show white matter changes	Rarely B-cell tumours	60% of cases respond to immunotherapy 20% had died Relapses may occur in 23% (Hara et al. 2017)

		rigidity and myoclonus (PERM) (Balint et al. 2014)		in 33% cases (Hara et al. 2017; Boronat et al. 2013)		
MgluR5	Depression, anxiety, delusions, hallucinations, anterograde amnesia	Limbic encephalitis	Median age 25.5 (range 15–46) M/F 1:1 (Lancaster et al. 2011c; Mat et al. 2013)	MRI brain: May show increased signal in mesiotemporal lobes or hyperintensities in posterior parietal-occipital cortex CSF: May show lymphocytosis (Lancaster et al. 2011c; Mat et al. 2013)	Hodgkin's lymphoma	May have complete recovery with immunotherapy/chemotherapy treatment

CSF cerebrospinal fluid, *EEG* electroencephalogram, *F* female, *M* male, *MRI* magnetic resonance imaging, *WCC* white cell count

3.1 Structure of the NMDA Receptor

The NMDA receptor is essential to the development and function of the nervous system and plays a central role in synaptic plasticity and memory formation (Bliss and Collingridge 1993). It is an ionotropic type of receptor with eight alternatively spliced GluN1 isoforms and two GluN3 subunits (A–B), which bind glycine, and four GluN2 subunits (A–D), which bind glutamate. Target epitopes are located in extracellular regions of NR1–NR2B NMDA receptors (Dalmau et al. 2007) with the crucial epitopes in NMDAR-Ab encephalitis present in the more widely expressed NR1 subunit (Dalmau et al. 2008).

In healthy NMDA receptors, glutamate is released from the presynaptic terminal into the synaptic cleft to act on postsynaptic glutamate receptors. Subsequent activation of α-amino-3-hydroxy-5-methyl-4-isoxazolepropionic (AMPA) receptors by glutamate depolarises the postsynaptic membrane and removes a magnesium block on the NMDA receptor (Mayer et al. 1984; Nowak et al. 1984). This allows cations, including calcium ions (Ca2+), to flow into the postsynaptic dendrite, leading to excitatory transmission and synaptic plasticity (Mayer and Westbrook 1987). The strengthening of the impulse between the two neurons is an essential process in synaptic plasticity, cognition and memory formation. The hippocampus contains the highest density of NMDARs which are important for shaping the strength of synaptic connections through involvement in long-term potentiation (LTP) and long-term depression (LTD) (Newcomer and Krystal 2001; Waxman and Lynch 2005).

3.2 Mechanism of NMDAR-Ab Encephalitis

Blockade of the NMDA receptor through NMDA antagonists such as phencyclidine (PCP) and ketamine is well known to cause both positive- and negative-type symptoms of schizophrenia (Javitt 2007) as well as agitation and dissociation (Javitt and Zukin 1991). Psychotic features due to ketamine are proportional to glutamate levels, suggesting a mechanism of action that may lead to psychosis (Stone et al. 2012).

Approximately 50% of adult women presenting with NMDAR-Ab encephalitis have an underlying teratoma, which may contain nervous tissue (Titulaer et al. 2013). In a paraneoplastic model, it is proposed that an antigen, released by tumour cells undergoing apoptosis, is taken up by antigen-presenting cells (Martinez-Hernandez et al. 2011; Moscato et al. 2014). This breaks down immune tolerance by a sequence of mechanisms including antigen presentation by T or dendritic cells generating memory B cells and antibody-producing plasma cells. Systemically synthesised antibodies can bind to NMDA receptors present in the tumour. It is postulated that antibodies can also pass through a disrupted blood-brain barrier

(BBB) or reach the brain through the choroid plexus. The memory B cells (also activated T cells) undergo restimulation (resident antigen-presenting cells or T cells), antigen-driven affinity maturation, clonal expansion and differentiation into NMDA receptor antibody-secreting plasma cells (Moscato et al. 2014). Antibodies bind to extracellular epitopes of the NMDA receptor causing subsequent dysfunction and internalisation.

Greater than 90% of young girls or males with the disorder have no identified trigger, although it may be associated with herpes simplex virus 1 (HSV-1) infection (Irani et al. 2010b; Titulaer et al. 2013; Armangue et al. 2014). Approximately 10–25% of patients with HSV-1 encephalitis have an immune-mediated relapse associated with GluN1-specific antibodies and NMDAR-Ab encephalitis symptoms.

NMDAR-Ab have been shown to cause reversible reduction in neuronal surface NMDA receptors without causing cell death (Hughes et al. 2010; Moscato et al. 2014) leading to reduction in NMDAR-mediated currents and synaptic plasticity. Removal of the pathogenic antibodies through tumour removal and treatment with immunosuppression (see below) leads to clinical improvement (Titulaer et al. 2013). Antibody titres may decline (months or even years) after recovery or in some cases remain detectable (Gresa-Arribas et al. 2014).

3.3 *Incidence and Clinical Characteristics of NMDAR-Ab Encephalitis*

The incidence of NMDAR-Ab encephalitis has been estimated at 0.85 per million children per year in the UK (Wright et al. 2015). While it can affect both men and women of any age, it appears to predominantly affect young women (median age 21 [range 1–95], 81% female) (Titulaer et al. 2013). As mentioned previously, in women over 18 years of age, ~50% have an underlying tumour (overwhelmingly found to be ovarian teratoma) compared to only 5% of men (Dalmau et al. 2011).

About 70% of cases have prodromal symptoms including headache, fever, nausea and respiratory tract infections (Dalmau et al. 2008). Within 2 weeks they can develop psychiatric symptoms characterised by psychotic features, e.g. delusions, hallucinations, paranoia and, less commonly, anxiety and mood symptoms. These are the presenting symptoms in over 65% of cases (Titulaer et al. 2013). Children are more likely to have abnormal movements, e.g. chorea earlier in the disease course, and to experience seizures compared to adults. Psychosis is less common compared to adults, but behavioural regression is frequently noted (Titulaer et al. 2013).

During the first month of the disease, the majority of patients (87%) developed four or more of eight categories of symptoms (see Table 3) (Dalmau et al. 2008). Short-term memory loss is common early in the disease process but can be underestimated because psychopathology and language deficits can dominate the

Table 3 Features of NMDAR-Ab encephalitis

Behavioural change
Memory dysfunction
Speech disorder/mutism
Seizures
Decrease in level of consciousness
Movement disorder
Autonomic dysfunction
Central hypoventilation

clinical presentation. Following this, abnormal movements such as catatonia, chorea, akathisia, dystonia and orofacial dyskinesia manifest (Baizabal-Carvallo et al. 2013; Duan et al. 2016; Mohammad et al. 2014). Features may progress and include autonomic instability leading to cardiac/respiratory arrest or refractory status epilepticus. Patients are often (75%) transferred to the intensive care unit (ICU) at this stage for ventilation, intravenous antiepileptic medication and inotropic support (Titulaer et al. 2013).

3.4 *Diagnosis of NMDAR-Ab Encephalitis*

Criteria to aid the early diagnosis of NMDAR-Ab encephalitis have been published by lead researchers in the field based primarily on clinical features and the results of investigations (Graus et al. 2016) (see Table 4). Techniques that identify NMDAR-Ab in serum and CSF are cell-based assays (CBAs) (used by most clinical laboratories) using live or fixed cells, immunohistochemistry of brain sections adapted to membrane proteins (commercially available; sometimes used as a confirmatory test) and immunocytochemistry of cultures of dissociated rodent live hippocampal neurons (used in research laboratories) (Graus et al. 2016).

Electroencephalogram (EEG) is abnormal in approximately 90% of patients, usually showing non-specific, slow and disorganised activity (Dalmau et al. 2008). Slow rhythmic activity in the delta-theta range (the so-called ‘delta brush’ sign) was initially thought to predominate in the catatonic-like stage (Schmitt et al. 2012); however this feature was later shown to appear in other neurological disorders (Baykan et al. 2018). CSF analysis is abnormal in about 80% of patients, demonstrating lymphocytic pleocytosis (Dalmau et al. 2008; Irani et al. 2010b). Other findings include normal or mildly increased protein concentration and, in 60% of patients, CSF-specific oligoclonal bands (Dalmau et al. 2011). Importantly, CSF abnormalities may be the only remarkable clinical finding (Scott et al. 2018), hence the importance of offering lumbar puncture to patients found to be seropositive for NMDAR-Ab.

Clinical routine brain magnetic resonance imaging (MRI) is abnormal in only 30% of patients approximately, despite patients being significantly unwell at the

Table 4 Diagnostic criteria for anti-NMDA receptor encephalitis (Graus et al. 2016)

Probable anti-NMDA receptor encephalitis Diagnosis can be made when all three of the following criteria have been met: Rapid onset (less than 3 months) of at least four of the six following major groups of symptoms: • Abnormal (psychiatric) behaviour or cognitive dysfunction • Speech dysfunction (pressured speech, verbal reduction, mutism) • Seizures • Movement disorder, dyskinesias or rigidity/abnormal postures • Decreased level of consciousness • Autonomic dysfunction or central hypoventilation At least one of the following laboratory study results: • Abnormal EEG (focal or diffuse slow or disorganised activity, epileptic activity or extreme delta brush) • CSF with pleocytosis or oligoclonal bands Reasonable exclusion of other disorders • Diagnosis can also be made in the presence of three of the above groups of symptoms accompanied by a systemic teratoma
Definite anti-NMDA receptor encephalitis Diagnosis can be made in the presence of one or more of the six major groups of symptoms and IgG anti-GluN1 antibodies, after reasonable exclusion of other disorders. Antibody testing should include testing of CSF. If only serum is available, confirmatory tests should be included (e.g., live neurons or tissue immunohistochemistry, in addition to cell-based assay)

time of imaging (Titulaer et al. 2013). Changes may include white matter hyperintensities. Follow-up brain MRI has been shown to remain normal or show minimum change despite the severity and duration of symptoms (Dalmau et al. 2011). Brain biopsy may show normal or non-specific signs of inflammation. These include perivascular lymphocytic cuffing (predominantly of B cells), sparse parenchymal T-cell infiltrates or microglial activation (Camdessanche et al. 2011).

Patients who have had NMDAR-Ab encephalitis identified should undergo screening for malignancies associated with these occasional paraneoplastic antibodies. In the largest study of patients with NMDAR-Ab encephalitis ($N = 571$) (Titulaer et al. 2013), 38% of patients were found to have underlying malignancy, invariably ovarian teratomas. While teratomas and carcinoma located elsewhere (lung, breast, testicular; ovarian carcinoma, thymic carcinomas and pancreatic cancer) can occur, they are uncommon.

Given the prominent psychiatric, especially psychotic, features, there is significant interest by the psychiatric community in what has been described as a new 'identifiable treatable subtype of psychosis' (Kayser and Dalmau 2016). Therefore, a key issue for psychiatrists is whether patients with NMDAR-Ab encephalitis may have psychiatric features alone and whether patients with psychosis and NMDAR-Ab require a different treatment pathway (Lennox et al. 2012).

3.5 Psychiatric Features Associated with NMDAR-Ab Encephalitis

As outlined above, psychiatric features appear prominently in the early course of NMDAR-Ab encephalitis, and psychiatrists are usually the first clinicians to assess such cases, who usually do not have a previous psychiatric history (Peery et al. 2012). Patients with psychotic disorders such as first episode psychosis (FEP) and schizophrenia would seem a natural cohort to assess for NMDAR antibodies. There has been much debate as to the most robust methodologies used to identify NMDAR-Ab, the relevance of NMDAR-Ab in serum only in the absence of CSF and the significance of IgM and IgA antibodies amongst others.

A study (Kayser et al. 2013) reviewed the presentations of 571 patients with NMDAR-Ab encephalitis, who were identified as having NMDAR-Ab both in serum and CSF with a fixed cell-based assay and immunohistochemistry. The authors found that 23 (~4%) of these had isolated psychiatric symptoms – at the time of presentation (0.9%, 5/571) or at relapse (3.2%, 18/571). When considering that only 64 individuals (11.2%) experienced a relapse, the proportion of patients with psychiatric features at relapse increases (28%, 18/64). Seventy-four percent had some form of delusional thinking, 43% had abnormal perceptions (auditory or visual hallucinations) and 57% had aggressive behaviour. Seventy percent had a mood element to their presentation usually mania, mood lability and impulsivity. Depressive symptoms were less common (Kayser et al. 2013). Psychotic phenomena in presenting cases have been described as fragmented in comparison with those typically found in functional psychoses, with delusions being poorly formed and non-systematised (Barry et al. 2015).

Several screening studies have examined NMDAR-Ab in psychosis populations. A landmark study (Zandi et al. 2011) identified that 3/46 (6.5%) patients with first episode psychosis but without traditional encephalopathic features were seropositive using a cell-based assay with live Human Embryonic Kidney (HEK) cells (Irani et al. 2010b). All three cases met DSM-IV criteria for schizophrenia, and one responded to immunotherapy treatment. Similar results were found in a later study (Pathmanandavel et al. 2015) of 43 children experiencing a first episode of psychosis (FEP) (median age 15 years), which identified antibodies to NMDAR using a flow cytometry live cell-based assay and none in a paediatric control cohort ($N = 43$). Studies utilising fixed and permeabilised CBAs found no difference in the seroprevalences of NMDAR antibodies in patients and in controls (Dahm et al. 2014; Hammer et al. 2014). While some studies (de Witte et al. 2015; Masopust et al. 2015) which used both fixed cell-based assay and immunohistochemistry did not identify any NMDAR-Ab in serum of cohorts of patients with schizophrenia and first episode psychosis, other studies have identified seropositive cases in post-partum psychosis, using this 'double validation' method (Bergink et al. 2015).

In 2014, a meta-analysis by Pollak and colleagues examined the seroprevalence of NMDAR-Ab in 1441 patients with psychotic illnesses and 1,598 healthy controls from several screening studies (Hammer et al. 2014; Haussleiter et al. 2012;

Masdeu et al. 2012; Rhoads et al. 2011; Steiner et al. 2013; Tsutsui et al. 2012; Zandi et al. 2011). This identified 115 individuals [7.98%, 95% confidence interval (CI) 6.69–9.50] who were anti-NMDA receptor antibody positive. Of these, 21 patients (1.46%, 95% CI 0.94–2.23) were positive for antibodies of the IgG subclass. Prevalence rates were greater in cases than controls only for IgG antibodies. A meta-analysis of NMDAR-Ab seropositivity in patients with schizophrenia, schizoaffective disorder, bipolar affective disorder (BPAD) or major depressive disorder (MDD) demonstrated a higher odds ratio of 3.1 compared to healthy controls (Pearlman and Najjar 2014).

Scott et al. (2018) screened all inpatients (children as well as adults up to age 45 years old) admitted with a first episode psychosis for autoantibodies. Importantly, they endeavoured to recruit every FEP case including those who had a drug-induced psychosis, and each participant had a clinical assessment and was investigated with EEG, MRI brain and antibodies both in serum and CSF. The authors identified that approximately 5% (6/113) of participants had autoimmune encephalitis. Four of these cases (3.5%, 4/113) were diagnosed with NMDAR-Ab encephalitis. Two other recent studies (Lennox et al. 2017; Gaughran et al. 2018) which identified an NMDAR-Ab seroprevalence rate of ~3% in first episode psychosis services, with the prevalence in controls varying between 0 and 1%, using a live cell-based assay (L-CBA) only. Although the results of clinical investigations were not described, none of the cases were reported as having developed NMDAR-Ab encephalitis at 6-month follow-up in one study (Lennox et al. 2017). The prevalence of autoimmune encephalitis may be underestimated in research cohorts or community cohorts (who are capacitous, prescreened for organic red flags and often in at least partial remission) (Pollak and Lennox 2018). Therefore, the study by Scott and colleagues is important as they recruited participants as early as possible, clinically investigated them and obtained CSF which is often challenging in psychosis populations and a criticism of studies that use serum only (Leypoldt et al. 2017).

NMDAR-Ab in serum has also been identified in patients with treatment-resistant psychosis with an identified point prevalence of 7% (3/43) using a live cell-based assay (Beck et al. 2015). NMDAR-Ab has also been identified in serum in cases of autism (Creten et al. 2011), bipolar affective disorder (Choe et al. 2013), eating disorders (Perogamvros et al. 2012) and post-partum psychosis (Bergink et al. 2015).

Differences between current assays suggest examining for NMDAR-Ab with at least two techniques, e.g. both a CBA and immunohistochemistry, or binding to cultured neurons, sampling both CSF and serum for NMDAR-Ab whenever possible, and cross-laboratory comparative assays should help understand the differences described (Varley et al. 2015). Recent consensus criteria by Graus et al. (2016) which utilise much of this will ideally lead to clarity for clinicians in the diagnosis of NMDAR-Ab encephalitis. However, such an approach does exclude those patients who have psychotic symptoms alone and are seropositive for NMDAR-Ab but who do not have abnormal investigative findings or who have declined investigations. Yet there are reports that such patients can respond to treatment with immunotherapy and research into this is ongoing (Zandi et al. 2014).

Table 5 Overview of immunotherapy treatment in NMDAR-Ab encephalitis

First-line immunotherapy	Oral/IV steroids (methylprednisolone)
	Intravenous immunoglobulins (IVIG)
	Plasmapheresis
	Plasma exchange
Second-line immunotherapy	Rituximab
	Cyclophosphamide
Other immunotherapy	Azathioprine, mycophenolate mofetil, tacrolimus, methotrexate

3.6 Treatment of NMDAR-Ab Encephalitis

A framework for treatment of NMDAR-Ab encephalitis has emerged (see Table 5) (Titulaer et al. 2013). Utilising this, approximately 50% of patients that are treated with first-line immunotherapy or tumour removal will experience improvement within 4 weeks of treatment, reaching an mRS [modified Rankin Scale, which runs from no symptoms (0) to death (6) (Quinn et al. 2008)] of 0–2 by 24 months (Titulaer et al. 2013). Notably patients with psychosis due to NMDAR-Ab encephalitis may experience worsening of autonomic dysfunction with neuroleptic treatment (Lejuste et al. 2016), and psychiatric symptoms can resolve in response to immunotherapy (Kayser et al. 2013; Zandi et al. 2011, 2014). 70% of those who do not improve with first-line treatment receive second-line immunotherapy (Dalmau et al. 2008; Irani et al. 2010b; Titulaer et al. 2013). Approximately 10–15% of patients relapse in a 2-year period (Titulaer et al. 2013). Relapses are more likely to be monosymptomatic (including psychiatric symptoms only) (Kayser et al. 2013) and result in fewer admissions to ICU (Titulaer et al. 2013). NMDAR-Ab can persist for a prolonged period following recovery (Gresa-Arribas et al. 2014).

3.7 Clinical Course and Relapse

Delays in treatment are associated with cognitive and functional impairment as well as death (Finke et al. 2012; Titulaer et al. 2013). The estimated mortality from NMDAR-Ab encephalitis is 4% with the median time from disease onset to death estimated at three and a half months (Dalmau et al. 2011). Female patients need to be monitored with yearly ultrasound scanning in case ovarian pathology develops. Screening protocols for males and prepubescent (<12) females are unknown (Titulaer et al. 2013). A recent systematic review of cognitive outcomes identified that, while intellectual functioning was more impaired within the acute recovery period than in the later phase of convalescence (Mckeon et al. 2017), rates of impaired processing speed, episodic memory and aspects of executive functioning were consistent across time points. Adverse neuropsychological outcomes occurred at higher frequency in patients where immunotherapy was delayed. Persistent cognitive deficits are reported up to several years post-clinical remission (Finke et al. 2012; Mckeon et al. 2016).

4 Voltage-Gated Potassium Channel Complex Antibodies (VGKCC-Ab)

In 2001, a type of limbic encephalitis characterised by neuropsychiatric features including psychosis, seizures, amnesia and hyponatremia was described in patients with antibodies to the voltage-gated potassium channel receptor (VGKC-Ab) (Buckley et al. 2001). Later work (Irani et al. 2010a; Lai et al. 2010) demonstrated that most VGKC-Ab were directed towards cell surface proteins complexed to the voltage-gated potassium channel subunits (Kv), predominantly leucine-rich glioma-inactivated-1 (anti-LGI1 antibody) and contactin-associated protein-2 (anti-CASPR2 antibody) and more rarely contactin-2. Antibodies to this entire complex were termed voltage-gated potassium channel *complex* antibodies (VGKCC-Ab). However patients positive for LGI1-Ab and CASPR2-Ab accounted for almost all of the immunotherapy-responsive cases, and many VGKC-Ab patients including those without anti-LGI1 antibody and anti-CASPR2 antibody (so-called 'double-negative' VGKC-Ab cases) were tested and were less clearly immunoresponsive (Varley et al. 2017; Grain et al. 2017). Recently, double-negative autoantibodies have been shown to be directed against intracellular aspects of the channel itself as well as against the non-neuronal protein dendrotoxin (DTX). These are both unlikely to be pathogenic autoantibodies (Lang et al. 2017). At present, direct testing for anti-LGI1 antibody and anti-CASPR2 antibody is recommended instead of testing for VGKC-Ab.

Anti-LGI1 antibodies are more often associated with limbic encephalitis and epilepsy, whereas anti-CASPR2 antibodies are associated with movement disorders such as peripheral nerve hyperexcitability, neuromyotonia and Morvan's syndrome. Neuromyotonia is a syndrome of peripheral nerve hyperexcitability (PNH) with fasciculations and cramps (Shillito et al. 1995). Morvan's syndrome is characterised by neuromyotonia, encephalopathy, autonomic dysfunction, insomnia and complex nocturnal behaviours (Klein et al. 2013; Liguori et al. 2001).

LGI1 and CASPR2 antibodies are identified through cell-based assays examining the binding of IgG immunoglobulins to human embryonic kidney (HEK) cells, transfected with complementary DNA encoding the relevant autoantigen. This binding is then visualised using a fluorescence-labelled secondary antibody (Irani et al. 2010a).

4.1 Leucine-Rich Glioma-Inactivated-1 Antibody (LGI1-Ab)

4.1.1 Structure of LGI1

LGI1 is a neuronal secreted synaptic protein expressed mainly in the hippocampus and neocortex (Irani et al. 2010a). It forms a complex with presynaptic disintegrin and metalloproteinase domain-containing protein 23 (ADAM23) and postsynaptic disintegrin and metalloproteinase domain-containing protein 22 (ADAM22). LGI1 may interact with postsynaptic AMPA receptors, postsynaptic density protein 95 (PSD95) and presynaptic shaker Kv1-potassium channels. Through interacting with pre- and postsynaptic proteins, LGI1 may have a role in the regulation of neurotransmitter release (Fukata et al. 2010). It is known that murine mutations of LGI1 disrupt the formation of this complex and alter AMPA-mediated signalling (Fukata et al. 2010; Varley et al. 2017). Mutations in the LGI1 gene are associated with lateral temporal lobe epilepsy and psychiatric features including psychosis (Striano et al. 2011).

4.1.2 Mechanism of Action of LGI1-Ab

LGI1 antibodies cause reversible CNS synaptic dysfunction by several mechanisms (Lancaster and Dalmau 2012). The antibodies may prevent binding of LGI1 to the receptors that it regulates, or they might act on the LGI1-ADAM protein complex. Alternatively, LGI1 antibodies could disrupt currents mediated by Kv1.1 and Kv1.2 and/or impair AMPAR function, either indirectly by blocking LGI1-mediated regulation of these proteins or directly by disrupting the entire protein complex. Binding of antibodies to LGI1 leads to secondary channel dysfunction, caused by a reversible reduction of synaptic AMPA receptors and a loss of function of inhibitory interneurons, with consequent excess neuronal network excitation and seizures (Ohkawa et al. 2013).

A study involving application of serum from a patient with LGI1 antibodies to a hippocampal slice preparation showed increased nerve hyperexcitability, effects similar to application of a Kv1.1 and Kv1.2 antagonist (Lalic et al. 2011). Antibody titres appear to correlate with clinical presentation, with immunotherapy treatment resulting in clinical improvement in patients (Malter et al. 2014). LGI1 antibodies are strongly associated with Human Leukocyte Antigen-DR 7 isotype (HLA-DR7) and HLA-DRB4 in nontumour patients, supporting the autoimmune hypothesis (Kim et al. 2017; Van Sonderen et al. 2017).

4.1.3 Clinical Characteristics of LGI1-Ab Encephalitis

The median age of onset of LGI1-Ab encephalitis is around 60 years and most often occurs in males (Irani et al. 2010a; Lai et al. 2010). Cases may develop neuropsychiatric features, including mood disturbances, psychosis, amnesia and disorientation (Zandi et al. 2011; Irani et al. 2013; Navarro et al. 2016). 80% of patients with limbic encephalitis due to LGI1 antibodies present with a variety of seizures, including faciobrachial dystonic seizures (FBDS) (Irani et al. 2013). These are brief (<3 s) unilateral contractions of the arm, often involving the ipsilateral face (or leg) and occurring up to 100 times a day. Subsequently, patients develop subacute onset amnesia and behavioural/psychiatric disturbances. Seizures may develop into tonic clonic type (Irani et al. 2010a; Lai et al. 2010).

Hyponatremia is present in at least 60% of patients which may be due to the syndrome of inappropriate antidiuretic hormone (SIADH) (Van Sonderen et al. 2016b, c). One possible mechanism is inflammation of the hypothalamic-pituitary neuraxis (Newey and Sarwal 2014). Brain MRI shows T2 high signal of the medial temporal lobe in approximately two-thirds of patients (Irani et al. 2010a; Lai et al. 2010; Shin et al. 2013). EEG can be abnormal in up to 90% of cases (Celicanin et al. 2017; Zandi et al. 2011; Van Sonderen et al. 2016c). CSF cell count and protein are unremarkable in 75% of cases (Van Sonderen et al. 2016c). Tumour screening is positive in up to 11% of the patients with thymoma and lung cancer being the most common; however other malignancies have been associated (Irani et al. 2010a; Lai et al. 2010; Malter et al. 2014).

4.1.4 Psychiatric Features Associated with LGI1-Ab

Both mood and psychotic symptoms are described in association with LGI1 antibodies (Pruss and Lennox 2016). Additional psychiatric features include apathy, disinhibition and compulsive behaviour (Van Sonderen et al. 2016c). The prevalence of LGI1 antibodies has been found to be 0.1% in patients with established schizophrenia in one study and none were identified in patients with affective disorders or borderline personality disorder (Dahm et al. 2014). A more recent study of patients with first episode psychosis ($N = 228$) and controls ($N = 105$) routinely screened for LGI1-Ab found no significant difference between both cases and controls (Lennox et al. 2017). Anti-LGI1 encephalitis is reported occurring post-partum with prominent anxiety features that responded to immunotherapy (Gotkine et al. 2011). Psychotic features with LGI1-Ab encephalitis show a response to treatment with immunotherapy (Klein et al. 2013). New psychotic illness have also been described in individuals following treatment for LGI1-Ab encephalitis (Pollak and Moran 2017).

4.1.5 Treatment of LGI1-Ab Encephalitis and Follow-Up

Treatment pathways are similar to those utilised in NMDAR-Ab encephalitis with first- and second-line immunotherapies. In refractory cases, rituximab and cyclophosphamide may be used (second-line therapies) (Lancaster et al. 2011b). Treatment with first-line immunotherapy in 32 patients with anti-LGI1 encephalitis was considered effective in 80% of cases (Van Sonderen et al. 2016c). Shorter time to start immunotherapy allows resolution of FBDS and recovery of basal functions (Irani et al. 2011).

At 2-year follow-up of 21 patients with LGI1-Ab encephalitis, 67% of cases had a favourable outcome (measured by mRS of 0–2). Six of seventeen (35%) patients with available data had a relapse of symptoms, with a median time to relapse of 35 months (range 21–98 months) (Van Sonderen et al. 2016c). Cognitive assessment of 11 immunotherapy treated individuals at median 44 months (range 25–95) follow-up showed reduced spatial recognition, but they were otherwise normal compared to normative data (Van Sonderen et al. 2016c). Other studies have reported that patients may develop hippocampal atrophy and memory deficits (Butler et al. 2014; Malter et al. 2014).

4.2 Contactin-Associated Protein-2 Antibody (CASPR2-Ab)

4.2.1 Structure of CASPR2

CASPR2 is a transmembrane axonal protein of the neurexin IV superfamily that is localised to the juxtaparanode of myelinated axons (Poliak et al. 1999). It's extracellular domain interacts with contactin-2, and it connects with the cytoskeleton via protein 4.1B. CASPR2, contactin-2 and protein 4.1B are necessary to concentrate Kv1.1 and Kv1.2 channels in the juxtaparanode, which is important for the proper electrical functioning of axons in both the CNS and PNS (Lancaster and Dalmau 2012; Zhou et al. 1999).

4.2.2 Mechanism of Action of CASPR2-Ab

CASPR2 antibodies are widely thought to act by disrupting axonal potassium currents. Factors such as differences in time to establish intrathecal antibody synthesis, or in the structure of tight, septate-like junctions of myelinating cells around the axons, may explain the variability of PNS and CNS symptoms in patients with CASPR2 antibodies (Lancaster and Dalmau 2012). Genetic variation in the human gene encoding CASPR2 (CNTNAP2) is associated with autism (Jackman et al. 2009; Whalley et al. 2011), epilepsy (Strauss et al. 2006), Tourette syndrome, obsessive-compulsive disorder (Verkerk et al. 2003), schizophrenia (Malhotra and Sebat 2012), Pitt-Hopkins syndrome (Zweier et al. 2009) and other mental disabilities (Gregor et al. 2011).

4.2.3 Clinical Characteristics of CASPR2-Ab-Related Disease

Patients typically are males with age of onset later in life, e.g. approximately 66 years (Lai et al. 2010; Lancaster et al. 2011a; Irani et al. 2010a; Van Sonderen et al. 2016a). Clinical syndromes include neuromyotonia alone, a purely CNS-based limbic encephalitis or symptoms of both in Morvan's syndrome (Varley et al. 2017). They can also experience peripheral nerve hyperexcitability (PNH). Limbic encephalitis with CASPR2-Ab can be associated with autonomic features, pain or cerebellar symptoms (Irani et al. 2010a) and can potentially mimic Creutzfeldt-Jakob disease (CJD). Indeed some patients with a positive CASPR2-Ab have been subsequently diagnosed with CJD (Rossi et al. 2015).

In a case series of 38 patients with CASPR2 antibody-associated disease (Van Sonderen et al. 2016a), the most common presenting symptoms were cognitive disturbance, seizures, PNH or neuropathic pain. During the disease course, approximately 80% of patients experienced cognitive dysfunction, and 53% had seizures. Paraclinical data is outlined in Table 2. CASPR2 antibodies are associated with malignancy in approximately 20% of cases (Van Sonderen et al. 2016a). These are usually thymomas, found in 40% of cases of Morvan's syndrome and 10% of other presentations (Vincent and Irani 2010; Lancaster et al. 2010).

4.2.4 Psychiatric Features Associated with CASPR2-Ab

Behavioural disturbances, persecutory delusions, hallucinations and psychosis have been reported in cases with CASPR2 antibodies (Irani et al. 2012; Lancaster et al. 2011a; Van Sonderen et al. 2016a). Rare mutations of the CNTNAP2 gene that codes for CASPR2 protein (Ottman et al. 2004) can cause a clinical presentation of auditory hallucinations and partial epilepsy. CASPR2-Ab have been estimated at 0.9% in one large study of 2,533 seropositive patients with a variety of neuropsychiatric illness including schizophrenia (Dahm et al. 2014), although the seroprevalence rate in FEP in another study was not significantly different to healthy controls (Lennox et al. 2017).

4.2.5 Treatment of CASPR-Ab-Related Disease and Follow-Up

Treatment of CASPR2-Ab disease includes identifying and treating any malignancy and treatment with first- and second-line immunotherapy. Full recovery can be achieved in ~40% of cases and partial recovery with 12%. Approximately 25% of cases with >1-year follow-up had relapses presenting at a median of 19 months post initial episode (Van Sonderen et al. 2016a).

5 α-Amino-3-hydroxy-5-methyl-4-isoxazolepropionic Acid Receptor Antibody (AMPAR-Ab)

5.1 *Structure and Mechanism of Action*

The α-amino-3-hydroxy-5-methyl-4-isoxazolepropionic acid receptor (AMPAR) is a gated ion channel which consists of GluA 1–4 tetrameric subunits and is part of the family of glutamate receptors. They mediate fast excitatory synaptic transmission, necessary for learning, memory and synaptic plasticity (Henley and Wilkinson 2016). Antibodies target the extracellular domains of GluA1 and GluA2 subunits of the AMPA receptor and were first described in ten patients with limbic encephalitis (Lai et al. 2009). They cause a selective reversible decrease in the total surface amount and synaptic location of GluA1 and GluA2 through increased internalisation and degradation. This leads to a decrease of AMPAR-mediated currents (Lai et al. 2009).

5.2 *Clinical Features of AMPAR-Ab-Related Disease*

AMPAR antibody-associated disease mostly affects older females in response to malignancy (lung, breast, thymus) in approximately 65% of cases (Hoftberger et al. 2015; Lai et al. 2009). Clinical features vary but may include those of limbic encephalitis (short-term memory deficits, confusion, abnormal behaviour and seizures). Seizures are frequently present. Insomnia, lethargy and decreased level of consciousness have also been described (Graus et al. 2010; Hoftberger et al. 2015). Paraclinical data is available in Table 2.

5.3 *Psychiatric Features Associated with AMPAR-Ab*

Almost one in three patients (27%) in a case series of AMPAR-Ab encephalitis ($N = 22$) had psychotic features as part of their presentation (Hoftberger et al. 2015). One of these cases had a background history of BPAD. Recent screening studies however have identified no differences in the seroprevalence of AMPAR-Ab in FEP compared to healthy controls (Lennox et al. 2017). An earlier study reported that AMPAR-Ab were not found in a cohort of schizophrenia, affective disorders or borderline personality disorders (Dahm et al. 2014). However an additional study described that 3 of 4,819 patients with neuropsychiatric presentations had GluR2 AMPAR antibodies and that all had memory deficits and mood symptoms (Dogan Onugoren et al. 2015). Additional case reports of atypical psychosis (Graus et al. 2010) and mood and psychotic symptoms (Elamin et al. 2015) responding to immunotherapy treatment are described. Antibodies to AMPAR have been identified in patients with Turner's syndrome with co-morbid bipolar disorder and psychotic features (Quaranta et al. 2015).

5.4 Treatment and Clinical Course

Treatment with immunotherapy or oncological treatment leads to an improvement in the majority (70–90%) of cases (Hoftberger et al. 2015; Lai et al. 2009). Relapses occur in approximately 16% of cases which may be reduced by the use of more aggressive forms of treatment (chemotherapy or rituximab).

6 γ-Aminobutyric Acid A Receptor ($GABA_AR$) and γ-Aminobutyric Acid B Receptor ($GABA_BR$) Antibody Disease

6.1 Structure and Mechanism of Action

GABA receptors act as inhibitory receptors in the central nervous system through mediation of GABA (γ-Aminobutyric acid), the major inhibitory neurotransmitter. $GABA_A$ receptor ($GABA_AR$) is a ligand-gated chloride channel that mediates fast inhibitory synaptic transmission in the CNS. At the synapse, most $GABA_AR$s contain 2 α subunits, 2 β subunits and 1 γ subunit, arranged as γ-β-α-β-α (Sieghart 2006). Genetic or pharmacological alteration of the $GABA_A$ receptor has been associated with seizures (Baulac et al. 2001; Sun et al. 2013). $GABA_B$ receptors are heterodimers comprising of the $GABA_BR$ subunits 1 and 2 with $GABA_B1$ subunits containing the ligand-binding domain (Pagano et al. 2001). $GABA_BR$s are G-protein-coupled receptors coupled indirectly to either calcium or potassium channels to produce prolonged inhibitory responses (Bowery 2010) and are found in both the CNS (cortex, cerebellum, thalamus and hippocampus) and the PNS (autonomic ganglia, visceral tissue) (Benarroch 2012). GABAB1R polymorphisms are associated with temporal lobe epilepsy (Xi et al. 2011), schizophrenia and obsessive-compulsive disorder (Zai et al. 2005, 2009).

6.2 Clinical and Psychiatric Features of GABA Antibody-Related Disease

Anti-$GABA_AR$ antibody encephalitis has been described relatively recently, and the spectrum of symptoms has not been fully defined. One study (Pettingill et al. 2015) identified autoantibodies against the $\alpha 1$ and/or $\gamma 2$ subunits of the GABA receptors in

40 of 2046 patients whose serum was negative for other neuronal surface antibodies. Features of 15 representative patients included seizures (47%), memory impairment (47%), hallucinations (33%) and anxiety (27%). Notably one of these patients had pre-existing obsessive-compulsive disorder, and diagnoses for other cases included paranoid schizophrenia. The majority of patients in this study were not treated with immunotherapy. Antibodies to the GABA $\alpha 1/\beta 3$ subunits were identified at high serum titres as well as in the CSF of patients with limbic encephalitis, status epilepticus or epilepsia partialis continua (Petit-Pedrol et al. 2014). Their presentations included affective problems, behavioural changes and/or psychotic features. All cases developed extensive cortical-subcortical MRI abnormalities and had a high mortality and a variable response to immunotherapy. Forty percent of patients with anti-$GABA_A R$ encephalitis have tumours, mostly thymomas and, less commonly, other neoplasms (e.g. Hodgkin lymphoma, multiple myeloma). See Table 2 for paraclinical data.

Anti-$GABA_B$ receptor encephalitis was first described in 15 cases with limbic encephalitis with a tendency towards severe seizures (Lancaster et al. 2010). Behavioural disturbances and psychotic features (delusions, paranoia, gustatory and visual hallucinations) formed part of the presenting complaint in 4 of these 15 cases (~25%) and in 5% (1/20) of another series (Hoftberger et al. 2013). Additional studies since then (Lancaster et al. 2010, 2011a; Hoftberger et al. 2013; Kim et al. 2014) show that the majority of patients present with seizures, confusion and memory deficits. Ataxia and opsoclonus-myoclonus have also been reported (Hoftberger et al. 2013). $GABA_B$ receptor antibodies have been identified in 0.3% of affective disorders (Dahm et al. 2014). About 50% of $GABA_B$ receptor encephalitis cases have lung cancer (mainly small cell lung cancer) (Hoftberger et al. 2013; Lancaster et al. 2010). Underlying carcinoid of the thymus (Lancaster et al. 2011a) and melanoma (Jarius et al. 2013) have also been reported.

6.3 Treatment and Clinical Course

Immunotherapy and treatment of malignancy resulted in substantial improvement in 18/21 (86%) patients with $GABA_A$ receptor encephalitis. Fourteen percent of cases died from status epilepticus or sepsis (Spatola et al. 2017). Approximately 75% of patients with $GABA_B$ receptor encephalitis show a partial or complete response to immunotherapy and oncological treatment (Hoftberger et al. 2013).

7 Dopamine 2 Receptor Antibody (D2R-Ab)

7.1 *Structure and Mechanism of Action*

Dopamine signalling is mediated through dopamine receptors. These are G-protein-coupled seven-transmembrane domain receptors. Five receptors are divided into two groups: D1- (D1R and D5R) and D2-class receptors (D2R, D3R and D4R) (Dale et al. 2012). The dopamine 2 receptor (D2R) has long isoforms (located on the postsynaptic membrane) and short isoforms (on presynaptic membrane) coded by alternative splicing of the same D2R gene (Usiello et al. 2000). D2R are highly expressed in the basal ganglia, cortex, hippocampus and substantia nigra and involved in synaptic plasticity and memory formation (Beaulieu and Gainetdinov 2011). D2R expression modulation has been associated with schizophrenia, depression and movement disorders (Beaulieu and Gainetdinov 2011).

7.2 *Clinical and Psychiatric Features*

Clinical features of D2R antibody disease include prominent movement disorders (parkinsonism, dystonia and chorea). Psychiatric features including psychotic features can occur in 25% of cases from one series (Dale et al. 2012). Where MRI brain was abnormal, findings were localised to the basal ganglia. EEG was abnormal in a minority of cases. D2R antibodies were also reported in 3 of 43 (7%) children with FEP (Pathmanandavel et al. 2015) using a live CBA. They were not identified in patients with schizophrenia, major depression or borderline personality disorder (Muller et al. 2014). D2R antibodies have been found in PANDAS (paediatric autoimmune neuropsychiatric disorders associated with streptococcal infection) (Cox et al. 2013).

7.3 *Treatment and Clinical Course*

There is limited data on treatment of patients with D2R antibodies. In a follow-up of 12 cases with D2R antibodies who received first-line immunotherapy, 5/12 are described as returning to baseline at 5-year follow-up; however residual motor, cognitive and psychiatric deficits were common in other cases (Dale et al. 2012).

8 Dipeptidyl-Peptidase-Like Protein 6 Antibody (DPPX-Ab)

8.1 *Structure*

DPPX (dipeptidyl-peptidase-like protein 6) is an auxiliary subunit of Kv4.2 potassium channels. It is a membrane glycoprotein involved in increasing the surface expression and channel conductance of Kv4.2 channels (Kaulin et al. 2009). Kv4.2 channels and DPPX are distributed throughout the nervous and enteric system (Boronat et al. 2013; Tobin et al. 2014).

8.2 *Clinical Features*

Hara et al. (2017) reviewed 39 known cases to date (median age 57 years); of these the majority (67%) developed weight loss/diarrhoea, behavioural disturbances and cognitive dysfunction and CNS hyperexcitability. DPPX antibodies have been associated with progressive encephalomyelitis with rigidity and myoclonus (PERM) (Balint et al. 2014). Psychiatric manifestations may occur with prominent psychotic features or behavioural disturbances that may lead to admission under psychiatry (Boronat et al. 2013) or neuropsychiatry (Hara et al. 2017). Forty-four percent of cases may have mood symptoms (appetite loss, depression and apathy) at the time of presentation (Hara et al. 2017). See Table 2 for paraclinical data.

Routine screening in a cohort of patients with first episode psychosis and high-risk psychosis did not reveal any seropositive cases (Mantere et al. 2017). The prevalence of DPPX antibodies in a cohort of patients with schizophrenia was reported as less than 1% (Dahm et al. 2014).

8.3 *Treatment and Clinical Course*

Sixty percent of cases had moderate-substantial improvement in response to treatment with immunotherapy. At follow-up, 20% of anti-DPPX encephalitis cases had died. Relapse occurred in 8 of 35 patients (23%) and was responsive to immunotherapy (Hara et al. 2017).

9 Rare Autoimmune Syndromes

Antibodies to the metabotropic glutamate receptor 5 (mGluR5), highly expressed in the hippocampus, are associated with Hodgkin's lymphoma (Carr 1982; Lancaster et al. 2011c; Mat et al. 2013; Pruss et al. 2014). Symptoms include depression, agitation, hallucinations and bizarre behaviour. Patient can have complete recovery from this rare syndrome following immunotherapy/oncological therapy (Leypoldt et al. 2015). Although glycine receptor and voltage-gated calcium channel (VGCC) antibodies are associated with characteristic neurological symptoms, they are rarely associated with psychiatric symptoms (Pollak et al. 2016).

10 Summary and Future Directions

The concept and emerging evidence of antibodies driving psychiatric symptoms particularly psychosis are important. This may lead to a better understanding of the mechanisms of schizophrenia and improved treatment pathways in a proportion of individuals found to have antibody-related disease. Early identification of such and treatment with immunotherapy lead to improved cognitive and functional outcomes. Autoimmune encephalitis frequently affects individuals with no previous psychiatric history, and behavioural change and psychosis are some of the earliest features. Careful history taking (including collateral history where available) and clinical examination for sometimes subtle neurological signs, CSF and paraclinical investigations, e.g. MRI brain and EEG, lead to diagnosis and guide treatment decisions (Midgley et al. 2018). Routine serum screening for such antibodies in patients with first episode psychosis is recommended (Lennox et al. 2017). For patients with isolated psychosis and NMDAR-Ab in serum, there appears to be no difference in psychopathology compared to other patients with psychosis (Lennox et al. 2017).

Examining for autoantibodies (in particular NMDAR-Ab) in several large cohorts of patients with first episode psychosis, treatment-resistant psychosis, established schizophrenia and healthy controls has shown that autoantibodies are identified in a small proportion of patients with psychosis. The lack of paired CSF samples in some studies has been criticised; however performing lumbar puncture in patients with new psychiatric symptoms can be challenging as patients may refuse and in those who are not neurologically ill it may be hard to justify. However, it is important that the seropositive patients are offered this investigation which may in some cases be the only abnormal finding (Scott et al. 2018).

This approach does however leave patients who present with psychotic symptoms and who have a positive serum neuronal autoantibody test result but who do not have EEG, neuroimaging or CSF abnormalities with diagnostic uncertainty. While rare, there is evidence that patients with psychosis and NMDAR-Ab in serum who are refractory to regular psychiatric treatment and receive immunotherapy may demonstrate clinical improvement (Zandi et al. 2014). Such patients have been designated 'synaptic and neuronal autoantibody-associated psychiatric syndromes' (SNAps) by Al-Diwani et al. (2017). Whether such antibodies identified in serum alone affect isolated psychotic symptoms is currently under investigation.

The SINNAPS2 trial (clinical trial number NCT03194815) is a randomised phase II double-blinded placebo-controlled trial designed to explore the utility of immunotherapy for patients with acute psychosis associated with anti-neuronal antibodies including NMDAR and LGI1 (Lennox et al. 2018). This trial may answer whether patients with psychosis and cell surface antibodies improve with immunotherapy treatment. If so, it could fundamentally change how we screen and treat patients with psychosis in the future.

For the present, awareness by psychiatrists of the features of autoimmune encephalitis is essential. Psychiatric features are early clinical signs of NMDAR-Ab encephalitis and neurological features may only evolve when the patient is already under psychiatric care (Barry et al. 2015). Hence, psychiatrists and psychiatry services need to be aware of emerging clinical signs which may be subtle or when the individual fails to respond to standard treatment or potentially has an adverse reaction to treatment with antipsychotics. Training and education of staff on the features of encephalitis is especially important in a time when psychiatric services are becoming increasingly demedicalised (Craddock et al. 2008; Oyebode and Humphreys 2011). In addition to considering the diagnosis, discussing such cases with clinical neurology and agreeing joint management strategies between psychiatrists and neurologists provide the best opportunity to enhance patient care at present.

References

Al-Diwani A, Pollak TA, Langford AE, Lennox BR (2017) Synaptic and neuronal autoantibody-associated psychiatric syndromes: controversies and hypotheses. Front Psych 8:13

Armangue T, Leypoldt F, Málaga I, Raspall-Chaure M, Marti I, Nichter C, Pugh J, Vicente-Rasoamalala M, Lafuente-Hidalgo M, Macaya A, Ke M, Titulaer MJ, Höftberger R, Sheriff H, Glaser C, Dalmau J (2014) Herpes simplex virus encephalitis is a trigger of brain autoimmunity. Ann Neurol 75(2):317–323

Baizabal-Carvallo JF, Stocco A, Muscal E, Jankovic J (2013) The spectrum of movement disorders in children with anti-NMDA receptor encephalitis. Mov Disord 28:543–547

Balint B, Jarius S, Nagel S, Haberkorn U, Probst C, Blöcker IM, Bahtz R, Komorowski L, Stöcker W, Kastrup A, Kuthe M, Meinck HM (2014) Progressive encephalomyelitis with rigidity and myoclonus: a new variant with DPPX antibodies. Neurology 82(17):1521–1528

Barry H, Byrne S, Barrett E, Murphy KC, Cotter DR (2015) Anti-N-methyl-d-aspartate receptor encephalitis: review of clinical presentation, diagnosis and treatment. BJPsych Bull 39:19–23

Barry H, Hardiman O, Healy DG, Keogan M, Moroney J, Molnar PP, Cotter DR, Murphy KC (2011) Anti-NMDA receptor encephalitis: an important differential diagnosis in psychosis. Br J Psychiatry 199(6):508–509

Baulac S, Huberfeld G, Gourfinkel-An I, Mitropoulou G, Beranger A, Prud'homme JF, Baulac M, Brice A, Bruzzone R, LeGuern E (2001) First genetic evidence of GABA(A) receptor dysfunction in epilepsy: a mutation in the gamma2-subunit gene. Nat Genet 28(1):46–48

Baykan B, Gungor Tuncer O, Vanli-Yavuz EN, Baysal Kirac L, Gundogdu G, Bebek N, Gurses C, Altindag E, Tuzun E (2018) Delta brush pattern is not unique to NMDAR encephalitis: evaluation of two independent long-term EEG cohorts. Clin EEG Neurosci 49(4):278–284
Beaulieu JM, Gainetdinov RR (2011) The physiology, signaling, and pharmacology of dopamine receptors. Pharmacol Rev 63:182–217
Beck K, Lally J, Shergill SS, Bloomfield MA, Maccabe JH, Gaughran F, Howes OD (2015) Prevalence of serum N-methyl-D-aspartate receptor autoantibodies in refractory psychosis. Br J Psychiatry 206:164–165
Benarroch EE (2012) GABAB receptors: structure, functions, and clinical implications. Neurology 78:578–584
Benros ME, Nielsen PR, Nordentoft M, Eaton WW, Dalton SO, Mortensen PB (2011) Autoimmune diseases and severe infections as risk factors for schizophrenia: a 30-year population-based register study. Am J Psychiatry 168:1303–1310
Bergink V, Armangue T, Titulaer MJ, Markx S, Dalmau J, Kushner SA (2015) Autoimmune encephalitis in postpartum psychosis. Am J Psychiatry 172:901–908
Bliss TV, Collingridge GL (1993) A synaptic model of memory: long-term potentiation in the hippocampus. Nature 361:31–39
Boronat A, Gelfand JM, Gresa-Arribas N, Jeong HY, Walsh M, Roberts K, Martinez-Hernandez E, Rosenfeld MR, Balice-Gordon R, Graus F, Rudy B, Dalmau J (2013) Encephalitis and antibodies to dipeptidyl-peptidase-like protein-6, a subunit of Kv4.2 potassium channels. Ann Neurol 73:120–128
Bowery NG (2010) Historical perspective and emergence of the GABAB receptor. Adv Pharmacol 58:1–18
Brierley JB, Corsellis JAN, Hierons R, Nevin S (1960) Subacute encephalitis of later adult life. Mainly affecting the limbic areas. Brain 83:357–368
Brown AS, Hooton J, Schaefer CA, Zhang H, Petkova E, Babulas V, Perrin M, Gorman JM, Susser ES (2004) Elevated maternal interleukin-8 levels and risk of schizophrenia in adult offspring. Am J Psychiatry 161:889–895
Buckley C, Oger J, Clover L, Tuzun E, Carpenter K, Jackson M, Vincent A (2001) Potassium channel antibodies in two patients with reversible limbic encephalitis. Ann Neurol 50:73–78
Butler CR, Miller TD, Kaur MS, Baker IW, Boothroyd GD, Illman NA, Rosenthal CR, Vincent A, Buckley CJ (2014) Persistent anterograde amnesia following limbic encephalitis associated with antibodies to the voltage-gated potassium channel complex. J Neurol Neurosurg Psychiatry 85:387–391
Camdessanche JP, Streichenberger N, Cavillon G, Rogemond V, Jousserand G, Honnorat J, Convers P, Antoine JC (2011) Brain immunohistopathological study in a patient with anti-NMDAR encephalitis. Eur J Neurol 18:929–931
Cannon M, Clarke MC, Cotter DR (2014) Priming the brain for psychosis: maternal inflammation during fetal development and the risk of later psychiatric disorder. Am J Psychiatry 171(9):901–905
Carr I (1982) The Ophelia syndrome: memory loss in Hodgkin's disease. Lancet 1:844–845
Celicanin M, Blaabjerg M, Maersk-Moller C, Beniczky S, Marner L, Thomsen C, Bach FW, Kondziella D, Andersen H, Somnier F, Illes Z, Pinborg LH (2017) Autoimmune encephalitis associated with voltage-gated potassium channels-complex and leucine-rich glioma-inactivated 1 antibodies – a national cohort study. Eur J Neurol 24:999–1005
Choe CU, Karamatskos E, Schattling B, Leypoldt F, Liuzzi G, Gerloff C, Friese MA, Mulert C (2013) A clinical and neurobiological case of IgM Nmda receptor antibody associated encephalitis mimicking bipolar disorder. Psychiatry Res 208:194–196
Corsellis JA, Goldberg GJ, Norton AR (1968) "Limbic encephalitis" and its association with carcinoma. Brain 91:481–496
Cotter D, Takei N, Farrell M, Sham P, Quinn P, Larkin C, Oxford J, Murray RM, O'Callaghan E (1995) Does prenatal exposure to influenza in mice induce pyramidal cell disarray in the dorsal hippocampus? Schizophr Res 16:233–241

Cox CJ, Sharma M, Leckman JF, Zuccolo J, Zuccolo A, Kovoor A, Swedo SE, Cunningham MW (2013) Brain human monoclonal autoantibody from sydenham chorea targets dopaminergic neurons in transgenic mice and signals dopamine D2 receptor: implications in human disease. J Immunol 191:5524–5541

Craddock N, Antebi D, Attenburrow MJ, Bailey A, Carson A, Cowen P, Craddock B, Eagles J, Ebmeier K, Farmer A, Fazel S, Ferrier N, Geddes J, Goodwin G, Harrison P, Hawton K, Hunter S, Jacoby R, Jones I, Keedwell P, Kerr M, Mackin P, Mcguffin P, Macintyre DJ, Mcconville P, Mountain D, O'donovan MC, Owen MJ, Oyebode F, Phillips M, Price J, Shah P, Smith DJ, Walters J, Woodruff P, Young A, Zammit S (2008) Wake-up call for British psychiatry. Br J Psychiatry 193:6–9

Creten C, Van Der Zwaan S, Blankespoor RJ, Maatkamp A, Nicolai J, Van Os J, Schieveld JN (2011) Late onset autism and anti-Nmda-receptor encephalitis. Lancet 378:98

Dahm L, Ott C, Steiner J, Stepniak B, Teegen B, Saschenbrecker S, Hammer C, Borowski K, Begemann M, Lemke S, Rentzsch K, Probst C, Martens H, Wienands J, Spalletta G, Weissenborn K, Stocker W, Ehrenreich H (2014) Seroprevalence of autoantibodies against brain antigens in health and disease. Ann Neurol 76:82–94

Dale RC, Merheb V, Pillai S, Wang D, Cantrill L, Murphy TK, Ben-Pazi H, Varadkar S, Aumann TD, Horne MK, Church AJ, Fath T, Brilot F (2012) Antibodies to surface dopamine-2 receptor in autoimmune movement and psychiatric disorders. Brain 135:3453–3468

Dalman C, Allebeck P, Gunnell D, Harrison G, Kristensson K, Lewis G, Lofving S, Rasmussen F, Wicks S, Karlsson H (2008) Infections in the CNS during childhood and the risk of subsequent psychotic illness: a cohort study of more than one million Swedish subjects. Am J Psychiatry 165:59–65

Dalmau J, Bataller L (2006) Clinical and immunological diversity of limbic encephalitis: a model for paraneoplastic neurologic disorders. Hematol Oncol Clin North Am 20:1319–1335

Dalmau J, Graus F, Rosenblum MK, Posner JB (1992) Anti-Hu-associated paraneoplastic encephalomyelitis/sensory neuronopathy. A clinical study of 71 patients. Medicine (Baltimore) 71:59–72

Dalmau J, Tuzun E, Wu HY, Masjuan J, Rossi JE, Voloschin A, Baehring JM, Shimazaki H, Koide R, King D, Mason W, Sansing LH, Dichter MA, Rosenfeld MR, Lynch DR (2007) Paraneoplastic anti-N-methyl-D-aspartate receptor encephalitis associated with ovarian teratoma. Ann Neurol 61:25–36

Dalmau J, Gleichman AJ, Hughes EG, Rossi JE, Peng X, Lai M, Dessain SK, Rosenfeld MR, Balice-Gordon R, Lynch DR (2008) Anti-NMDA-receptor encephalitis: case series and analysis of the effects of antibodies. Lancet Neurol 7:1091–1098

Dalmau J, Lancaster E, Martinez-Hernandez E, Rosenfeld MR, Balice-Gordon R (2011) Clinical experience and laboratory investigations in patients with anti-Nmdar encephalitis. Lancet Neurol 10:63–74

Davies G, Welham J, Chant D, Torrey EF, Mcgrath J (2003) A systematic review and meta-analysis of Northern Hemisphere season of birth studies in schizophrenia. Schizophr Bull 29:587–593

De Witte LD, Hoffmann C, Van Mierlo HC, Titulaer MJ, Kahn RS, Martinez-Martinez P, European Consortium of Autoimmune Mental D (2015) Absence of N-methyl-D-aspartate receptor IgG autoantibodies in schizophrenia: the importance of cross-validation studies. JAMA Psychiatry 72:731–733

Deakin J, Lennox BR, Zandi MS (2014) Antibodies to the N-methyl-D-aspartate receptor and other synaptic proteins in psychosis. Biol Psychiatry 75:284–291

Denny-Brown D (1948) Primary sensory neuropathy with muscular changes associated with carcinoma. J Neurol Neurosurg Psychiatry 11:73–87

Dogan Onugoren M, Deuretzbacher D, Haensch CA, Hagedorn HJ, Halve S, Isenmann S, Kramme C, Lohner H, Melzer N, Monotti R, Presslauer S, Schabitz WR, Steffanoni S, Stoeck K, Strittmatter M, Stogbauer F, Trinka E, Von Oertzen TJ, Wiendl H, Woermann FG, Bien CG (2015) Limbic encephalitis due to GABAB and AMPA receptor antibodies: a case series. J Neurol Neurosurg Psychiatry 86:965–972

Duan BC, Weng WC, Lin KL, Wong LC, Li ST, Hsu MH, Lin JJ, Fan PC, Lin MI, Chiu NC, Lin YC, Wang HS, Hung KL, Lee WT (2016) Variations of movement disorders in anti-N-methyl-D-aspartate receptor encephalitis: a nationwide study in Taiwan. Medicine (Baltimore) 95: e4365

Elamin M, Lonergan R, Killeen RP, O'riordan S, Tubridy N, Mcguigan C (2015) Posterior cortical and white matter changes on Mri in anti-Ampa receptor antibody encephalitis. Neurol Neuroimmunol Neuroinflamm 2:e118

Farrelly L, Focking M, Piontkewitz Y, Dicker P, English J, Wynne K, Cannon M, Cagney G, Cotter DR (2015) Maternal immune activation induces changes in myelin and metabolic proteins, some of which can be prevented with risperidone in adolescence. Dev Neurosci 37:43–55

Finke C, Kopp UA, Pruss H, Dalmau J, Wandinger KP, Ploner CJ (2012) Cognitive deficits following anti-NMDA receptor encephalitis. J Neurol Neurosurg Psychiatry 83:195–198

Fukata Y, Lovero KL, Iwanaga T, Watanabe A, Yokoi N, Tabuchi K, Shigemoto R, Nicoll RA, Fukata M (2010) Disruption of LGI1-linked synaptic complex causes abnormal synaptic transmission and epilepsy. Proc Natl Acad Sci U S A 107:3799–3804

Gaughran F, Lally J, Beck K, McCormack R, Gardner-Sood P, Coutinho E, Jacobson L, Lang B, Sainz-Fuertes R, Papanastasiou E, Di Forti M, Nicholson T, Vincent A, Murray RM (2018) Brain-relevant antibodies in first-episode psychosis: a matched case-control study. Psychol Med 48(8):1257–1263

Gotkine M, Ben-Hur T, Vincent A, Vaknin-Dembinsky A (2011) Limbic encephalitis presenting as a post-partum psychiatric condition. J Neurol Sci 308:152–154

Grain R, Lally J, Stubbs B, Malik S, Lemince A, Nicholson TR, Murray RM, Gaughran F (2017) Autoantibodies against voltage-gated potassium channel and glutamic acid decarboxylase in psychosis: a systematic review, meta-analysis, and case series. Psychiatry Clin Neurosci 71:678–689

Graus F, Cordon-Cardo C, Posner JB (1985) Neuronal antinuclear antibody in sensory neuronopathy from lung cancer. Neurology 35:538–543

Graus F, Boronat A, Xifro X, Boix M, Svigelj V, Garcia A, Palomino A, Sabater L, Alberch J, Saiz A (2010) The expanding clinical profile of anti-AMPA receptor encephalitis. Neurology 74:857–859

Graus F, Titulaer MJ, Balu R, Benseler S, Bien CG, Cellucci T, Cortese I, Dale RC, Gelfand JM, Geschwind M, Glaser CA, Honnorat J, Hoftberger R, Iizuka T, Irani SR, Lancaster E, Leypoldt F, Pruss H, Rae-Grant A, Reindl M, Rosenfeld MR, Rostasy K, Saiz A, Venkatesan A, Vincent A, Wandinger KP, Waters P, Dalmau J (2016) A clinical approach to diagnosis of autoimmune encephalitis. Lancet Neurol 15:391–404

Gregor A, Albrecht B, Bader I, Bijlsma EK, Ekici AB, Engels H, Hackmann K, Horn D, Hoyer J, Klapecki J, Kohlhase J, Maystadt I, Nagl S, Prott E, Tinschert S, Ullmann R, Wohlleber E, Woods G, Reis A, Rauch A, Zweier C (2011) Expanding the clinical spectrum associated with defects in CNTNAP2 and NRXN1. BMC Med Genet 12:106

Gresa-Arribas N, Titulaer MJ, Torrents A, Aguilar E, Mccracken L, Leypoldt F, Gleichman AJ, Balice-Gordon R, Rosenfeld MR, Lynch D, Graus F, Dalmau J (2014) Antibody titres at diagnosis and during follow-up of anti-NMDA receptor encephalitis: a retrospective study. Lancet Neurol 13:167–177

Hammer C, Stepniak B, Schneider A, Papiol S, Tantra M, Begemann M, Siren AL, Pardo LA, Sperling S, Mohd Jofrry S, Gurvich A, Jensen N, Ostmeier K, Luhder F, Probst C, Martens H, Gillis M, Saher G, Assogna F, Spalletta G, Stocker W, Schulz TF, Nave KA, Ehrenreich H (2014) Neuropsychiatric disease relevance of circulating anti-NMDA receptor autoantibodies depends on blood-brain barrier integrity. Mol Psychiatry 19:1143–1149

Hara M, Arino H, Petit-Pedrol M, Sabater L, Titulaer MJ, Martinez-Hernandez E, Schreurs MW, Rosenfeld MR, Graus F, Dalmau J (2017) DPPX antibody-associated encephalitis: main syndrome and antibody effects. Neurology 88:1340–1348

Haussleiter IS, Emons B, Schaub M, Borowski K, Brune M, Wandinger KP, Juckel G (2012) Investigation of antibodies against synaptic proteins in a cross-sectional cohort of psychotic patients. Schizophr Res 140:258–259

Henley JM, Wilkinson KA (2016) Synaptic AMPA receptor composition in development, plasticity and disease. Nat Rev Neurosci 17:337–350

Hoftberger R, Titulaer MJ, Sabater L, Dome B, Rozsas A, Hegedus B, Hoda MA, Laszlo V, Ankersmit HJ, Harms L, Boyero S, De Felipe A, Saiz A, Dalmau J, Graus F (2013) Encephalitis and GABAB receptor antibodies: novel findings in a new case series of 20 patients. Neurology 81:1500–1506

Hoftberger R, Van Sonderen A, Leypoldt F, Houghton D, Geschwind M, Gelfand J, Paredes M, Sabater L, Saiz A, Titulaer MJ, Graus F, Dalmau J (2015) Encephalitis and AMPA receptor antibodies: novel findings in a case series of 22 patients. Neurology 84:2403–2412

Hughes EG, Peng X, Gleichman AJ, Lai M, Zhou L, Tsou R, Parsons TD, Lynch DR, Dalmau J, Balice-Gordon RJ (2010) Cellular and synaptic mechanisms of anti-NMDA receptor encephalitis. J Neurosci 30:5866–5875

Irani SR, Alexander S, Waters P, Kleopa KA, Pettingill P, Zuliani L, Peles E, Buckley C, Lang B, Vincent A (2010a) Antibodies to Kv1 potassium channel-complex proteins leucine-rich, glioma inactivated 1 protein and contactin-associated protein-2 in limbic encephalitis, Morvan's syndrome and acquired neuromyotonia. Brain 133:2734–2748

Irani SR, Bera K, Waters P, Zuliani L, Maxwell S, Zandi MS, Friese MA, Galea I, Kullmann DM, Beeson D, Lang B, Bien CG, Vincent A (2010b) N-methyl-D-aspartate antibody encephalitis: temporal progression of clinical and paraclinical observations in a predominantly non-paraneoplastic disorder of both sexes. Brain 133:1655–1667

Irani SR, Pettingill P, Kleopa KA, Schiza N, Waters P, Mazia C, Zuliani L, Watanabe O, Lang B, Buckley C, Vincent A (2012) Morvan syndrome: clinical and serological observations in 29 cases. Ann Neurol 72:241–255

Irani SR, Michell AW, Lang B, Pettingill P, Waters P, Johnson MR, Schott JM, Armstrong RJ, S Zagami A, Bleasel A, Somerville ER, Smith SM, Vincent A (2011) Faciobrachial dystonic seizures precede Lgi1 antibody limbic encephalitis. Ann Neurol 69(5):892–900

Irani SR, Stagg CJ, Schott JM, Rosenthal CR, Schneider SA, Pettingill P, Pettingill R, Waters P, Thomas A, Voets NL, Cardoso MJ, Cash DM, Manning EN, Lang B, Smith SJ, Vincent A, Johnson MR (2013) Faciobrachial dystonic seizures: the influence of immunotherapy on seizure control and prevention of cognitive impairment in a broadening phenotype. Brain 136:3151–3162

Jackman C, Horn ND, Molleston JP, Sokol DK (2009) Gene associated with seizures, autism, and hepatomegaly in an Amish girl. Pediatr Neurol 40:310–313

Jarius S, Steinmeyer F, Knobel A, Streitberger K, Hotter B, Horn S, Heuer H, Schreiber SJ, Wilhelm T, Trefzer U, Wildemann B, Ruprecht K (2013) GABAB receptor antibodies in paraneoplastic cerebellar ataxia. J Neuroimmunol 256:94–96

Javitt DC (2007) Glutamate and schizophrenia: phencyclidine, N-methyl-D-aspartate receptors, and dopamine-glutamate interactions. Int Rev Neurobiol 78:69–108

Javitt DC, Zukin SR (1991) Recent advances in the phencyclidine model of schizophrenia. Am J Psychiatry 148:1301–1308

Kappelmann N, Khandaker GM, Dal H, Stochl J, Kosidou K, Jones PB, Dalman C, Karlsson H (2019) Systemic inflammation and intelligence in early adulthood and subsequent risk of schizophrenia and other non-affective psychoses: a longitudinal cohort and co-relative study. Psychol Med 49:295–302

Kaulin YA, De Santiago-Castillo JA, Rocha CA, Nadal MS, Rudy B, Covarrubias M (2009) The dipeptidyl-peptidase-like protein DPP6 determines the unitary conductance of neuronal Kv4.2 channels. J Neurosci 29:3242–3251

Kayser MS, Dalmau J (2016) Anti-NMDA receptor encephalitis, autoimmunity, and psychosis. Schizophr Res 176:36–40

Kayser MS, Titulaer MJ, Gresa-Arribas N, Dalmau J (2013) Frequency and characteristics of isolated psychiatric episodes in anti-N-methyl-d-aspartate receptor encephalitis. JAMA Neurol 70:1133–1139

Kelleher E, Corvin A (2013) In: Mitchell KJE (ed) The genetics of neurodevelopmental disorder. Wiley, Cambridge

Khandaker GM, Zimbron J, Dalman C, Lewis G, Jones PB (2012) Childhood infection and adult schizophrenia: a meta-analysis of population-based studies. Schizophr Res 139:161–168

Khandaker GM, Pearson RM, Zammit S, Lewis G, Jones PB (2014) Association of serum interleukin 6 and C-reactive protein in childhood with depression and psychosis in young adult life: a population-based longitudinal study. JAMA Psychiatry 71:1121–1128

Khandaker GM, Cousins L, Deakin J, Lennox BR, Yolken R, Jones PB (2015) Inflammation and immunity in schizophrenia: implications for pathophysiology and treatment. Lancet Psychiatry 2:258–270

Khandaker GM, Dalman C, Kappelmann N, Stochl J, Dal H, Kosidou K, Jones PB, Karlsson H (2018) Association of childhood infection with IQ and adult nonaffective psychosis in Swedish men: a population-based longitudinal cohort and co-relative study. JAMA Psychiatry 75:356–362

Kim TJ, Lee ST, Shin JW, Moon J, Lim JA, Byun JI, Shin YW, Lee KJ, Jung KH, Kim YS, Park KI, Chu K, Lee SK (2014) Clinical manifestations and outcomes of the treatment of patients with GABAB encephalitis. J Neuroimmunol 270:45–50

Kim TJ, Lee ST, Moon J, Sunwoo JS, Byun JI, Lim JA, Shin YW, Jun JS, Lee HS, Lee WJ, Yang AR, Choi Y, Park KI, Jung KH, Jung KY, Kim M, Lee SK, Chu K (2017) Anti-LGI1 encephalitis is associated with unique Hla subtypes. Ann Neurol 81:183–192

Klein CJ, Lennon VA, Aston PA, Mckeon A, O'toole O, Quek A, Pittock SJ (2013) Insights from LGI1 and CASPR2 potassium channel complex autoantibody subtyping. JAMA Neurol 70:229–234

Kraepelin E (1899) Psychiatrie: Ein lehrbuch fur studierende und aerzte. JA Barth, Leipzig

Lai M, Hughes EG, Peng X, Zhou L, Gleichman AJ, Shu H, Mata S, Kremens D, Vitaliani R, Geschwind MD, Bataller L, Kalb RG, Davis R, Graus F, Lynch DR, Balice-Gordon R, Dalmau J (2009) AMPA receptor antibodies in limbic encephalitis alter synaptic receptor location. Ann Neurol 65:424–434

Lai M, Huijbers MG, Lancaster E, Graus F, Bataller L, Balice-Gordon R, Cowell JK, Dalmau J (2010) Investigation of LGI1 as the antigen in limbic encephalitis previously attributed to potassium channels: a case series. Lancet Neurol 9:776–785

Lalic T, Pettingill P, Vincent A, Capogna M (2011) Human limbic encephalitis serum enhances hippocampal mossy fiber-CA3 pyramidal cell synaptic transmission. Epilepsia 52:121–131

Lancaster E, Dalmau J (2012) Neuronal autoantigens–pathogenesis, associated disorders and antibody testing. Nat Rev Neurol 8:380–390

Lancaster E, Lai M, Peng X, Hughes E, Constantinescu R, Raizer J, Friedman D, Skeen MB, Grisold W, Kimura A, Ohta K, Iizuka T, Guzman M, Graus F, Moss SJ, Balice-Gordon R, Dalmau J (2010) Antibodies to the GABA(B) receptor in limbic encephalitis with seizures: case series and characterisation of the antigen. Lancet Neurol 9:67–76

Lancaster E, Huijbers MG, Bar V, Boronat A, Wong A, Martinez-Hernandez E, Wilson C, Jacobs D, Lai M, Walker RW, Graus F, Bataller L, Illa I, Markx S, Strauss KA, Peles E, Scherer SS, Dalmau J (2011a) Investigations of caspr2, an autoantigen of encephalitis and neuromyotonia. Ann Neurol 69:303–311

Lancaster E, Martinez-Hernandez E, Dalmau J (2011b) Encephalitis and antibodies to synaptic and neuronal cell surface proteins. Neurology 77:179–189

Lancaster E, Martinez-Hernandez E, Titulaer MJ, Boulos M, Weaver S, Antoine JC, Liebers E, Kornblum C, Bien CG, Honnorat J, Wong S, Xu J, Contractor A, Balice-Gordon R, Dalmau J (2011c) Antibodies to metabotropic glutamate receptor 5 in the Ophelia syndrome. Neurology 77:1698–1701

Lang B, Makuch M, Moloney T, Dettmann I, Mindorf S, Probst C, Stoecker W, Buckley C, Newton CR, Leite MI, Maddison P, Komorowski L, Adcock J, Vincent A, Waters P, Irani SR (2017) Intracellular and non-neuronal targets of voltage-gated potassium channel complex antibodies. J Neurol Neurosurg Psychiatry 88:353–361

Lejuste F, Thomas L, Picard G, Desestret V, Ducray F, Rogemond V, Psimaras D, Antoine JC, Delattre JY, Groc L, Leboyer M, Honnorat J (2016) Neuroleptic intolerance in patients with anti-NMDAR encephalitis. Neurol Neuroimmunol Neuroinflamm 3:e280

Lennox BR, Coles AJ, Vincent A (2012) Antibody-mediated encephalitis: a treatable cause of schizophrenia. Br J Psychiatry 200(2):92–94

Lennox BR, Palmer-Cooper EC, Pollak T, Hainsworth J, Marks J, Jacobson L, Lang B, Fox H, Ferry B, Scoriels L, Crowley H, Jones PB, Harrison PJ, Vincent A (2017) Prevalence and clinical characteristics of serum neuronal cell surface antibodies in first-episode psychosis: a case-control study. Lancet Psychiatry 4:42–48

Lennox BR, Tomei G, Vincent SA, Yeeles K, Pollard R, Palmer-Cooper E, Jones P, Zandi MS, Coles A (2018) Study of immunotherapy in antibody positive psychosis: feasibility and acceptability (SINAPPS1). J Neurol Neurosurg Psychiatry. pii: jnnp-2018-318124. https://doi.org/10.1136/jnnp-2018-318124

Leypoldt F, Armangue T, Dalmau J (2015) Autoimmune encephalopathies. Ann N Y Acad Sci 1338:94–114

Leypoldt F, de Witte L, Lancaster E, Titulaer MJ (2017) Serum neuronal cell-surface antibodies in first-episode psychosis. Lancet Psychiatry 4(3):186–187

Liguori R, Vincent A, Clover L, Avoni P, Plazzi G, Cortelli P, Baruzzi A, Carey T, Gambetti P, Lugaresi E, Montagna P (2001) Morvan's syndrome: peripheral and central nervous system and cardiac involvement with antibodies to voltage-gated potassium channels. Brain 124:2417–2426

Malhotra D, Sebat J (2012) CNVS: harbingers of a rare variant revolution in psychiatric genetics. Cell 148:1223–1241

Malter MP, Frisch C, Schoene-Bake JC, Helmstaedter C, Wandinger KP, Stoecker W, Urbach H, Surges R, Elger CE, Vincent AV, Bien CG (2014) Outcome of limbic encephalitis with VGKC-complex antibodies: relation to antigenic specificity. J Neurol 261:1695–1705

Mantere O, Saarela M, Kieseppa T, Raij T, Mantyla T, Lindgren M, Rikandi E, Stoecker W, Teegen B, Suvisaari J (2017) Anti-neuronal anti-bodies in patients with early psychosis. Schizophr Res 192:404–407

Martinez-Hernandez E, Horvath J, Shiloh-Malawsky Y, Sangha N, Martinez-Lage M, Dalmau J (2011) Analysis of complement and plasma cells in the brain of patients with anti-NMDAR encephalitis. Neurology 77:589–593

Masdeu JC, Gonzalez-Pinto A, Matute C, Ruiz De Azua S, Palomino A, De Leon J, Berman KF, Dalmau J (2012) Serum IgG antibodies against the NR1 subunit of the NMDA receptor not detected in schizophrenia. Am J Psychiatry 169:1120–1121

Masopust J, Andrys C, Bazant J, Vysata O, Kuca K, Valis M (2015) Anti-NMDA receptor antibodies in patients with a first episode of schizophrenia. Neuropsychiatr Dis Treat 11:619–623

Mat A, Adler H, Merwick A, Chadwick G, Gullo G, Dalmau JO, Tubridy N (2013) Ophelia syndrome with metabotropic glutamate receptor 5 antibodies in CSF. Neurology 80:1349–1350

Mayer ML, Westbrook GL (1987) Permeation and block of N-methyl-D-aspartic acid receptor channels by divalent cations in mouse cultured central neurones. J Physiol 394:501–527

Mayer ML, Westbrook GL, Guthrie PB (1984) Voltage-dependent block by Mg2+ of NMDA responses in spinal cord neurones. Nature 309:261–263

Mcgrath JJ, Welham JL (1999) Season of birth and schizophrenia: a systematic review and meta-analysis of data from the Southern Hemisphere. Schizophr Res 35:237–242

Mckeon GL, Scott JG, Spooner DM, Ryan AE, Blum S, Gillis D, Langguth D, Robinson GA (2016) Cognitive and social functioning deficits after anti-N-methyl-D-aspartate receptor encephalitis: an exploratory case series. J Int Neuropsychol Soc 22:828–838

Mckeon GL, Robinson GA, Ryan AE, Blum S, Gillis D, Finke C, Scott JG (2017) Cognitive outcomes following anti-N-methyl-D-aspartate receptor encephalitis: a systematic review. J Clin Exp Neuropsychol 40:1–19
Metcalf SA, Jones PB, Nordstrom T, Timonen M, Maki P, Miettunen J, Jaaskelainen E, Jarvelin MR, Stochl J, Murray GK, Veijola J, Khandaker GM (2017) Serum C-reactive protein in adolescence and risk of schizophrenia in adulthood: a prospective birth cohort study. Brain Behav Immun 59:253–259
Midgley L, Kelleher E, Zandi MS (2018) Autoimmune encephalitis. Br J Hosp Med (Lond) 79(9): C134–C137. https://doi.org/10.12968/hmed.2018.79.9.C134
Mohammad SS, Sinclair K, Pillai S, Merheb V, Aumann TD, Gill D, Dale RC, Brilot F (2014) Herpes simplex encephalitis relapse with chorea is associated with autoantibodies to N-methyl-D-aspartate receptor or dopamine-2 receptor. Mov Disord 29:117–122
Mortensen PB, Norgaard-Pedersen B, Waltoft BL, Sorensen TL, Hougaard D, Yolken RH (2007) Early infections of Toxoplasma gondii and the later development of schizophrenia. Schizophr Bull 33:741–744
Moscato EH, Peng X, Jain A, Parsons TD, Dalmau J, Balice-Gordon RJ (2014) Acute mechanisms underlying antibody effects in anti-N-methyl-D-aspartate receptor encephalitis. Ann Neurol 76:108–119
Muller UJ, Teegen B, Probst C, Bernstein HG, Busse S, Bogerts B, Schiltz K, Stoecker W, Steiner J (2014) Absence of dopamine receptor serum autoantibodies in schizophrenia patients with an acute disease episode. Schizophr Res 158:272–274
Navarro V, Kas A, Apartis E, Chami L, Rogemond V, Levy P, Psimaras D, Habert MO, Baulac M, Delattre JY, Honnorat J (2016) Motor cortex and hippocampus are the two main cortical targets in LGI1-antibody encephalitis. Brain 139:1079–1093
Newcomer JW, Krystal JH (2001) NMDA receptor regulation of memory and behavior in humans. Hippocampus 11:529–542
Newey CR, Sarwal A (2014) Hyponatremia and voltage gated potassium channel antibody associated limbic encephalitis. J Neurol Neurophysiol 5:195
Nokura K, Yamamoto H, Okawara Y, Koga H, Osawa H, Sakai K (1997) Reversible limbic encephalitis caused by ovarian teratoma. Acta Neurol Scand 95:367–373
Noll R (2004) Historical review: autointoxication and focal infection theories of dementia praecox. World J Biol Psychiatry 5(2):66–72
Nowak L, Bregestovski P, Ascher P, Herbet A, Prochiantz A (1984) Magnesium gates glutamate-activated channels in mouse central neurones. Nature 307:462–465
O'Callaghan E, Sham P, Takei N, Glover G, Murray RM (1991) Schizophrenia after prenatal exposure to 1957 A2 influenza epidemic. Lancet 337:1248–1250
Ohkawa T, Fukata Y, Yamasaki M, Miyazaki T, Yokoi N, Takashima H, Watanabe M, Watanabe O, Fukata M (2013) Autoantibodies to epilepsy-related Lgi1 in limbic encephalitis neutralize LGI1-ADAM22 interaction and reduce synaptic AMPA receptors. J Neurosci 33:18161–18174
Okamura H, Oomori N, Uchitomi Y (1997) An acutely confused 15-year-old girl. Lancet 350:488
Ottman R, Winawer MR, Kalachikov S, Barker-Cummings C, Gilliam TC, Pedley TA, Hauser WA (2004) LGI1 mutations in autosomal dominant partial epilepsy with auditory features. Neurology 62:1120–1126
Oyebode F, Humphreys M (2011) The future of psychiatry. Br J Psychiatry 199:439–440
Pagano A, Rovelli G, Mosbacher J, Lohmann T, Duthey B, Stauffer D, Ristig D, Schuler V, Meigel I, Lampert C, Stein T, Prezeau L, Blahos J, Pin J, Froestl W, Kuhn R, Heid J, Kaupmann K, Bettler B (2001) C-terminal interaction is essential for surface trafficking but not for heteromeric assembly of GABA(b) receptors. J Neurosci 21:1189–1202
Pathmanandavel K, Starling J, Merheb V, Ramanathan S, Sinmaz N, Dale RC, Brilot F (2015) Antibodies to surface dopamine-2 receptor and N-methyl-D-aspartate receptor in the first episode of acute psychosis in children. Biol Psychiatry 77:537–547

Pearlman DM, Najjar S (2014) Meta-analysis of the association between N-methyl-d-aspartate receptor antibodies and schizophrenia, schizoaffective disorder, bipolar disorder, and major depressive disorder. Schizophr Res 157:249–258

Peery HE, Day GS, Dunn S, Fritzler MJ, Pruss H, De Souza C, Doja A, Mossman K, Resch L, Xia C, Sakic B, Belbeck L, Foster WG (2012) Anti-NMDA receptor encephalitis. The disorder, the diagnosis and the immunobiology. Autoimmun Rev 11:863–872

Perogamvros L, Schnider A, Leemann B (2012) The role of NMDA receptors in human eating behavior: evidence from a case of anti-NMDA receptor encephalitis. Cogn Behav Neurol 25:93–97

Petit-Pedrol M, Armangue T, Peng X, Bataller L, Cellucci T, Davis R, Mccracken L, Martinez-Hernandez E, Mason WP, Kruer MC, Ritacco DG, Grisold W, Meaney BF, Alcala C, Sillevis-Smitt P, Titulaer MJ, Balice-Gordon R, Graus F, Dalmau J (2014) Encephalitis with refractory seizures, status epilepticus, and antibodies to the GABAA receptor: a case series, characterisation of the antigen, and analysis of the effects of antibodies. Lancet Neurol 13:276–286

Pettingill P, Kramer HB, Coebergh JA, Pettingill R, Maxwell S, Nibber A, Malaspina A, Jacob A, Irani SR, Buckley C, Beeson D, Lang B, Waters P, Vincent A (2015) Antibodies to GABAA receptor alpha1 and gamma2 subunits: clinical and serologic characterization. Neurology 84:1233–1241

Poliak S, Gollan L, Martinez R, Custer A, Einheber S, Salzer JL, Trimmer JS, Shrager P, Peles E (1999) Caspr2, a new member of the neurexin superfamily, is localized at the juxtaparanodes of myelinated axons and associates with K+ channels. Neuron 24:1037–1047

Pollak TA, Lennox BR (2018) Time for a change of practice: the real-world value of testing for neuronal autoantibodies in acute first-episode psychosis. BJPsych Open 4(4):262–264

Pollak TA, Moran N (2017) Emergence of new-onset psychotic disorder following recovery from LGI1 antibody-associated limbic encephalitis. BMJ Case Rep 2017. https://doi.org/10.1136/bcr-2016-218328

Pollak TA, Beck K, Irani SR, Howes OD, David AS, Mcguire PK (2016) Autoantibodies to central nervous system neuronal surface antigens: psychiatric symptoms and psychopharmacological implications. Psychopharmacology (Berl) 233:1605–1621

Pruss H, Lennox BR (2016) Emerging psychiatric syndromes associated with antivoltage-gated potassium channel complex antibodies. J Neurol Neurosurg Psychiatry 87:1242–1247

Pruss H, Rothkirch M, Kopp U, Hamer HM, Hagge M, Sterzer P, Saschenbrecker S, Stocker W, Harms L, Endres M (2014) Limbic encephalitis with mGluR5 antibodies and immunotherapy-responsive prosopagnosia. Neurology 83:1384–1386

Purcell SM, Wray NR, Stone JL, Visscher PM, O'donovan MC, Sullivan PF, Sklar P (2009) Common polygenic variation contributes to risk of schizophrenia and bipolar disorder. Nature 460:748–752

Quaranta G, Maremmani AG, Perugi G (2015) Anti-AMPA-receptor encephalitis presenting as a rapid-cycling bipolar disorder in a young woman with turner syndrome. Case Rep Psychiatry 2015:273192

Quinn TJ, Dawson J, Walters M (2008) Dr John Rankin; his life, legacy and the 50th anniversary of the Rankin Stroke Scale. Scott Med J 53:44–47

Rhoads J, Guirgis H, Mcknight C, Duchemin AM (2011) Lack of anti-NMDA receptor autoantibodies in the serum of subjects with schizophrenia. Schizophr Res 129:213–214

Rossi M, Mead S, Collinge J, Rudge P, Vincent A (2015) Neuronal antibodies in patients with suspected or confirmed sporadic Creutzfeldt-Jakob disease. J Neurol Neurosurg Psychiatry 86:692–694

Schmitt SE, Pargeon K, Frechette ES, Hirsch LJ, Dalmau J, Friedman D (2012) Extreme delta brush: a unique EEG pattern in adults with anti-NMDA receptor encephalitis. Neurology 79:1094–1100

Schulz P, Pruss H (2015) "Hirnsymptome bei Carcinomatose" – Hermann oppenheim and an early description of a paraneoplastic neurological syndrome. J Hist Neurosci 24:371–377

Scott JG, Gillis D, Ryan AE, Hargovan H, Gundarpi N, McKeon G et al (2018) The prevalence and treatment outcomes of antineuronal antibody-positive patients admitted with first episode of psychosis. BJPsych Open 4(2):69–74

Shi J, Levinson DF, Duan J, Sanders AR, Zheng Y, Pe'er I, Dudbridge F, Holmans PA, Whittemore AS, Mowry BJ, Olincy A, Amin F, Cloninger CR, Silverman JM, Buccola NG, Byerley WF, Black DW, Crowe RR, Oksenberg JR, Mirel DB, Kendler KS, Freedman R, Gejman PV (2009) Common variants on chromosome 6p22.1 are associated with schizophrenia. Nature 460:753–757

Shillito P, Molenaar PC, Vincent A, Leys K, Zheng W, Van Den Berg RJ, Plomp JJ, Van Kempen GT, Chauplannaz G, Wintzen AR et al (1995) Acquired neuromyotonia: evidence for autoantibodies directed against K+ channels of peripheral nerves. Ann Neurol 38:714–722

Shin YW, Lee ST, Shin JW, Moon J, Lim JA, Byun JI, Kim TJ, Lee KJ, Kim YS, Park KI, Jung KH, Lee SK, Chu K (2013) VGKC-complex/LGI1-antibody encephalitis: clinical manifestations and response to immunotherapy. J Neuroimmunol 265:75–81

Sieghart W (2006) Structure, pharmacology, and function of Gabaa receptor subtypes. Adv Pharmacol 54:231–263

Spatola M, Petit-Pedrol M, Simabukuro MM, Armangue T, Castro FJ, Barcelo Artigues MI, Julià Benique MR, Benson L, Gorman M, Felipe A, Caparó Oblitas RL, Rosenfeld MR, Graus F, Dalmau J (2017) Investigations in GABAA receptor antibody-associated encephalitis. Neurology 88(11):1012–1020

Stefansson H, Ophoff RA, Steinberg S, Andreassen OA, Cichon S, Rujescu D, Werge T, Pietilainen OP, Mors O, Mortensen PB, Sigurdsson E, Gustafsson O, Nyegaard M, Tuulio-Henriksson A, Ingason A, Hansen T, Suvisaari J, Lonnqvist J, Paunio T, Borglum AD, Hartmann A, Fink-Jensen A, Nordentoft M, Hougaard D, Norgaard-Pedersen B, Bottcher Y, Olesen J, Breuer R, Moller HJ, Giegling I, Rasmussen HB, Timm S, Mattheisen M, Bitter I, Rethelyi JM, Magnusdottir BB, Sigmundsson T, Olason P, Masson G, Gulcher JR, Haraldsson M, Fossdal R, Thorgeirsson TE, Thorsteinsdottir U, Ruggeri M, Tosato S, Franke B, Strengman E, Kiemeney LA, Melle I, Djurovic S, Abramova L, Kaleda V, Sanjuan J, De Frutos R, Bramon E, Vassos E, Fraser G, Ettinger U, Picchioni M, Walker N, Toulopoulou T, Need AC, GE D, Yoon JL, Shianna KV, Freimer NB, Cantor RM, Murray R, Kong A, Golimbet V, Carracedo A, Arango C, Costas J, Jonsson EG, Terenius L, Agartz I, Petursson H, Nothen MM, Rietschel M, Matthews PM, Muglia P, Peltonen L, ST Clair D, Goldstein DB, Stefansson K, Collier DA (2009) Common variants conferring risk of schizophrenia. Nature 460:744–747

Steiner J, Walter M, Glanz W, Sarnyai Z, Bernstein HG, Vielhaber S, Kastner A, Skalej M, Jordan W, Schiltz K, Klingbeil C, Wandinger KP, Bogerts B, Stoecker W (2013) Increased prevalence of diverse N-methyl-D-aspartate glutamate receptor antibodies in patients with an initial diagnosis of schizophrenia: specific relevance of IgG NR1a antibodies for distinction from N-methyl-D-aspartate glutamate receptor encephalitis. JAMA Psychiatry 70:271–278

Stone JM, Dietrich C, Edden R, Mehta MA, De Simoni S, Reed LJ, Krystal JH, Nutt D, Barker GJ (2012) Ketamine effects on brain GABA and glutamate levels with 1H-MRS: relationship to ketamine-induced psychopathology. Mol Psychiatry 17:664–665

Strauss KA, Puffenberger EG, Huentelman MJ, Gottlieb S, Dobrin SE, Parod JM, Stephan DA, Morton DH (2006) Recessive symptomatic focal epilepsy and mutant contactin-associated protein-like 2. N Engl J Med 354:1370–1377

Striano P, Busolin G, Santulli L, Leonardi E, Coppola A, Vitiello L, Rigon L, Michelucci R, Tosatto SC, Striano S, Nobile C (2011) Familial temporal lobe epilepsy with psychic auras associated with a novel LGI1 mutation. Neurology 76:1173–1176

Sun Y, Wu Z, Kong S, Jiang D, Pitre A, Wang Y, Chen G (2013) Regulation of epileptiform activity by two distinct subtypes of extrasynaptic GABAA receptors. Mol Brain 6(1):21. https://doi.org/10.1186/1756-6606-6-21

Titulaer MJ, Mccracken L, Gabilondo I, Armangue T, Glaser C, Iizuka T, Honig LS, Benseler SM, Kawachi I, Martinez-Hernandez E, Aguilar E, Gresa-Arribas N, Ryan-Florance N, Torrents A, Saiz A, Rosenfeld MR, Balice-Gordon R, Graus F, Dalmau J (2013) Treatment and prognostic factors for long-term outcome in patients with anti-NMDA receptor encephalitis: an observational cohort study. Lancet Neurol 12:157–165

Tobin WO, Lennon VA, Komorowski L, Probst C, Clardy SL, Aksamit AJ, Appendino JP, Lucchinetti CF, Matsumoto JY, Pittock SJ, Sandroni P, Tippmann-Peikert M, Wirrell EC, Mckeon A (2014) DPPX potassium channel antibody: frequency, clinical accompaniments, and outcomes in 20 patients. Neurology 83:1797–1803

Tsutsui K, Kanbayashi T, Tanaka K, Boku S, Ito W, Tokunaga J, Mori A, Hishikawa Y, Shimizu T, Nishino S (2012) Anti-NMDA-receptor antibody detected in encephalitis, schizophrenia, and narcolepsy with psychotic features. BMC Psychiatry 12:37

Usiello A, Baik JH, Rouge-Pont F, Picetti R, Dierich A, Lemeur M, Piazza PV, Borrelli E (2000) Distinct functions of the two isoforms of dopamine D2 receptors. Nature 408:199–203

Van Sonderen A, Arino H, Petit-Pedrol M, Leypoldt F, Kortvelyessy P, Wandinger KP, Lancaster E, Wirtz PW, Schreurs MW, Sillevis Smitt PA, Graus F, Dalmau J, Titulaer MJ (2016a) The clinical spectrum of Caspr2 antibody-associated disease. Neurology 87:521–528

Van Sonderen A, Schreurs MW, Wirtz PW, Sillevis Smitt PA, Titulaer MJ (2016b) From VGKC to LGI1 and Caspr2 encephalitis: the evolution of a disease entity over time. Autoimmun Rev 15:970–974

Van Sonderen A, Thijs RD, Coenders EC, Jiskoot LC, Sanchez E, De Bruijn MA, Van Coevorden-Hameete MH, Wirtz PW, Schreurs MW, Sillevis Smitt PA, Titulaer MJ (2016c) Anti-LGI1 encephalitis: clinical syndrome and long-term follow-up. Neurology 87:1449–1456

Van Sonderen A, Roelen DL, Stoop JA, Verdijk RM, Haasnoot GW, Thijs RD, Wirtz PW, Schreurs MW, Claas FH, Sillevis Smitt PA, Titulaer MJ (2017) Anti-LGI1 encephalitis is strongly associated with HLA-DR7 and HLA-DRB4. Ann Neurol 81:193–198

Varley J, Vincent A, Irani SR (2015) Clinical and experimental studies of potentially pathogenic brain-directed autoantibodies: current knowledge and future directions. J Neurol 262:1081–1095

Varley J, Taylor J, Irani SR (2017) Autoantibody-mediated diseases of the CNS: structure, dysfunction and therapy. Neuropharmacology 132:71–82

Verkerk AJ, Mathews CA, Joosse M, Eussen BH, Heutink P, Oostra BA (2003) CNTNAP2 is disrupted in a family with Gilles de la Tourette syndrome and obsessive compulsive disorder. Genomics 82:1–9

Vincent A, Irani SR (2010) Caspr2 antibodies in patients with thymomas. J Thorac Oncol 5:S277–S280

Vitaliani R, Mason W, Ances B, Zwerdling T, Jiang Z, Dalmau J (2005) Paraneoplastic encephalitis, psychiatric symptoms, and hypoventilation in ovarian teratoma. Ann Neurol 58:594–604

Wagner-Jauregg J, Bruetsch WL (1946) The history of the malaria treatment of general paralysis. Am J Psychiatry 102:577–582

Waxman EA, Lynch DR (2005) N-methyl-D-aspartate receptor subtypes: multiple roles in excitotoxicity and neurological disease. Neuroscientist 11:37–49

Whalley HC, O'connell G, Sussmann JE, Peel A, Stanfield AC, Hayiou-Thomas ME, Johnstone EC, Lawrie SM, Mcintosh AM, Hall J (2011) Genetic variation in CNTNAP2 alters brain function during linguistic processing in healthy individuals. Am J Med Genet B Neuropsychiatr Genet 156b:941–948

Wright S, Hacohen Y, Jacobson L, Agrawal S, Gupta R, Philip S, Smith M, Lim M, Wassmer E, Vincent A (2015) N-methyl-D-aspartate receptor antibody-mediated neurological disease: results of a UK-based surveillance study in children. Arch Dis Child 100(6):521–526

Xi B, Chen J, Yang L, Wang W, Fu M, Wang C (2011) GABBR1 gene polymorphism(G1465A)is associated with temporal lobe epilepsy. Epilepsy Res 96:58–63

Zai G, Arnold P, Burroughs E, Barr CL, Richter MA, Kennedy JL (2005) Evidence for the gamma-amino-butyric acid type B receptor 1 (GABBR1) gene as a susceptibility factor in obsessive-compulsive disorder. Am J Med Genet B Neuropsychiatr Genet 134b:25–29

Zai CC, Tiwari AK, King N, De Luca V, Mueller DJ, Shaikh S, Wong GW, Meltzer HY, Lieberman JA, Kennedy JL (2009) Association study of the gamma-aminobutyric acid type a receptor gamma2 subunit gene with schizophrenia. Schizophr Res 114:33–38

Zandi MS, Irani SR, Lang B, Waters P, Jones PB, Mckenna P, Coles AJ, Vincent A, Lennox BR (2011) Disease-relevant autoantibodies in first episode schizophrenia. J Neurol 258:686–688

Zandi MS, Deakin JB, Morris K, Buckley C, Jacobson L, Scoriels L, Cox AL, Coles AJ, Jones PB, Vincent A, Lennox BR (2014) Immunotherapy for patients with acute psychosis and serum N-methyl d-aspartate receptor (NMDAR) antibodies: a description of a treated case series. Schizophr Res 160:193–195

Zhou L, Messing A, Chiu SY (1999) Determinants of excitability at transition zones in Kv1.1-deficient myelinated nerves. J Neurosci 19:5768–5781

Zweier C, De Jong EK, Zweier M, Orrico A, Ousager LB, Collins AL, Bijlsma EK, Oortveld MA, Ekici AB, Reis A, Schenck A, Rauch A (2009) CNTNAP2 and NRXN1 are mutated in autosomal-recessive Pitt-Hopkins-like mental retardation and determine the level of a common synaptic protein in Drosophila. Am J Hum Genet 85:655–666

Herpes Simplex Virus Type-1 Infection: Associations with Inflammation and Cognitive Aging in Relation to Schizophrenia

Vishwajit L. Nimgaonkar, Triptish Bhatia, Abdelaziz Mansour, Maribeth A. Wesesky, and Smita Deshpande

Contents

V. L. Nimgaonkar (✉)
Department of Psychiatry, University of Pittsburgh School of Medicine, Pittsburgh, PA, USA

Department of Human Genetics, University of Pittsburgh, Pittsburgh, PA, USA
e-mail: nimga@pitt.edu

T. Bhatia
Training Program for Psychiatric Genetics in India, Post-graduate Institute for Medical Education and Research, Dr Ram Manohar Lohia Hospital, Delhi, India

A. Mansour and M. A. Wesesky
Department of Psychiatry, University of Pittsburgh School of Medicine, Pittsburgh, PA, USA

S. Deshpande
Post-graduate Institute for Medical Education and Research, Dr Ram Manohar Lohia Hospital, New Delhi, India

Curr Topics Behav Neurosci (2020) 44: 125–140
DOI 10.1007/7854_2018_86
Published Online: 3 May 2019

Abstract Most persons experience cognitive decline as they grow older. The term "cognitive aging," coined to describe milder varieties of cognitive decline, is likely to be due to multiple causes. Persistent or repeated infections of the central nervous system (whether subclinical or diagnosable) can cause damage to neurons directly or indirectly through inflammation resulting in incremental neuronal damage, thus eroding cognitive reserve. This possibility has not been considered widely. We evaluated the data linking persistent infection with herpes simplex virus type 1 (HSV-1) and cognitive aging by applying the Bradford Hill criteria. Despite inherent problems in establishing causal relations for chronic disorders, our analyses suggest plausible links. These studies are pertinent for patients with schizophrenia, who are particularly vulnerable due to disorder-related cognitive impairment. Further investigations are warranted to test a causal hypothesis, particularly prospective studies and intervention studies.

Keywords Bradford Hill criteria · Causality · Cognitive aging · Herpes virus · HSV-1 · Schizophrenia

1 Introduction

Cognitive dysfunction, whether mild or severe, extracts a heavy public health burden (Zhu et al. 2013). The type of cognitive dysfunction spans the continuum from mild changes to severe dementia. The Institute of Medicine recently published a report on "Cognitive Aging" to draw attention to incremental cognitive dysfunction that is noticeable as we age (Institute of Medicine 2015; Carbone et al. 2014). The cognitive dysfunction spans a spectrum of changes and severity. How and why the dysfunction occurs is a matter of conjecture, but it is reasonable to assume that it is multifactorial; in other words, no one cause needs be necessary or sufficient (Gould and Gottesman 2006). If mild cognitive impairment (MCI) or dementia occurs in even a minority of individuals with cognitive aging, the burden is likely to be much higher (Zhu et al. 2013; Paradise et al. 2015; Springate and Tremont 2013; Lin and Neumann 2013). Age-associated cognitive dysfunction will demand even more resources in the next decade, because the number of persons aged 60 years and older is expected to increase to 1.2 billion across the world by 2025 (http://www.who.int/ageing/en/index.html).

Numerous genetic factors have been associated with severe cognitive decline and dementias. While genetic factors certainly play a role in age-related cognitive decline, preventable and potentially treatable environmental factors need to be investigated too. We postulate that a portion of the risk could be contributed by chronic viral infections. Like cognitive aging, the prevalence of many viral infections increases with age (Smith and Robinson 2002). Furthermore, individuals with chronic infections perform less efficiently on cognitive tests compared with uninfected individuals in several cross-sectional studies, many of them focusing on herpes viruses (Aiello et al. 2006; Carbone et al. 2014; DeVaughn et al. 2015;

Barrientos et al. 2012; Bucks et al. 2008). Such infections have the potential to affect cognitive functions if they afflict the brain directly or indirectly. Mild or moderate levels of dysfunction in the cognitive domains of attention, working, and verbal memory were reported repeatedly among HSV-1 seropositive schizophrenia and bipolar patients and even healthy individuals in 19 studies (Dickerson et al. 2003a, 2004, 2012; Strandberg et al. 2003; Aiello et al. 2006; Prasad et al. 2012b; Schretlen et al. 2010; Prasad et al. 2011; Yolken et al. 2011; Gerber et al. 2012; Watson et al. 2013; Katan et al. 2013; Thomas et al. 2013; Tarter et al. 2014; Jonker et al. 2014; Wright et al. 2015; Fruchter et al. 2015; Nimgaonkar et al. 2016; Bhatia et al. 2017; Hamdani et al. 2017; Vanyukov et al. 2017). We and others have systematically investigated cognitive functions among individuals with chronic infections caused by herpes simplex virus type 1 (HSV-1) and explored whether such observations could explain a portion of cognitive aging (Watson et al. 2013; Thomas et al. 2013; Bhatia et al. 2016, 2017; Hamdani et al. 2017). Henceforth, for the sake of brevity, we refer to the putative association as HSV-1 – cognitive decline/dysfunction (HSVCD). In the following sections, we describe aspects of HSV-1 infections in humans relevant to a causal hypothesis, followed by possible mechanism for the HSVCD. In the final section, we evaluate the published HSV-1-related data in relation to Bradford Hill criteria, the current gold standard for investigating causal connections between chronic risk factors and noncommunicable human diseases. Our review includes cross-sectional and prospective studies relating HSV-1 infection to cognitive dysfunction, brain imaging studies, the effects of highly specific antiviral drugs on cognitive dysfunction, and efforts to model HSV-1 infections in vitro in human neuronal cells.

2 The Nature of HSV-1 Infections

HSV-1 is a double-stranded DNA virus that causes human-specific infections. It initially infects mucosal surfaces. It migrates to sensory ganglia from the primary infection site, where it can lie dormant in a latent state for the lifetime of the host (Steiner et al. 2007). When latent, viral DNA assumes a circular form and exists without replication in the neuronal nucleus; it only produces relatively few untranslated viral transcripts that are not translated into viral proteins (Harkness et al. 2014; Steiner et al. 2007). Reactivation from latency can be induced by stress or immunosuppression (Steiner et al. 2007; Shimomura and Higaki 2011); reactivated virions migrate through sensory nerves to the original site of infection, where recurrent infection flares (Steiner et al. 2007). It is the reactivated state that causes acute lytic mucosal lesions. Viral particles are typically detectable at sites of reactivation and in mucosal fluids bathing the lytic mucosal lesions. Oral transmission of mucosal fluids from an infected individual is the typical route of primary infection, but transmission through sexual infection is an increasingly frequent form of primary infection, and infection of neonates during childbirth is thus a mounting public health concern (Kriebs 2008) (Fig. 1).

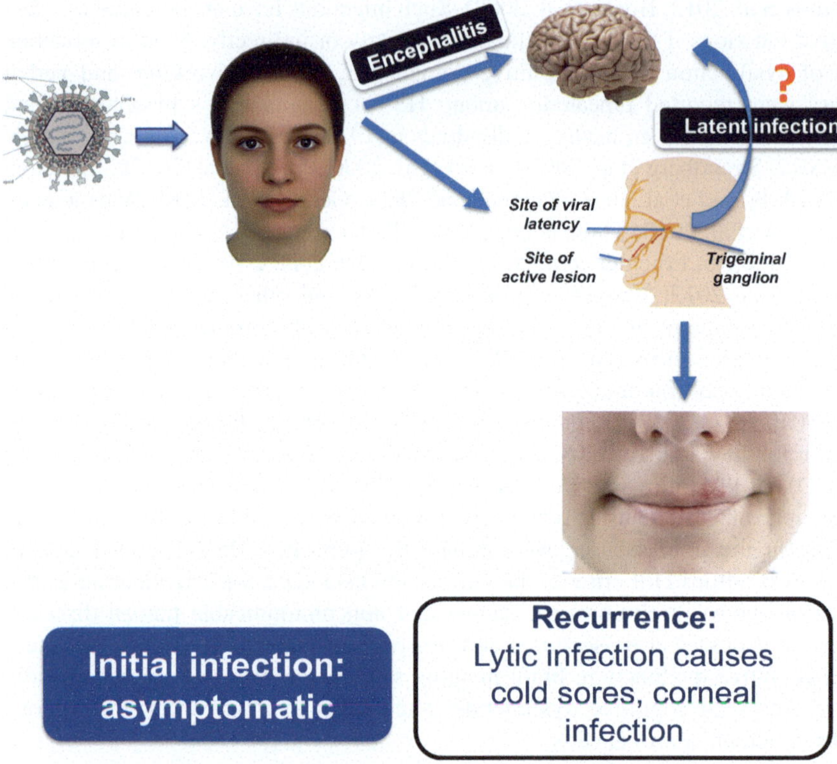

Fig. 1 Herpes simplex virus, type 1 (HSV-1) infection natural history and the human brain

More than 3.7 billion individuals aged 0–49 years are infected with HSV-1 worldwide, with rates exceeding 87% in Africa (Looker et al. 2015). The prevalence of HSV-1 seropositivity increases with age, being approximately 40% among US children/teens and 40–70% among adults less than 70 years of age (http://www.cdc.gov/nchs/nhanes.htm). The primary infection can be asymptomatic or it may be associated with mild fever. As noted above, HSV-1 causes recurrent mucosal infections, but the most damaging infections involve the eye and the brain. Approximately 500,000 persons have ocular HSV-1 infection across the world at any one time (Liesegang et al. 1989). Recurrent herpes stromal keratitis (HSK) can cause corneal scarring and blindness (Rowe et al. 2013). Ocular herpes is one of the chief causes of blindness worldwide, with a global incidence of approximately 1.5 million cases per year, and it is estimated that there are 40,000 new cases of visual impairment or blindness annually (Farooq and Shukla 2012). HSV-1 can also cause encephalitis, leading to death or severe residual cognitive impairment in survivors (Steiner et al. 2007). Thankfully, it is rare (~0.004%), and it typically occurs only in adults with compromised immune functions or increasingly in neonates.

A conclusive diagnosis of HSV-1 infection requires detection of the viral particles at reactivated sites or in mucosal fluids, but reactivation is unpredictable. Therefore, it is difficult to detect viral particles in the blood or saliva during the latent, persistent phases (Corstjens et al. 2012). Consequently, specific host-generated IgG antibodies in the serum (which can remain elevated for prolonged periods) are typically used as indirect indicators of infection with HSV-1.

HSV-1 can be effectively treated with nucleoside analogues such as valacyclovir (VAL) and its derivatives. Once these drugs are converted into their active forms by viral thymidine kinase, they can inhibit DNA polymerase and even terminate its activity (Kimberlin and Whitley 2007). Because the conversion into active metabolites can only occur in cells with actively replicating virions, the drugs are inactive in uninfected human cells and thus have relatively few side effects. Thus, VAL-like drugs are highly specific antiviral agents. On the other hand, the drugs are ineffective against the latent viral form. Thus, the antiviral drugs can abort productive infection but cannot eliminate virions from the human body. No effective vaccines have been found for HSV-1. Thus, HSV-1 infection is essentially incurable at present even though highly efficacious and specific antiviral drugs are available.

3 Plausible Mechanisms for Cognitive Aging Attributed to HSV-1

Acute encephalitis caused by HSV-1 is potentially lethal but can be alleviated if it is detected in time and is treated with VAL. Individuals who survive encephalitis can have severe, enduring cognitive impairment. A similar mechanism could operate in the absence of severe encephalitis. If subacute or latent HSV-1 infection occurs in the brain, it is possible that localized recrudescence could occur without detection or overt symptoms, yet it could potentially culminate in cognitive impairment. This possibility was presciently suggested by Becker over a decade ago (Becker 2002). He suggested that HSV-1 virions could infect the olfactory regions following nasal infection and thence track to the temporal or frontal cortical regions. Even if such infection does not occur in the brain, another possibility is indirect damage to the brain following repeated infection in the periphery, with the release of cytokines that can cross the blood brain barrier and thus impair neuronal function (Yarlagadda et al. 2009). Infection with other viruses, such as cytomegalovirus (CMV), herpes simplex virus, type 2 (HSV-2), and Epstein-Barr virus (EBV), may result in cognitive deterioration in older individuals, independent of age-related variables (Nimgaonkar et al. 2016). It should be noted that the majority of studies linking HSV-1 infections with cognitive dysfunction have been conducted among persons with schizophrenia, who have disorder-related cognitive dysfunction (Gur et al. 2007). It is plausible that HSV-1 infection acts additively or even interacts with such factors.

4 Bradford Hill Viewpoints for Etiological Links for Chronic Diseases

Early in the twentieth century, Koch enunciated criteria that he felt must be fulfilled before a putative risk factor could be accepted as an etiological agent for a human disease (Evans 1976). These "postulates" were articulated in the era when infectious agents were being linked putatively to specific infections. Koch required that the infectious agent be identified from infected tissues, be cultured in vitro, and be shown to cause the disease in question in animal models. Though Koch's postulates definitively linked many infectious agents to deadly diseases such as cholera, it is difficult to apply the postulates to many noninfectious diseases or to diseases caused by genetic mutations. It is particularly difficult to utilize them for infectious diseases of the central nervous system (CNS), because of difficulties in obtaining CNS tissues from live individuals. Therefore, it may be prudent to seek other types of evidence. In the wake of uncertainties about links between cigarette smoking and lung cancer, Sir Austin Bradford Hill articulated a set of nine "viewpoints" that could be evaluated to test for causal links (Hill 1965). Unlike Koch, who required that all his postulates must be fulfilled before etiological links could be accepted, Bradford Hill emphasized that his viewpoints should not be considered as hard-and-fast rules; rather he suggested them as guideposts that could be used to evaluate the body of evidence to enable consensus. The Bradford Hill viewpoints have been used extensively in the past 50 years. Recent advances in genetics, statistics, and toxicology enable more sophisticated analyses and tests of these viewpoints (Fedak et al. 2015). In the following sections, we evaluate each of the Bradford Hill viewpoints (criteria) in relation to persistent HSV-1 infection and cognitive dysfunction.

4.1 *Strength*

Bradford Hill suggested that the magnitude of the association between the suspected risk factor and the outcome of interest could be used as a gauge, provided potential confounding factors were taken into account. With regard to smoking and lung cancer, he wrote: "prospective inquiries into smoking have shown that the death rate from cancer of the lung in cigarette smokers is nine to ten times the rate in non-smokers and the rate in heavy cigarette smokers is twenty to thirty times as great." By using carefully matched cases and controls, he suggested that the substantial odds ratios provided a persuasive pointer implicating smoking in the etiology of lung cancer. With regard to HSV-1 infections and cognitive dysfunction, the estimated odds ratios are 1.25–3.2, i.e., in the small to medium effect size range (Prasad et al. 2012b). Our analyses indicate population attributable risks (PAR) between 15.2 and 59.3% (Prasad et al. 2012b), making this a potentially important public health hazard. Further, HSV-1, like many infections, is common among individuals with lower socioeconomic status (SES). Thus, HSV-1 infection could serve as a proxy for low SES. Most of the cross-sectional studies have attempted to control for SES, though this is admittedly difficult; for example, respondents can provide misleading information regarding household income. Another concern – whether the cognitive dysfunction is a residuum of prior encephalitis – is unlikely, because acute encephalitis is so rare.

4.2 *Consistency*

Bradford Hill emphasized the importance of attempting replication not only from the viewpoint of scientific rigor but also because it provides more confidence in a causative link. He suggested that the association under scrutiny was unlikely to be due to chance if it could be detected at different sites, at different times, and using different study designs. He pointed out that such variation could serve as a check against factors that might not be obvious, such as disease severity. He also admonished against mistaking statistical significance for consistency. From the viewpoint of this guideline, HSVCD has been detected in 19 studies that were conducted in Europe, India, and the USA. The studies were conducted over a 25-year period. The majority of studies incorporated the conventional case-control design, while one study also included a family-based design (Watson et al. 2013). The association was investigated in healthy individuals, as well as clinic-based samples that included individuals with schizophrenia (SZ) and bipolar disorders (BD). Because a variety of cognitive tests were employed, there was variation in the cognitive domains that were associated, e.g., attention, immediate memory, language, verbal memory, and executive function. Most of the associations occurred with the cognitive domains of attention or the domains of working memory. As there is substantial correlation for performance in these domains, it is difficult to identify the domains with the primary association. In contrast, four groups of investigators did not report significantly impaired cognitive functions among seropositive individuals (Aiello et al. 2008; Barnes et al. 2014; Katan et al. 2013; Nimgaonkar et al. 2016). Some of these samples were larger than the samples in which a significant association was detected, e.g., (Nimgaonkar et al. 2016), and thus should have appropriate statistical power. The reasons for lack of replication are not obvious. It is notable all the studies with nonsignificant associations comprised individuals aged 60 or over (Aiello et al. 2008; Katan et al. 2013; Nimgaonkar et al. 2016). Thus, the HSVCD is largely consistent.

4.3 *Specificity*

Bradford Hill recognized that specificity is not a *sine qua non* for demonstrating etiology, though he pointed out that associations detected in specific situations or in groups of individuals were more likely to indicate causality. Pertinent to HSVCD, he also recognized that the outcome of interest could have a multifactorial etiology. Moreover, the same etiological factor could have multiple effects. HSV-1 can infect the brain, causing encephalitis and many postencephalitic sequelae resembling those observed with HSVCD, thus lending credibility to HSVCD. Yet, the cognitive dysfunction predicted by HSVCD is also observed in relation to many other putative etiological agents, including herpes viruses such as cytomegalovirus (Wright et al. 2015; Dickerson et al. 2014; Shirts et al. 2008; Hamdani et al. 2017).

4.4 Temporality

Bradford Hill pointed out that the putative etiological agent must predate the outcome of interest. This requirement is a sine qua non for postulating a cause and effect relationship. As noted earlier, it is difficult to detect HSV-1 virions in serum or saliva; they are only detected locally in lytic lesions. Though HSV-1 virions are detectable in the cerebrospinal fluid (CSF) during acute encephalitis, they are not detectable in CSF during persistent, quiescent infection. Therefore, we have to rely on antibody titers as a sensitive and specific mark of prior infection. Even so, antibody titers do not indicate the timing of infection. Hence, temporality would be difficult to test unless prospective cohorts of birth populations were conducted. Instead, many investigators have conducted prospective studies in adults, following up individuals who were seronegative and seropositive at study entry and simultaneously evaluating changes in cognitive functions. Four studies have reported significantly worse cognitive function over time when participants seropositive at baseline were followed over 1–2 years and their temporal profile was compared with participants who were HSV-1 seronegative at study entry (Strandberg et al. 2003; Prasad et al. 2012b; Fruchter et al. 2015; Bhatia et al. 2017). No cognitive decline was associated with HSV-1 in three studies of older individuals (Nimgaonkar et al. 2016; Aiello et al. 2008; Barnes et al. 2014). However, temporal cognitive improvement was reported after acute encephalitis in one study (Hokkanen and Launes 1997). Thus, longitudinal follow-up studies of older individuals – similar to the pattern observed in cross-sectional studies – do not support HSVCD. Thus, temporality, an important tenet for Bradford Hill, is not observed consistently for HSVCD.

4.5 Biological Gradient

A linear relationship between the exposure to the risk factor or the "dose" of the risk factor and the outcome of interest provides persuasive evidence supporting causality, though the absence of such a relationship need not disprove causality. Bradford Hill cited the example of increased risk of lung cancer being related to the quantity of cigarettes consumed in support of this contention. This guideline is difficult to examine with regard to viral infection, as the "dose" of initial infection, the extent of replication, and the residual infection are clearly impossible to quantify in the clinical research setting. Still, it has been reported that memory functions and executive functioning are associated with level of HSV1 exposure (Jonker et al. 2014). Though antibodies to HSV-1 are used routinely as a proxy for viral exposure, the antibody titers fluctuate with time and thus cannot be used to estimate "dosage" of exposure.

4.6 *Plausibility*

Bradford Hill suggested biological plausibility for the putative etiological link as another viewpoint, but he recognized that it depends on the level of knowledge available – and its unpredictable dependence on future scientific advances. In the case of HSV-1, the natural history of infection provides strong plausible links. It is well known that HSV-1 can infect the brain, causing acute encephalitis. Survivors of encephalitis are very likely to suffer from residual, lifetime cognitive impairment (Hokkanen and Launes 2007). The pertinent question is whether subacute (recurrent) encephalitis can occur and whether it is associated with cognitive impairment. Though replicating HSV-1 virions have not been detected in brains of humans, sans acute encephalitis, several studies have documented the presence of viral DNA among postmortem brain tissues from individuals who died from causes other than HSV-1 encephalitis (Baringer and Pisani 1994; Karatas et al. 2008; Hill et al. 2008). Thus, a plausible biological explanation is available for HSVCD.

4.7 *Coherence*

This viewpoint is related to plausibility. Bradford Hill suggested that the interpretation of cause-effect relationships should not conflict with what is known while recognizing that an absence of "coherence" should not be viewed as evidence against causality. With regard to smoking and lung cancer, Bradford Hill cited epidemiologic data relating to temporal changes in the incidence of smoking and lung cancer, as well as gender differences in the prevalence of both variables. Both sets of data are "coherent" with the proposed links between smoking and lung cancer. In the case of HSVCD, it may be instructive to draw on several brain imaging studies. All published studies of structural brain MRI scans have detected reduced gray matter volume in temporal-frontal brain regions among individuals without a history of encephalitis (Prasad et al. 2007, 2011; Schretlen et al. 2010; Pandurangi et al. 1994). While these cross-sectional studies are susceptible to some of the confounds discussed above with regard to the cross-sectional studies relating HSV-1 exposure to cognitive impairment, it is remarkable that the brain regions with significant differences among HSV-1-exposed and HSV-1-nonexposed individuals coincide with the regions predominantly affected in acute encephalitis due to HSV-1.

Furthermore, progressive cognitive impairments and gray matter loss were also reported among HSV-1 seropositive individuals with SZ, but not among seropositive nonpsychotic control individuals (Prasad et al. 2011).

4.8 Experiment

Bradford Hill recommended that experimental manipulation of the risk factor ought to provide valid tests of the putative hypothesis. This is not feasible for HSV-1, as effective vaccines are currently not available. On the other hand, randomized, controlled trials (RCTs) could provide relevant information, as could systematic reviews of such trials. As noted earlier, efficacious antiviral drugs, comprising acyclovir (ACV) and its prodrug valacyclovir (VAL), are widely available for treating HSV-1 infections. These drugs effectively halt the replicating stage and can be used prophylactically to reduce reactivation (Miserocchi et al. 2007; Acyclovir for the prevention of recurrent herpes simplex virus eye disease. Herpetic Eye Disease Study Group 1998; Wilhelmus et al. 1998), but they do not affect the latent state. Dickerson and colleagues initially reported improvement of psychotic symptoms following treatment with valacyclovir (VAL, an antiviral drug for HSV-1 infection) among individuals with SZ who were seropositive for cytomegalovirus (CMV) in a non-blinded study (Dickerson et al. 2003b), but the improvement could not be detected in randomized controlled trial (RCT) (Dickerson et al. 2009). Earlier, DeLisi and colleagues tested acyclovir in eight patients with schizophrenia and did not detect significant improvement (DeLisi et al. 1987). However, these studies were conducted among individuals with chronic SZ, and seropositivity for HSV-1 was not assessed. In contrast, our initial RCT indicated beneficial effects of adjunctive VAL over placebo among early-course HSV-1 seropositive SZ patients (Prasad et al. 2012a). A subsequent, larger RCT also detected improvement in two cognitive domains, although the cognitive domains in which improvement differed from the earlier RCT (Bhatia et al. 2017). A systematic review of these studies may provide worthwhile information, though it may be difficult to "harmonize" the differing sets of cognitive variables used in these treatment studies. Another caveat is the lack of efficacy of currently available antivirals against latent HSV-1 infection. If the cognitive impairment accrues from recurring cycles of latency and reactivation in small regions of the brain or it stems from immunological reactions to the infection (e.g., release of cytokines that can cross the blood-brain barrier to cause neuronal damage (Yarlagadda et al. 2009)), then the RCTs would not test the causal links rigorously.

4.9 Analogy

Bradford Hill suggested that persuasive evidence of a causal relationship between another agent and a specific disease could be marshalled to garner support for the agent of interest, even if the evidence was weaker. In the case of HSV-1, links between infection and cognitive impairment have been suggested for cytomegalovirus (CMV), although the evidence is also inconclusive (Shirts et al. 2008; Nimgaonkar et al. 2016). Stronger evidence has emerged for cognitive impairment even among individuals treated effectively for human immunodeficiency virus/ acquired immunodeficiency syndrome (HIV/AIDS) (Watkins and Treisman 2015; Shirts et al. 2008). Thus, analogy, an admittedly weaker viewpoint, can still be marshalled in support of the HSVCD hypothesis.

5 Discussion

Based on these analyses, six out of nine Bradford Hill criteria for causation are fulfilled for HSVCD, thus linking chronic, asymptomatic HSV-1 infection to cognitive dysfunction *sans* encephalitis (Prasad et al. 2012b). A major proportion of the relevant studies have been conducted among patients with schizophrenia, though the associations have also been reported among otherwise healthy individuals. On balance, the accumulating evidence continues to incriminate HSV-1, though it is by no means unequivocal. The main criticism of the HSVCD hypothesis is that infective virions have not been identified in human brain tissues in the absence of acute encephalitis. Arguably, another unidentified coincidental infection or even low socioeconomic status could explain the observed cross-sectional associations.

Suggestions for future studies. More convincing evidence will only come from additional treatment trials, as well as cohort-based longitudinal studies. Furthermore, other modern advances could be marshalled to examine the HSVCD (Fedak et al. 2015). For example, large archival data, including datasets with DNA sequences from postmortem tissues, could be examined for the presence of HSV-1 sequences. Cognitive dysfunction related to persistent HSV-1 infection has been detected among otherwise healthy individuals, as well as persons with schizophrenia. It is uncertain whether the effect size of the associations is greater among the latter group, though an interaction between HSV-1 exposure and psychiatric diagnosis has not been reported. Still this question needs to be addressed further.

6 Conclusions

Our work suggests plausible causal links with enormous public health consequences, based on fulfillment of a majority of Bradford Hill viewpoints. In our view, the bulk of evidence points to moderate to strong evidence for causality. We recommend specific additional studies to test the hypothesis further.

References

Acyclovir for the prevention of recurrent herpes simplex virus eye disease. Herpetic Eye Disease Study Group (1998) N Engl J Med 339(5):300–306

Aiello AE, Haan M, Blythe L, Moore K, Gonzalez JM, Jagust W (2006) The influence of latent viral infection on rate of cognitive decline over 4 years. J Am Geriatr Soc 54(7):1046–1054

Aiello AE, Haan MN, Pierce CM, Simanek AM, Liang J (2008) Persistent infection, inflammation, and functional impairment in older Latinos. J Gerontol A Biol Sci Med Sci 63(6):610–618

Baringer JR, Pisani P (1994) Herpes simplex virus genomes in human nervous system tissue analyzed by polymerase chain reaction. Ann Neurol 36(6):823–829

Barnes LL, Capuano AW, Aiello AE, Turner AD, Yolken RH, Torrey EF, Bennett DA (2014) Cytomegalovirus infection and risk of alzheimer disease in older black and white individuals. J Infect Dis 211(2):230–237

Barrientos RM, Frank MG, Watkins LR, Maier SF (2012) Aging-related changes in neuroimmune-endocrine function: implications for hippocampal-dependent cognition. Horm Behav 62 (3):219–227

Becker Y (2002) Herpes simplex virus evolved to use the human defense mechanisms to establish a lifelong infection in neurons – a review and hypothesis. Virus Genes 24(2):187–196

Bhatia T, Wood J, Iyenger S, Narayanan S, Prasad KM, Chen K, Gur RE, Gur RC, Nimgaonkar VL, Deshpande SN (2016) Dysfunctional emotion discrimination in schizophrenia is associated with HSV-1 infection and improves with antiviral treatment. 55th annual meeting of American College of Neuropsychiatry, American College of Neuropsychiatry, Hollywood, FL

Bhatia T, Wood J, Iyengar S, Narayanan SS, Beniwal RP, Prasad KM, Chen K, Yolken RH, Dickerson F, Gur RC, Gur RE, Deshpande SN, Nimgaonkar VL (2017) Emotion discrimination in humans: its association with HSV-1 infection and its improvement with antiviral treatment. Schizophr Res 193:161–167

Bucks RS, Gidron Y, Harris P, Teeling J, Wesnes KA, Perry VH (2008) Selective effects of upper respiratory tract infection on cognition, mood and emotion processing: a prospective study. Brain Behav Immun 22(3):399–407

Carbone I, Lazzarotto T, Ianni M, Porcellini E, Forti P, Masliah E, Gabrielli L, Licastro F (2014) Herpes virus in Alzheimer's disease: relation to progression of the disease. Neurobiol Aging 35(1):122–129

Corstjens PL, Abrams WR, Malamud D (2012) Detecting viruses by using salivary diagnostics. J Am Dent Assoc 143(Suppl 10):12S–18S

DeLisi LE, Goldin LR, Nurnberger JI, Simmons-Alling S, Hamovit J, Dingman CW (1987) Failure to alleviate symptoms of schizophrenia with the novel use of an antiviral agent, acyclovir (Zovirax). Biol Psychiatry 22(2):216–220

DeVaughn S, Müller-Oehring EM, Markey B, Brontë-Stewart HM, Schulte T (2015) Aging with HIV-1 infection: motor functions, cognition, and attention – a comparison with Parkinson's disease. Neuropsychol Rev 25(4):424–438

Dickerson FB, Boronow JJ, Stallings C, Origoni AE, Ruslanova I, Yolken RH (2003a) Association of serum antibodies to herpes simplex virus 1 with cognitive deficits in individuals with schizophrenia. Arch Gen Psychiatry 60(5):466–472

Dickerson FB, Boronow JJ, Stallings CR, Origoni AE, Yolken RH (2003b) Reduction of symptoms by valacyclovir in cytomegalovirus-seropositive individuals with schizophrenia. Am J Psychiatry 160(12):2234–2236

Dickerson FB, Boronow JJ, Stallings C, Origoni AE, Cole S, Krivogorsky B, Yolken RH (2004) Infection with herpes simplex virus type 1 is associated with cognitive deficits in bipolar disorder. Biol Psychiatry 55(6):588–593

Dickerson FB, Stallings CR, Boronow JJ, Origoni AE, Sullens A, Yolken RH (2009) Double blind trial of adjunctive valacyclovir in individuals with schizophrenia who are seropositive for cytomegalovirus. Schizophr Res 107(2-3):147–149

Dickerson F, Stallings C, Origoni A, Vaughan C, Khushalani S, Yolken R (2012) Additive effects of elevated C-reactive protein and exposure to herpes simplex virus type 1 on cognitive impairment in individuals with schizophrenia. Schizophr Res 134(1):83–88

Dickerson F, Stallings C, Origoni A, Katsafanas E, Schweinfurth LA, Savage CL, Yolken R (2014) Association between cytomegalovirus antibody levels and cognitive functioning in non-elderly adults. PLoS One 9(5):e95510

Evans AS (1976) Causation and disease: the Henle-Koch postulates revisited. Yale J Biol Med 49(2):175–195

Farooq AV, Shukla D (2012) Herpes simplex epithelial and stromal keratitis: an epidemiologic update. Surv Ophthalmol 57(5):448–462

Fedak KM, Bernal A, Capshaw ZA, Gross S (2015) Applying the Bradford Hill criteria in the 21st century: how data integration has changed causal inference in molecular epidemiology. Emerg Themes Epidemiol 12:14

Fruchter E, Goldberg S, Fenchel D, Grotto I, Ginat K, Weiser M (2015) The impact of Herpes simplex virus type 1 on cognitive impairments in young, healthy individuals – a historical prospective study. Schizophr Res 168(1–2):292–296

Gerber SI, Krienke UJ, Biedermann NC, Grunze H, Yolken RH, Dittmann S, Langosch JM (2012) Impaired functioning in euthymic patients with bipolar disorder – HSV-1 as a predictor. Prog Neuropsychopharmacol Biol Psychiatry 36(1):110–116

Gould TD, Gottesman II (2006) Psychiatric endophenotypes and the development of valid animal models. Genes Brain Behav 5(2):113–119

Gur RE, Nimgaonkar VL, Almasy L, Calkins ME, Ragland JD, Pogue-Geile MF, Kanes S, Blangero J, Gur RC (2007) Neurocognitive endophenotypes in a multiplex multigenerational family study of schizophrenia. Am J Psychiatry 164:813–819

Hamdani N, Daban-Huard C, Godin O, Laouamri H, Jamain S, Attiba D, Delavest M, Lépine JP, Le Corvoisier P, Houenou J, Richard JR, Yolken RH, Krishnamoorthy R, Tamouza R, Leboyer M, Dickerson FB (2017) Effects of cumulative herpesviridae and Toxoplasma gondii infections on cognitive function in healthy, bipolar, and schizophrenia subjects. J Clin Psychiatry 78(1): e18–e27

Harkness JM, Kader M, DeLuca NA (2014) Transcription of the herpes simplex virus 1 genome during productive and quiescent infection of neuronal and nonneuronal cells. J Virol 88(12):6847–6861

Hill A (1965) The environment and disease: association or causation? Proc R Soc Med 58:295–300

Hill JM, Ball MJ, Neumann DM, Azcuy AM, Bhattacharjee PS, Bouhanik S, Clement C, Lukiw WJ, Foster TP, Kumar M, Kaufman HE, Thompson HW (2008) The high prevalence of herpes simplex virus type 1 DNA in human trigeminal ganglia is not a function of age or gender. J Virol 82(16):8230–8234

Hokkanen L, Launes J (1997) Cognitive recovery instead of decline after acute encephalitis: a prospective follow up study. J Neurol Neurosurg Psychiatry 63(2):222–227

Hokkanen L, Launes J (2007) Neuropsychological sequelae of acute-onset sporadic viral encephalitis. Neuropsychol Rehabil 17(4-5):450–477

Institute of Medicine (2015) Cognitive aging: progress in understanding and opportunities for action. The National Academies Press, Washington

Jonker I, Klein HC, Duivis HE, Yolken RH, Rosmalen JG, Schoevers RA (2014) Association between exposure to HSV1 and cognitive functioning in a general population of adolescents. The TRAILS study. PLoS One 9(7):e101549

Karatas H, Gurer G, Pinar A, Soylemezoglu F, Tezel GG, Hascelik G, Akalan N, Tuncer S, Ciger A, Saygi S (2008) Investigation of HSV-1, HSV-2, CMV, HHV-6 and HHV-8 DNA by real-time PCR in surgical resection materials of epilepsy patients with mesial temporal lobe sclerosis. J Neurol Sci 264(1–2):151–156

Katan M, Moon YP, Paik MC, Sacco RL, Wright CB, Elkind MS (2013) Infectious burden and cognitive function: the Northern Manhattan Study. Neurology 80(13):1209–1215

Kimberlin D, Whitley R (2007) Antiviral therapy of HSV-1 and -2. In: Arvin A, Campadelli-Fiume G, Mocarski E, Moore PS, Roizman B, Whitley R, Yamanishi K (eds) Human herpesviruses: biology, therapy, and immunoprophylaxis. Cambridge University Press, Cambridge

Kriebs JM (2008) Understanding herpes simplex virus: transmission, diagnosis, and considerations in pregnancy management. J Midwifery Womens Health 53(3):202–208

Liesegang TJ, Melton LJ, Daly PJ, Ilstrup DM (1989) Epidemiology of ocular herpes simplex. Incidence in Rochester, Minn, 1950 through 1982. Arch Ophthalmol 107(8):1155–1159

Lin PJ, Neumann PJ (2013) The economics of mild cognitive impairment. Alzheimers Dement 9(1):58–62

Looker KJ, Magaret AS, May MT, Turner KM, Vickerman P, Gottlieb SL, Newman LM (2015) Global and regional estimates of prevalent and incident herpes simplex virus type 1 infections in 2012. PLoS One 10(10):e0140765

Miserocchi E, Modorati G, Galli L, Rama P (2007) Efficacy of valacyclovir vs acyclovir for the prevention of recurrent herpes simplex virus eye disease: a pilot study. Am J Ophthalmol 144(4):547–551

Nimgaonkar VL, Yolken RH, Wang T, Chang CC, McClain L, McDade E, Snitz BE, Ganguli M (2016) Temporal cognitive decline associated with exposure to infectious agents in a population-based, aging cohort. Alzheimer Dis Assoc Disord 30(3):216–222

Pandurangi AK, Pelonero AL, Nadel L, Calabrese VP (1994) Brain structure changes in schizophrenics with high serum titers of antibodies to herpes virus. Schizophr Res 11(3):245–250

Paradise M, McCade D, Hickie IB, Diamond K, Lewis SJ, Naismith SL (2015) Caregiver burden in mild cognitive impairment. Aging Ment Health 19(1):72–78

Prasad KM, Shirts BH, Yolken RH, Keshavan MS, Nimgaonkar VL (2007) Brain morphological changes associated with exposure to HSV1 in first-episode schizophrenia. Mol Psychiatry 12(1):105–13, 1

Prasad KM, Eack SM, Goradia D, Pancholi KM, Keshavan MS, Yolken RH, Nimgaonkar VL (2011) Progressive gray matter loss and changes in cognitive functioning associated with exposure to herpes simplex virus 1 in schizophrenia: a longitudinal study. Am J Psychiatry 168(8):822–830

Prasad KM, Eack SM, Keshavan MS, Yolken RH, Iyengar S, Nimgaonkar VL (2012a) Antiherpes virus-specific treatment and cognition in schizophrenia: a test-of-concept randomized double-blind placebo-controlled trial. Schizophr Bull 39(4):857–66

Prasad KM, Watson AM, Dickerson FB, Yolken RH, Nimgaonkar VL (2012b) Exposure to Herpes simplex virus type 1 and cognitive impairments in individuals with schizophrenia. Schizophr Bull 38(6):1137–1148

Rowe AM, St Leger AJ, Jeon S, Dhaliwal DK, Knickelbein JE, Hendricks RL (2013) Herpes keratitis. Prog Retin Eye Res 32:88–101

Schretlen DJ, Vannorsdall TD, Winicki JM, Mushtaq Y, Hikida T, Sawa A, Yolken RH, Dickerson FB, Cascella NG (2010) Neuroanatomic and cognitive abnormalities related to herpes simplex virus type 1 in schizophrenia. Schizophr Res 118(1-3):224–231

Shimomura Y, Higaki S (2011) The kinetics of herpes virus on the ocular surface and suppression of its reactivation. Cornea 30(Suppl 1):S3–S7

Shirts BH, Prasad KM, Pogue-Geile MF, Dickerson F, Yolken RH, Nimgaonkar VL (2008) Antibodies to cytomegalovirus and Herpes simplex virus 1 associated with cognitive function in schizophrenia. Schizophr Res 106(2-3):268–274

Smith JS, Robinson NJ (2002) Age-specific prevalence of infection with herpes simplex virus types 2 and 1: a global review. J Infect Dis 186(Suppl 1):S3–S28

Springate B, Tremont G (2013) Caregiver burden and depression in mild cognitive impairment. J Appl Gerontol 32(6):765–775

Steiner I, Kennedy PG, Pachner AR (2007) The neurotropic herpes viruses: herpes simplex and varicella-zoster. Lancet Neurol 6(11):1015–1028

Strandberg TE, Pitkala KH, Linnavuori KH, Tilvis RS (2003) Impact of viral and bacterial burden on cognitive impairment in elderly persons with cardiovascular diseases. Stroke 34(9):2126–2131

Tarter KD, Simanek AM, Dowd JB, Aiello AE (2014) Persistent viral pathogens and cognitive impairment across the life course in the third national health and nutrition examination survey. J Infect Dis 209(6):837–844

Thomas P, Bhatia T, Gauba D, Wood J, Long C, Prasad K, Dickerson FB, Gur RE, Gur RC, Yolken RH, Nimgaonkar VL, Deshpande SN (2013) Exposure to herpes simplex virus, type 1 and reduced cognitive function. J Psychiatr Res 47(11):1680–1685

Vanyukov MM, Nimgaonkar VL, Kirisci L, Kirillova GP, Reynolds MD, Prasad K, Tarter RE, Yolken RH (2017) Association of cognitive function and liability to addiction with childhood herpesvirus infections: a prospective cohort study-ERRATUM. Dev Psychopathol:1–2

Watkins CC, Treisman GJ (2015) Cognitive impairment in patients with AIDS – prevalence and severity. HIV AIDS (Auckl) 7:35–47

Watson AM, Prasad KM, Klei L, Wood JA, Yolken RH, Gur RC, Bradford LD, Calkins ME, Richard J, Edwards N, Savage RM, Allen TB, Kwentus J, McEvoy JP, Santos AB, Wiener HW, Go RC, Perry RT, Nasrallah HA, Gur RE, Devlin B, Nimgaonkar VL (2013) Persistent infection with neurotropic herpes viruses and cognitive impairment. Psychol Med 43(5):1023–1031

Wilhelmus KR, Beck RW, Moke PS, Dawson CR, Barron BA, Jones DB, Kaufman HE, Kurinij N, Stulting D, Sugar J, Cohen JA, Hyndiuk RA, Asbell PA (1998) Acyclovir for the prevention of recurrent herpes simplex virus eye disease. Herpetic Eye Disease Study Group. N Engl J Med 339(5):300–306

Wright CB, Gardener H, Dong C, Yoshita M, DeCarli C, Sacco RL, Stern Y, Elkind MS (2015) Infectious burden and cognitive decline in the Northern Manhattan Study. J Am Geriatr Soc 63(8):1540–1545

Yarlagadda A, Alfson E, Clayton AH (2009) The blood brain barrier and the role of cytokines in neuropsychiatry. Psychiatry (Edgmont) 6(11):18–22

Yolken RH, Torrey EF, Lieberman JA, Yang S, Dickerson FB (2011) Serological evidence of exposure to Herpes Simplex Virus type 1 is associated with cognitive deficits in the CATIE schizophrenia sample. Schizophr Res 128(1-3):61–65

Zhu CW, Sano M, Ferris SH, Whitehouse PJ, Patterson MB, Aisen PS (2013) Health-related resource use and costs in elderly adults with and without mild cognitive impairment. J Am Geriatr Soc 61(3):396–402

Role of Infection, Autoimmunity, Atopic Disorders, and the Immune System in Schizophrenia: Evidence from Epidemiological and Genetic Studies

Michael E. Benros and Preben B. Mortensen

Contents

Abstract An immunologic component to schizophrenia has been increasingly recognized, where infections and chronic inflammatory diseases as atopic disorders and autoimmune diseases could be involved in the pathogenesis of schizophrenia. Psychotic symptoms can be directly triggered by infections reaching the CNS, or be secondary to systemic inflammation indirectly affecting the brain through immune components, such as brain-reactive antibodies and cytokines. Large-scale epidemiological studies have consistently displayed that infections, autoimmune diseases,

M. E. Benros (✉)
Mental Health Centre Copenhagen, Copenhagen University Hospital, Copenhagen, Denmark

National Centre for Register Based Research, Aarhus University, Aarhus, Denmark
e-mail: Benros@dadlnet.dk

P. B. Mortensen
National Centre for Register Based Research, Aarhus University, Aarhus, Denmark

Curr Topics Behav Neurosci (2020) 44: 141–160
DOI 10.1007/7854_2019_93
Published Online: 21 March 2019

and atopic disorders are associated with increased risk of schizophrenia and that schizophrenia is associated with increased levels of immune markers at diagnosis. However, since there is also an increased risk of immune-related diseases after the diagnosis with schizophrenia and in family members of individuals with schizophrenia, parts of the association could also be due to heritable factors. Shared genetic factor might account for some of this increased prevalence of immune-related diseases among individuals with schizophrenia, and indeed the most pronounced genetic association with schizophrenia lies within the HLA region, which is one of the most important regions for the immune system. However, genetic studies have shown that the common genetic variants associated with schizophrenia do not seem to increase the susceptibility for acquiring infections. Nonetheless, shared genes with the susceptibility for acquiring infections not captured by the polygenic risk score for schizophrenia could still influence the association.

Keywords Autoimmune diseases · Epidemiology · Genetics · Immunology · Infection · Inflammation · Register-based · Schizophrenia

1 Introduction

Immune-related hypotheses to the etiology of schizophrenia have become increasingly prominent (Meyer et al. 2011a), suggesting that inflammation and autoimmunity could be involved in the pathogenesis of some patients with schizophrenia (Benros et al. 2012). Moreover, genetic studies have consistently shown associations between schizophrenia and specific immune parameters (Ripke et al. 2014), indicating that particularly genetically vulnerable individuals might be at risk of developing schizophrenia as a consequence of inflammation and immune components affecting the brain. The brain has traditionally been regarded as an immune-privileged site, protected by the blood-brain barrier, but it has become increasingly clear within the recent decades that the immune system does play an important role within the central nervous system (CNS). Inflammatory mechanisms can in fact affect the brain through many different pathways. During periods with infections and/or inflammation, the permeability of the blood-brain barrier might be increased, making the brain vulnerable to immune components in the blood, such as antibodies and cytokines (Margutti et al. 2006; Irani and Lang 2008). Several infectious agents can also penetrate the blood-brain barrier and invade the CNS directly (Kim 2008). Moreover, infectious agents, such as *toxoplasma* and certain viruses like Borna disease virus, HIV, and hepatitis C virus, can persist in the CNS, and even if not directly involved in destruction of CNS tissue, it might trigger CNS immune responses and thereby indirectly cause damage (Wilkinson et al. 2010; Shankar et al. 1992; Fishman et al. 2008). Peripheral infections and inflammation can furthermore affect the brain through proinflammatory cytokines activating the tryptophan-kynurenine pathway regulating NMDA glutamate receptor activity together with serotonin production (Dantzer et al. 2008), which can also indirectly affect dopamine regulation (Muller and Schwarz 2010). Moreover, peripheral inflammation and microbiome alterations can affect the brain through stimulation of peripheral nerves such as the vagal nerve (Dantzer et al. 2008; Cryan and Dinan 2012). Infections can additionally induce the development of autoimmune diseases (Rose 1998), characterized by autoantibodies and T-cells that can react against the body's own tissues, and general dysfunction of the immune system which can affect the brain and induce psychiatric symptoms (Margutti et al. 2006; Kayser et al. 2010). Inflammation reaching the brain might act as a priming event on microglia, inducing a long-term development of abnormal signal patterns possibly involved in schizophrenia. Furthermore, experimental animal studies have found that psychotic-like symptoms can be induced by inflammation or brain-reactive antibodies (Katzav et al. 2007; Kowal et al. 2004).

2 Immune Alterations in Individuals with Schizophrenia

Many diverse immune alterations have been observed in persons with schizophrenia, such as elevated levels of cytokines and inflammation markers (Goldsmith et al. 2016; Nikkilä et al. 2001; Potvin et al. 2008). Abnormalities of the blood-brain barrier have been indicated in studies of patients with schizophrenia, together with signs of CNS inflammation (Bechter et al. 2010). Additionally, increased prevalence of autoimmune diseases has been observed, and studies have indicated increased autoantibody reactivity and elevated autoantibody levels even in the patients with no known autoimmune diseases (Laske et al. 2008). Moreover, a Danish large-scale prospective population study has shown that elevated C-reactive protein levels in the general population are associated with an increased risk of late-onset schizophrenia (Wium-Andersen et al. 2014), which have been supported by subsequent studies from Finland showing the same association (Metcalf et al. 2017). Furthermore, schizophrenia has also been associated with genetic markers related to the immune system (Ripke et al. 2014).

3 Associations Between Infections and Schizophrenia

Bacterial infections have been suggested to be causally related to psychoses as early as 1896 (Noll 2007), and since the 1918 influenza pandemic was followed by multiple reports of postinfluenza psychoses and schizophrenia-like symptoms, virus infections have also been associated with schizophrenia (Torrey et al. 2006). Later, when antibiotics were introduced, the discovery that antibiotics can effectively treat neurosyphilis was one of the greatest successes within treatment of

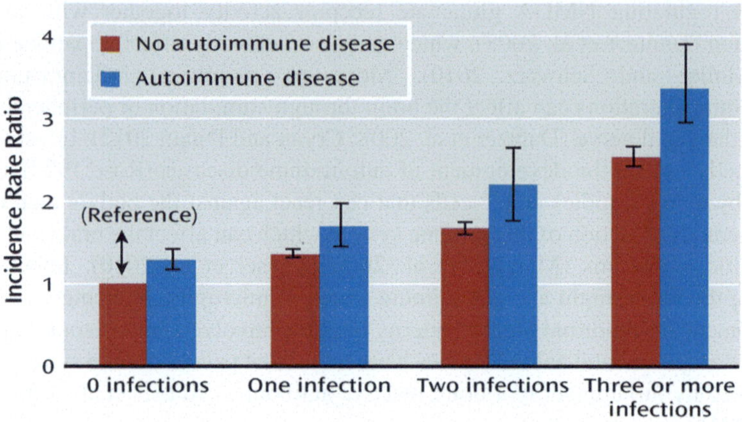

Fig. 1 Incidence rate ratios (IRRs), with confidence intervals, of schizophrenia spectrum disorders in persons with autoimmune disease and infections in Denmark 1977–2006. The linear trend between number of infections is significant (0,1,2,3+). Source: Benros et al. (2011)

mental disorders and had huge impact since significant proportions of patients with psychosis in state mental hospitals suffered from neurosyphilis in the pre-antibiotic era and were also cured for their psychiatric symptoms (Sullivan et al. 2012). However, after this remarkable treatment success, interest in possible infectious causes of psychosis waned, but during the last decades, research on the relationship between infections and psychiatric disorders has re-emerged (Yolken and Torrey 2008).

The largest study to date utilized the nationwide Danish registers to show that infections requiring hospital contacts increased the subsequent risk of schizophrenia spectrum disorders by 60% ($N = 39{,}076$ with schizophrenia spectrum disorders) (Benros and Nielsen 2011). The number of prior infections requiring hospital contacts increased the risk of schizophrenia in a dose-response relationship (Fig. 1) and was increased the most with the temporal proximity of the last infection. Hospital contacts due to infection had occurred in 23.6% prior to the diagnosis with schizophrenia spectrum disorders, resulting in a population-attributable risk of 9% associated with hospital contacts due to infections. A subsequent Swedish nationwide study confirmed the associations with previous hospital contacts with infections and increased risk of subsequent schizophrenia spectrum disorders (Blomstrom et al. 2014). A later Danish nationwide study on a younger cohort only, including individuals with complete follow-up of all hospital contacts from birth, showed that 45% of persons with schizophrenia had a previous hospital contact with infection, and in this younger cohort with shorter follow-up than the previous, hospital contacts increased the risk of schizophrenia by 41%, with bacterial infections increasing the risk by 63% (Nielsen et al. 2014). A new study investigating the effect of the infections not requiring hospitalization showed that also the less severe infections treated by mainly the general practitioners increased the risk of schizophrenia in a dose-response association but to a lesser extent than the more severe infections requiring hospital contacts (Köhler et al. 2017). It was primarily bacterial infections treated with antibiotics at the general practitioner that conferred the increased risk, particularly if the infection was more severe needing treatment with broad-spectrum antibiotics.

CNS infections have in meta-analysis shown significant associations with schizophrenia (Khandaker et al. 2012). Most studies on individuals with hospitalizations for CNS infections have found an increased risk of schizophrenia, including population-based studies from Denmark, Sweden, Finland, and Australia (Benros and Nielsen 2011; Dalman et al. 2008; Liang and Chikritzhs 2012; Koponen et al. 2004; Abrahao et al. 2005); however, this association with CNS infections was not shown in all studies (Weiser et al. 2010; Suvisaari et al. 2003).

Many different infectious agents have been associated with increased risk of schizophrenia, and a meta-analysis showed significant associations between schizophrenia with *Toxoplasma gondii*, human herpesvirus 2, Borna disease virus, human endogenous retrovirus W, *Chlamydophila psittaci*, and *Chlamydophila* pneumonia (Arias et al. 2012). Several studies have associated *Toxoplasma gondii* infection with schizophrenia (Torrey et al. 2007, 2012), and a large-scale population-based study showed that increased serum titer levels of *Toxoplasma* were associated with increased subsequent risk of schizophrenia in a dose-response association (Pedersen et al. 2011). Also herpes simplex virus infection, detected by both serum antibodies (Dickerson et al. 2003; Niebuhr et al. 2008) and CSF antibodies (Bartova et al. 1987), has been associated with increased risk of schizophrenia. Cytomegalovirus (CMV) antibody titers have been found to be higher in the serum of patients with schizophrenia (Torrey et al. 2006; Leweke et al. 2004), with stronger associations found in the serum of newly diagnosed and untreated patients (Leweke et al. 2004). CMV have also been found elevated in the CSF (Torrey et al. 1982), but to date, no neuropathologic evidence of CMV in the brains of patients with schizophrenia has been found (Torrey et al. 2006). Also retroviral antigens and products have been identified in patients with schizophrenia (Karlsson et al. 2001; Hart et al. 1999) and increased serum prevalence of Borna virus (Chen et al. 1999). Increased prevalence of *Chlamydophila* infection has also been found in patients with schizophrenia, particularly when linked to genetic markers of the immune system (Fellerhoff et al. 2007). This is supported by postmortem studies finding increased prevalence of *Chlamydophila* DNA in brains from patients with schizophrenia (Fellerhoff and Wank 2011). Psychotic disorders have also in population-based studies been associated with higher rates of subsequent infections, such as pneumonia and pneumococcal disease (Crump et al. 2013). Nonetheless, the increased prevalence of various infections in individuals with schizophrenia could also be influenced by unmeasured confounding.

4 Associations Between Autoimmune Diseases and Schizophrenia

Since the 1950s, investigators have been puzzled by the apparent protective effect of schizophrenia with a smaller than expected co-occurrence of the autoimmune disease rheumatoid arthritis (Trevathan and Tatum 1953). On the other hand, in the 1950s and 1960s, an unusually high occurrence of celiac disease in persons with schizophrenia was observed (Graff and Handford 1961). Also, as early as the 1960s, a variety of autoantibodies with cross-reactivity against brain antigens were described in the blood and CSF of patients with schizophrenia (Fessel 1962; Heath and Krupp 1967a, b). Since then, a wider range of autoimmune diseases and autoantibodies has been associated with increased risk for schizophrenia.

Large-scale Danish population-based studies with up to 20,317 patients with schizophrenia and 39,076 patients with non-affective psychosis showed that schizophrenia are associated with a nearly 50% higher lifetime prevalence of autoimmune disorders (Benros and Nielsen 2011; Eaton et al. 2006). A cross-sectional analysis of a national sample from Taiwan on 10,811 individuals with schizophrenia similarly showed that a range of autoimmune diseases were associated with schizophrenia, including specific positive associations with celiac disease, Graves' disease, psoriasis, pernicious anemia, hypersensitivity vasculitis, and again the negative association with rheumatoid arthritis (Chen et al. 2012).

Based on the Danish register data, hospital contacts due to autoimmune diseases had occurred in 2.4% of the patients before a schizophrenia diagnosis and in 3.6% of patients with schizophrenia after the diagnosis, totaling in 6% of people with schizophrenia that had a hospital contact with autoimmune diseases during follow-up (Benros and Nielsen 2011; Benros et al. 2014). Based on data from the study in Taiwan (Chen et al. 2012), 3.4% of persons with a hospital contact for autoimmune diseases also had a hospital contact with schizophrenia during a shorter follow-up period than the Danish studies. These prevalence estimates are based on hospital contacts only, and the actual prevalence of autoimmune diseases in people with schizophrenia is likely much higher if one were to screen the individuals, as exemplified by the clinical studies investigating the prevalence of celiac disease.

Screening studies of persons with schizophrenia have found antibodies to the self-antigen tissue transglutaminase, indicative of celiac disease, in about five times as many persons as expected (5.4% vs 0.8% in the CATIE study, $n = 1{,}401$) (Cascella et al. 2011). Furthermore, antibodies to gliadin, indicating sensitivity to wheat not necessarily associated with autoimmune disease, are also found in much higher proportion in patients with schizophrenia (23.1% vs 3.1% in the CATIE study), which is interesting since a wide range of neurological complications are associated with antibodies to gliadin, even in the absence of autoimmune disease (Cascella et al. 2011; Samaroo et al. 2010; Dickerson et al. 2010). Clinical studies have estimated the prevalence of celiac disease to be 2.1–2.6% in patients with schizophrenia compared to 0.3–1% in the general population (Cascella et al. 2011; Kalaydjian et al. 2006).

4.1 Temporal Associations Between Diagnosis of Schizophrenia and Autoimmune Disease Diagnosis

A Danish population-based study on 7,704 patients with schizophrenia showed that a prior history of autoimmune disease increased the risk of schizophrenia with about 45% (Eaton et al. 2006). Subsequent larger nationwide Danish studies including 20,317 individuals with schizophrenia and a total of 39,076 individuals with non-affective psychosis confirmed that the risk of autoimmune diseases was increased by 45% after an autoimmune disease diagnosis (Benros and Nielsen 2011). However, when only including individuals without a history of infection, the increased risk of schizophrenia diminished from 45% to 29%. Additionally, the study found that when autoimmune diseases and severe infections occurred together, they interacted in synergy and increased the risk of schizophrenia by 2.25 times.

On the reverse association, individuals diagnosed with schizophrenia had a 53% increased risk of a subsequent diagnosis with autoimmune diseases (Benros et al. 2014). Moreover, there was an interaction between having both a schizophrenia diagnosis and a hospital contact due to infections, which was associated with an increased risk of subsequent autoimmune diseases with 2.7 times. In individuals with schizophrenia, but no hospital contacts due to infections, the risk of autoimmune diseases was elevated with 32% and diminished with time to a nonsignificant level, whereas for persons with schizophrenia and infections, the risk remained elevated. Hence, the relationship between schizophrenia and autoimmune diseases seem to be bidirectional; however, the increased incidence of autoimmune diseases following a diagnosis of schizophrenia might in some cases reflect neuropsychiatric manifestations from the not-yet diagnosed autoimmune disease.

5 The Combined Effect of Infections and Autoimmune Diseases as Risk Factors for Schizophrenia

Experimental animal studies have shown that when brain-reactive antibodies are present in the blood and agents that increase the permeability of the blood-brain barrier are given, they can cause a temporary opening of the blood-brain barrier, with influx of brain-reactive antibodies into the brain leading to a subsequent development of a neuropsychiatric syndrome (Kowal et al. 2004). This indicates that brain-reactive antibodies in the circulation might not have pathological consequences until there is a breach of blood-brain barrier integrity (Diamond et al. 2009).

Therefore, studies have investigated the combined effect of hospital contacts due to autoimmune diseases and infections on the risk of developing schizophrenia, finding that the combined effect was larger than predicted by a combination of the single effects of the two disease groups, indicating the presence of a synergistic effect of the two exposures that increased the risk of schizophrenia by 2.3 times (Benros and Nielsen 2011). The risk of developing schizophrenia was elevated to the greatest degree in the group of autoimmune diseases with a suspected presence of brain-reactive antibodies and infections. The findings could support that CNS autoimmune disorders may require a "double insult" of circulating pathogenic serum antibodies present at the time when the blood-brain barrier is compromised by, for instance, infection, fever, or stress (Irani and Lang 2008). In persons with an autoimmune disease, three or more hospital contacts due to infections increased the risk of schizophrenia by 3.4 times (Benros and Nielsen 2011). Sepsis is the type of infection most likely to increase the blood-brain barrier permeability the most, and persons with a sepsis infection and an autoimmune disease had a fivefold increased risk of schizophrenia. The most elevated risk was observed in individuals that both had a hepatitis infection and an autoimmune disease, which was associated with an increased risk of schizophrenia by 8.9 times (Benros and Nielsen 2011). Autoimmune hepatitis is also associated with brain-reactive antibodies (Kimura et al. 2010), and in patients with severe affection of the liver, as seen in coma hepaticum in the initial phases, psychiatric symptoms are dominating (Butterworth 2011).

6 Associations Between Atopic Diseases and Schizophrenia

In a Danish nationwide study with 3,539 individuals with schizophrenia, the combined effect of being exposed to any atopic disorder increased the risk of schizophrenia by 35% (Pedersen et al. 2012). This increased risk was primarily driven by asthma diagnosis requiring hospital contact that was associated with a 46% increased risk of schizophrenia. When excluding asthma from the combined category of atopic disorders, the risk of schizophrenia was still associated with a 27% increased risk of schizophrenia, where urticaria were associated with a 34% increased risk, whereas allergic rhinitis was not significantly associated with increased risk of schizophrenia (Pedersen et al. 2012). Moreover, the atopic disorders asthma and eczema have in a UK birth cohort been associated with childhood psychotic experiences (Khandaker et al. 2014). In a subsequent Taiwanese study including 238 cases of schizophrenia utilizing health insurance records, asthma was also associated with a 40% increased risk of schizophrenia (Wang et al. 2017).

7 A Possible Genetic Association

7.1 Evidence from Epidemiological Studies Indicating Heritability

Several epidemiological studies have shown association with maternal infection during pregnancy and increased risk for schizophrenia in the offspring (Yolken and Torrey 2008; Brown et al. 2005; Brown and Derkits 2010; Mednick et al. 1988; Mortensen et al. 2007). It has been suggested that maternal immune response common to various infectious agents can influence fetal brain development leading to schizophrenia in later life, most likely in genetically susceptible individuals (Ozawa et al. 2006). As part of the immune response, elevated levels of antibodies are found in mothers of schizophrenic offspring at the end of pregnancy (Buka et al. 2001). The largest study to date is a Danish nationwide study on 3,722 individuals with schizophrenia, showing that maternal infections were associated with a 39% elevated risk of schizophrenia in the child, and the risk was increased by 23% after adjustments for parental mental illness (Nielsen et al. 2013). However, there was no significant difference in the elevated risk of schizophrenia if the infections occurred during or outside the pregnancy period, and the incidence of schizophrenia was similarly elevated when comparing infections in the mother or father during pregnancy, which could indicate that the increased risk of acquiring infections might be heritable.

A parental diagnosis with autoimmune diseases is associated with a 10% increased risk of schizophrenia, whereas a parental diagnosis with schizophrenia is associated with a 6% increased risk of autoimmune diseases in nationwide Danish studies (Benros et al. 2014; Eaton et al. 2010). A parental diagnosis with the following specific autoimmune diseases was associated with increased risk of schizophrenia: autoimmune hepatitis, type 1 diabetes, Sjögren's syndrome, iridocyclitis, multiple sclerosis, psoriasis vulgaris, and dermatopolymyositis (Eaton et al. 2010). The association with a family history of diabetes type 1 and autoimmune thyrotoxicosis with schizophrenia has been confirmed in other populations as well (Wright et al. 1996; Gilvarry et al. 1996). A parental diagnosis with schizophrenia were associated with increased risks of pernicious anemia, diabetes type 1, iridocyclitis, autoimmune hepatitis, systemic lupus erythematosus, Sjögren's syndrome, and primary biliary cirrhosis (Benros et al. 2014).

For a parental history of atopic disorders, only the mother's history of asthma was associated with an increased risk of schizophrenia in the offspring by 50%, which remained significantly elevated with 46% also after adjustment for the offspring's own history of asthma. However, again the risk of schizophrenia was not more elevated if the mother's hospital contact with asthma was prenatal compared with postnatal (Pedersen et al. 2012).

7.2 *Genetic Evidence*

The increased risk of schizophrenia in offspring to parents with hospital contacts for infections, asthma, and autoimmune diseases could reflect a genetic susceptibility toward acquiring infections and autoimmune diseases in individuals with schizophrenia. Studies have found associations between some inflammation-related genes and susceptibility to schizophrenia (Ripke et al. 2014), in which environmental influences, such as infections and autoimmune diseases, may interact with genetic factors. Furthermore, genetic markers within the HLA (human leukocyte antigen) region, which contains genes related to immune function, have been associated with the occurrence of both autoimmune diseases and schizophrenia (Ripke et al. 2014).

GWAS studies of schizophrenia patients have consistently implicated chromosome 6 in the HLA region to be the most significant genome-wide association with schizophrenia, and this has been shown in large-scale GWAS studies (Ripke et al. 2014). Nevertheless, when studies of autoimmune diseases and infections as risk factors for schizophrenia were stratified by a psychiatric family history, there was no additional increase in risk added in persons with a psychiatric family history but only the added effect of these risk factors (Benros and Nielsen 2011; Benros et al. 2014). However, several of the autoimmune diseases have been associated with, for instance, different markers in the HLA region, and these markers might be differently associated with schizophrenia. This could, for instance, explain the negative association between schizophrenia and rheumatoid arthritis, which has been consistently shown, a fact that could be due to the interplay of genetic influences (Benros et al. 2012; Eaton et al. 1992). However, ascertainment bias or anti-inflammatory and analgesic effects of antipsychotics might also be involved in this negative association (Mors et al. 1999; Sellgren et al. 2014; Torrey and Yolken 2001). Furthermore, the genetic association between the HLA region and schizophrenia has been shown to, in substantial part, be driven by the complement component 4 (C4) genes in the complement system, which also mediates synapse elimination during postnatal development (Sekar et al. 2016).

Moreover, the common genetic variants associated with schizophrenia, measured by the polygenic risk score for schizophrenia, were not associated with an increased risk of acquiring infections (Benros et al. 2016), which is a paradox since there is increased occurrence of immune-related diseases also outside of the pregnancy period in parents to individuals with schizophrenia, which could indicate a genetic association. Nonetheless, shared genes with the susceptibility for acquiring infections not captured by the polygenic risk score for schizophrenia could still influence the association. The polygenic risk score for schizophrenia increased the risk additively with infections, and individuals in the quartile with the highest load of the polygenic risk score for schizophrenia displayed an increased risk of 2.30-fold, and the individuals that had also had infections had a 3.39-fold increased risk of schizophrenia (Benros et al. 2016). In another study, the polygenic risk score for schizophrenia predicted the risk of the autoimmune diseases: type 1 diabetes, rheumatoid arthritis, and Crohn's disease (Stringer et al. 2014).

A few studies of patients with schizophrenia have profiled gene expression using microarrays focusing on analyzing inflammation-related genes in blood (Drexhage et al. 2010), lymphoblastoid cell lines (Sanders et al. 2017), and postmortem brains (Saetre et al. 2007). However, the largest transcriptome study of postmortem brains did not detect significant differential expression of the approximately 600 inflammation-related genes being studied (Gandal et al. 2018; Birnbaum et al. 2017). Nonetheless, other studies have investigated unbiased transcriptome expression profiles in individuals with schizophrenia finding enrichment of differential gene expression in immune-related pathways measured in the blood (Gardiner et al. 2013; Sainz et al. 2012; Xu et al. 2012) and brain (Karim et al. 2016; Hwang et al. 2013; Darby et al. 2016) of individuals with schizophrenia.

However, in a large genetic study, there was not found any enrichment of immune loci outside of the MHC region based on the findings from the largest GWAS of schizophrenia so far (Pouget et al. 2016).

Mendelian randomization studies have provided conflicting results on a potential causal effect of C-reactive protein in itself on schizophrenia (Prins et al. 2016; Hartwig et al. 2017), but Mendilian randomization studies of IL-6 indicated causal effect with schizophrenia (Khandaker et al. 2018; Hartwig et al. 2017). Moreover, epigenetic modifications after exposures to infections or other environmental factors could potentially also induce dysregulation of the immune system, but the evidence on this is still lacking.

Genetic studies have thus far offered some support for the roles of immune-related genes in schizophrenia but not on an individual level and only in large cohorts.

8 Confounding Factors Potentially Influencing the Association Between Immune-Related Diseases and Schizophrenia

Early-life exposure to infection and/or immune activation, such as maternal immune responses during pregnancy, may permanently alter the peripheral immune system of the child resulting in exacerbate reactions to subsequent immunological challenges later in life or increased vulnerability of acquiring infections (Patterson 2009; Meyer et al. 2011b).

Schizophrenia are associated with psychological stress particularly preceding the onset of the diagnosis, which can affect the immune system functioning and might increase the risk of acquiring infections and enhancing immunological responses in the individual (Pedersen et al. 2010; Ader et al. 1995). Thus, the observed co-occurrence of immune-related diseases and altered inflammatory markers might simply be a parallel finding and not a causal relationship. The relationship might also be bidirectional with immune responses leading to psychological stress preceding the more severe psychiatric symptoms. Moreover, several of the chronic autoimmune diseases did not by themselves increase the risk, which indicates that the associations are not only due to the psychological stress of living with a chronic disease or being hospitalized.

An iatrogenic effect of medical treatment seems unlikely to explain the major associations, since only some of the autoimmune diseases would be treated with, for instance, steroids or interferon, which may increase the risk of psychosis, and in fact even a decreased risk of psychosis has been reported associated with the use of steroids (Laan et al. 2009). Certain types of antibiotics might alter the gut microflora and thereby have positive or negative effects on the immune system and possibly also on schizophrenia (Cryan and Dinan 2012). Additionally, some antipsychotics can also affect immune responses possibly affecting the risk of subsequent autoimmune diseases and infections (Goldsmith and Rogers 2008).

Social and lifestyle factors of persons not yet diagnosed with schizophrenia can increase the probability of smoking and alcohol and drug abuse that can suppress the immune system and thereby increase the vulnerability to infection or have immune-activating effects resulting in autoimmunity (Margutti et al. 2006; Sperner-Unterweger 2005; Sopori 2002). Furthermore, social and lifestyle factors associated with schizophrenia may influence the help-seeking behavior and compliance of treatment initiated by the general practitioner, which could necessitate hospital treatment for autoimmune diseases and infections. Nonetheless, the added risk from infections or autoimmune diseases was not more pronounced in persons with a psychiatric family history or a personal history with substance abuse, which could be used as proxy variables for social and lifestyle factors (Benros et al. 2011).

8.1 Potential Etiological Mechanisms for the Associations

The potential underlying etiological mechanisms for parts of the associations between infections, atopic disorders, and autoimmune diseases with schizophrenia are numerous and not necessarily mutually exclusive and may in fact be interconnected. Both innate and adaptive immune responses might be involved, and there are many different routes of communication between the peripheral immune system and the brain (Margutti et al. 2006). Psychiatric symptoms can be directly triggered by immune components, such as brain-reactive antibodies and cytokines, or infections reaching the CNS, possibly through increased permeability of the blood-brain barrier, or be secondary to systemic inflammation indirectly affecting the brain (Dantzer et al. 2008; Muller and Schwarz 2010; Kayser et al. 2010; Diamond et al. 2009; Bechter 2012). The associations with a range of autoimmune diseases and infections may reflect inflammatory processes as a common pathway to schizophrenia. Systemic inflammation can increase the blood-brain barrier permeability, leaving the brain vulnerable to immune components from the periphery such as autoantibodies and cytokines, or possibly the effect of specific T-cell subsets that are involved in immune surveillance of the brain (Goverman 2009). Furthermore, an imbalance between the Th1 and Th2 systems has also been proposed as an etiological component (Muller and Schwarz 2010), which would fit with the increased prevalence of autoimmune diseases and atopic disorders in people with schizophrenia (Benros and Nielsen 2011; Benros et al. 2014; Pedersen et al. 2012). Inflammation can additionally affect the brain without passing the blood-brain barrier through stimulation of peripheral nerves (Dantzer et al. 2008) or proinflammatory cytokines activating the tryptophan-kynurenine pathway involved in regulation of the glutamate and serotonin system (Dantzer et al. 2008) and probably also indirectly the dopamine system (Muller and Schwarz 2010). Low-grade brain inflammation, which could be triggered by infections and autoimmune diseases, has also been proposed as an underlying causal mechanism of subgroups of patients with schizophrenia (Bechter 2012). Inflammation reaching the brain might act as a priming event on microglia, inducing a long-term development of abnormal signal patterns (Hickie et al. 2009). Some infectious agents escape surveillance by the immune system, but after an acute infection or inflammation, symptoms may flare up from the latent infection (Fellerhoff and Wank 2011). Additionally, alterations of the gut microbiota can also influence brain function and behavior possibly through neural, endocrine, and immune pathways (Cryan and Dinan 2012). Moreover, schizophrenia could prove to be of a more systemic character, with inflammation as a common etiological mechanism that, besides the neuropsychiatric symptoms, also manifest itself with increased prevalence of infections, atopic disorders, autoimmune disease, cardiovascular disease, and diabetes.

Genetically vulnerable individuals might be at a particular risk of developing schizophrenia as a consequence of inflammation and immune components affecting the brain. Subgroups of people with schizophrenia may demonstrate features of an autoimmune process, and the hypothesis is strengthened by the findings of an increased familial association between autoimmune diseases and schizophrenia (Benros and Nielsen 2011; Benros et al. 2014). Additionally, it is interesting that etiological mechanisms hypothesized to be involved in the initiation of autoimmunity, where genetic susceptibility is required along with triggering events such as infections, are very similar to the ones now proposed for schizophrenia (Rose 1998).

8.2 *Knowledge Gaps*

The observed associations between infections, atopic disorders, and autoimmune diseases with schizophrenia from the large-scale population-based register seem robust and support a possible immunological contribution in subgroups of individuals with schizophrenia. However, despite the temporal and a dose-response relationship particularly for infections, which could indicate causal associations, register-based studies cannot identify specific etiological pathways. Nonetheless, the associations seem biologically plausible based on experimental animal and human studies. Immune exposures are rather common, and, in particular, infections and initiation of the inflammatory cascade might prove to be important risk factors for schizophrenia, but direct evidence for it to play a major role in the etiology of schizophrenia is still lacking.

Screening for somatic diseases potentially affecting the brain, such as autoimmune diseases and infections, preferentially with material closer to the brain, such as CSF in addition to blood, could in the near future prove to be helpful in diagnosing and treatment planning for persons with first-onset symptoms of schizophrenia. However, the CNS is well protected and still difficult to access in vivo with regard to investigating signs of inflammation or relevant immune components not necessarily present even in CSF. Nonetheless, CSF studies of patients with schizophrenia, and no known autoimmune diseases or infection, have detected autoantibodies or antibodies against infectious agents in the CSF of 3.2–6% of patients with schizophrenia, leading to altered treatment (Meyer et al. 2011a). Furthermore, an increasing number of previously unknown antibodies with reactivity against the central nervous system (CNS) are being discovered these years (Graus et al. 2010). However, large-scale well-conducted CSF studies of individuals with schizophrenia accounting for confounding factors are still lacking. Some of the strongest evidence for the potential for autoimmunity and immune components to cause psychiatric symptoms comes from the autoimmune encephalitis and most strikingly for NMDA-receptor antibody-induced encephalitis where psychosis and psychiatric symptoms are often dominant in the initial and the remission phase of the disorder in up to 70% of the cases (Kayser and Dalmau 2011) and which have been demonstrated to be treatable with immune therapies (Graus et al. 2010; Kayser and Dalmau 2011). However, this is a very rare condition, and the prevalence of CNS-reactive antibodies in the CSF of patients with schizophrenia has not yet been determined and is very likely to be rare.

9 Conclusion

In summary, there seems to be a solid association between autoimmune diseases, asthma, and infections with schizophrenia based on large-scale nationwide studies. Experimental animal studies and human studies indicate that many diverse immune challenges can induce symptoms of schizophrenia. If the immunological alterations are confirmed to play a role in the pathogenesis of schizophrenia, it could provide an interesting and promising target of future prevention and treatment. Moreover, whether or not the co-occurrence of somatic diseases in people with psychiatric symptoms is causally related to the psychiatric symptoms, the individuals would benefit from treatment for their somatic comorbidity to reduce mortality and improve quality of life.

References

Abrahao AL, Focaccia R, Gattaz WF (2005) Childhood meningitis increases the risk for adult schizophrenia. World J Biol Psychiatry 6:44–48

Ader R, Cohen N, Felten D (1995) Psychoneuroimmunology: interactions between the nervous system and the immune system. Lancet 345:99–103

Arias I, Sorlozano A, Villegas E et al (2012) Infectious agents associated with schizophrenia: a meta-analysis. Schizophr Res 136:128–136

Bartova L, Rajcani J, Pogady J (1987) Herpes simplex virus antibodies in the cerebrospinal fluid of schizophrenic patients. Acta Virol 31:443–446

Bechter K (2012) Updating the mild encephalitis hypothesis of schizophrenia. Prog Neuro-Psychopharmacol Biol Psychiatry. https://doi.org/10.1016/j.pnpbp.2012.06.019

Bechter K, Reiber H, Herzog S, Fuchs D, Tumani H, Maxeiner HG (2010) Cerebrospinal fluid analysis in affective and schizophrenic spectrum disorders: identification of subgroups with immune responses and blood-CSF barrier dysfunction. J Psychiatr Res 44:321–330

Benros M, Nielsen P (2011) Autoimmune diseases and severe infections as risk factors for schizophrenia: a 30-year population-based register study. Am J Psychiatry 168:1303–1310

Benros ME, Nielsen PR, Nordentoft M, Eaton WW, Dalton SO, Mortensen PB (2011) Autoimmune diseases and severe infections as risk factors for schizophrenia: a 30-year population-based register study. Am J Psychiatry 168:1303–1310

Benros ME, Mortensen PB, Eaton WW (2012) Autoimmune diseases and infections as risk factors for schizophrenia. Ann N Y Acad Sci 1262:56–66

Benros ME, Pedersen MG, Rasmussen H, Eaton WW, Nordentoft M, Mortensen PB (2014) A nationwide study on the risk of autoimmune diseases in individuals with a personal or a family history of schizophrenia and related psychosis. Am J Psychiatry 171:218–226

Benros ME, Trabjerg BB, Meier S et al (2016) Influence of polygenic risk scores on the association between infections and schizophrenia. Biol Psychiatry 80:609–616

Birnbaum R, Jaffe AE, Chen Q et al (2017) Investigating the neuroimmunogenic architecture of schizophrenia. Nat Publ Group 23:1251–1260

Blomstrom A, Karlsson H, Svensson A et al (2014) Hospital admission with infection during childhood and risk for psychotic illness – a population-based cohort study. Schizophr Bull 40:1518–1525

Brown AS, Derkits EJ (2010) Prenatal infection and schizophrenia: a review of epidemiologic and translational studies. Am J Psychiatry 167:261–280

Brown AS, Schaefer CA, Quesenberry CP Jr, Liu L, Babulas VP, Susser ES (2005) Maternal exposure to toxoplasmosis and risk of schizophrenia in adult offspring. Am J Psychiatry 162:767–773

Buka SL, Tsuang MT, Torrey EF, Klebanoff MA, Bernstein D, Yolken RH (2001) Maternal infections and subsequent psychosis among offspring. Arch Gen Psychiatry 58:1032–1037

Butterworth RF (2011) Hepatic encephalopathy: a central neuroinflammatory disorder? Hepatology 53:1372–1376

Cascella NG, Kryszak D, Bhatti B et al (2011) Prevalence of celiac disease and gluten sensitivity in the United States clinical antipsychotic trials of intervention effectiveness study population. Schizophr Bull 37:94–100

Chen CH, Chiu YL, Wei FC et al (1999) High seroprevalence of Borna virus infection in schizophrenic patients, family members and mental health workers in Taiwan. Mol Psychiatry 4:33–38

Chen S-J, Chao Y-L, Chen C-Y et al (2012) Prevalence of autoimmune diseases in in-patients with schizophrenia: nationwide population-based study. Br J Psychiatry 200:374–380

Crump C, Winkleby MA, Sundquist K, Sundquist J (2013) Comorbidities and mortality in persons with schizophrenia: a Swedish national cohort study. Am J Psychiatry 170:324–333

Cryan JF, Dinan TG (2012) Mind-altering microorganisms: the impact of the gut microbiota on brain and behaviour. Nat Rev Neurosci 13:701–712

Dalman C, Allebeck P, Gunnell D et al (2008) Infections in the CNS during childhood and the risk of subsequent psychotic illness: a cohort study of more than one million Swedish subjects. Am J Psychiatry 165:59–65

Dantzer R, O'Connor JC, Freund GG, Johnson RW, Kelley KW (2008) From inflammation to sickness and depression: when the immune system subjugates the brain. Nat Rev Neurosci 9:46–56

Darby MM, Yolken RH, Sabunciyan S (2016) Consistently altered expression of gene sets in postmortem brains of individuals with major psychiatric disorders. Nat Publ Group 6:e890

Diamond B, Huerta PT, Mina-Osorio P, Kowal C, Volpe BT (2009) Losing your nerves? Maybe it's the antibodies. Nat Rev Immunol 9:449–456

Dickerson FB, Boronow JJ, Stallings C, Origoni AE, Ruslanova I, Yolken RH (2003) Association of serum antibodies to herpes simplex virus 1 with cognitive deficits in individuals with schizophrenia. Arch Gen Psychiatry 60:466–472

Dickerson F, Stallings C, Origoni A et al (2010) Markers of gluten sensitivity and celiac disease in recent-onset psychosis and multi-episode schizophrenia. Biol Psychiatry 68:100–104

Drexhage RC, van der Heul-Nieuwenhuijsen L, Padmos RC, Drexhage HA (2010) Inflammatory gene expression in monocytes of patients with schizophrenia: overlap and difference with bipolar disorder. A study in naturalistically treated patients. Int J Neuropsychopharmacol 13:1369–1381

Eaton WW, Hayward C, Ram R (1992) Schizophrenia and rheumatoid arthritis: a review. Schizophr Res 6:181–192

Eaton WW, Byrne M, Ewald H et al (2006) Association of schizophrenia and autoimmune diseases: linkage of Danish national registers. Am J Psychiatry 163:521–528

Eaton WW, Pedersen MG, Nielsen PR, Mortensen PB (2010) Autoimmune diseases, bipolar disorder, and non-affective psychosis. Bipolar Disord 12:638–646

Fellerhoff B, Wank R (2011) Increased prevalence of Chlamydophila DNA in post-mortem brain frontal cortex from patients with schizophrenia. Schizophr Res 129:191–195

Fellerhoff B, Laumbacher B, Mueller N, Gu S, Wank R (2007) Associations between Chlamydophila infections, schizophrenia and risk of HLA-A10. Mol Psychiatry 12:264–272

Fessel WJ (1962) Autoimmunity and mental illness. A preliminary report. Arch Gen Psychiatry 6:320–323

Fishman SL, Murray JM, Eng FJ, Walewski JL, Morgello S, Branch AD (2008) Molecular and bioinformatic evidence of hepatitis C virus evolution in brain. J Infect Dis 197:597–607

Gandal MJ, Haney JR, Parikshak NN et al (2018) Shared molecular neuropathology across major psychiatric disorders parallels polygenic overlap. Science 359:693–697

Gardiner EJ, Cairns MJ, Liu B et al (2013) Gene expression analysis reveals schizophrenia-associated dysregulation of immune pathways in peripheral blood mononuclear cells. J Psychiatr Res 47:425–437

Gilvarry CM, Sham PC, Jones PB et al (1996) Family history of autoimmune diseases in psychosis. Schizophr Res 19:33–40

Goldsmith C-AW, Rogers DP (2008) The case for autoimmunity in the etiology of schizophrenia. Pharmacotherapy 28:730–741

Goldsmith DR, Rapaport MH, Miller BJ (2016) A meta-analysis of blood cytokine network alterations in psychiatric patients: comparisons between schizophrenia, bipolar disorder and depression. Mol Psychiatry 21:1696–1709

Goverman J (2009) Autoimmune T cell responses in the central nervous system. Nat Rev Immunol 9:393–407

Graff H, Handford A (1961) Celiac syndrome in the case histories of five schizophrenics. Psychiatry Q 35:306–313

Graus F, Saiz A, Dalmau J (2010) Antibodies and neuronal autoimmune disorders of the CNS. J Neurol 257:509–517

Hart DJ, Heath RG, Sautter FJ Jr et al (1999) Antiretroviral antibodies: implications for schizophrenia, schizophrenia spectrum disorders, and bipolar disorder. Biol Psychiatry 45:704–714

Hartwig FP, Borges MC, Horta BL, Bowden J, Smith GD (2017) Inflammatory biomarkers and risk of schizophrenia: a 2-sample mendelian randomization study. JAMA Psychiatry 74(12):1226–1233

Heath RG, Krupp IM (1967a) Schizophrenia as an immunologic disorder. I. Demonstration of antibrain globulins by fluorescent antibody techniques. Arch Gen Psychiatry 16:1–9

Heath RG, Krupp IM (1967b) Catatonia induced in monkeys by antibrain antibody. Am J Psychiatry 123:1499–1504

Hickie IB, Banati R, Stewart CH, Stewart CH, Lloyd AR (2009) Are common childhood or adolescent infections risk factors for schizophrenia and other psychotic disorders? Med J Aust 190:S17–S21

Hwang Y, Kim J, Shin J, Kim J (2013) Gene expression profiling by mRNA sequencing reveals increased expression of immune/inflammation-related genes in the hippocampus of individuals with schizophrenia. Transl Psychiatry 3:e321–e329

Irani S, Lang B (2008) Autoantibody-mediated disorders of the central nervous system. Autoimmunity 41:55–65

Kalaydjian AE, Eaton W, Cascella N, Fasano A (2006) The gluten connection: the association between schizophrenia and celiac disease. Acta Psychiatr Scand 113:82–90

Karim S, Kamal MA, Iqbal Z et al (2016) Global expression studies of schizophrenic brain: a meta-analysis study linking neurological immune system with psychological disorders. CNS Neurol Disord Drug Targets 15:477–488

Karlsson H, Bachmann S, Schroder J, McArthur J, Torrey EF, Yolken RH (2001) Retroviral RNA identified in the cerebrospinal fluids and brains of individuals with schizophrenia. Proc Natl Acad Sci U S A 98:4634–4639

Katzav A, Solodeev I, Brodsky O et al (2007) Induction of autoimmune depression in mice by anti-ribosomal P antibodies via the limbic system. Arthritis Rheum 56:938–948

Kayser MS, Dalmau J (2011) Anti-NMDA receptor encephalitis in psychiatry. Curr Psychiatr Rev 7:189–193

Kayser MS, Kohler CG, Dalmau J (2010) Psychiatric manifestations of paraneoplastic disorders. Am J Psychiatry 167:1039–1050

Khandaker GM, Zimbron J, Dalman C, Lewis G, Jones PB (2012) Childhood infection and adult schizophrenia: a meta-analysis of population-based studies. Schizophr Res 139:161–168

Khandaker GM, Zammit S, Lewis G, Jones PB (2014) A population-based study of atopic disorders and inflammatory markers in childhood before psychotic experiences in adolescence. Schizophr Res 152:139–145

Khandaker GM, Zammit S, Burgess S, Lewis G, Jones PB (2018) Association between a functional interleukin 6 receptor genetic variant and risk of depression and psychosis in a population-based birth cohort. Brain Behav Immun 69:264–272

Kim KS (2008) Mechanisms of microbial traversal of the blood-brain barrier. Nat Rev Microbiol 6:625–634

Kimura A, Sakurai T, Koumura A et al (2010) High prevalence of autoantibodies against phosphoglycerate mutase 1 in patients with autoimmune central nervous system diseases. J Neuroimmunol 219:105–108

Köhler O, Petersen L, Mors O et al (2017) Infections and exposure to anti-infective agents and the risk of severe mental disorders: a nationwide study. Acta Psychiatr Scand 135:97–105

Koponen H, Rantakallio P, Veijola J, Jones P, Jokelainen J, Isohanni M (2004) Childhood central nervous system infections and risk for schizophrenia. Eur Arch Psychiatry Clin Neurosci 254:9–13

Kowal C, Degiorgio LA, Nakaoka T et al (2004) Cognition and immunity; antibody impairs memory. Immunity 21:179–188

Laan W, Smeets H, de Wit NJ, Kahn RS, Grobbee DE, Burger H (2009) Glucocorticosteroids associated with a decreased risk of psychosis. J Clin Psychopharmacol 29:288–290

Laske C, Zank M, Klein R et al (2008) Autoantibody reactivity in serum of patients with major depression, schizophrenia and healthy controls. Psychiatry Res 158:83–86

Leweke FM, Gerth CW, Koethe D et al (2004) Antibodies to infectious agents in individuals with recent onset schizophrenia. Eur Arch Psychiatry Clin Neurosci 254:4–8

Liang W, Chikritzhs T (2012) Early childhood infections and risk of schizophrenia. Psychiatry Res 200:214–217

Margutti P, Delunardo F, Ortona E (2006) Autoantibodies associated with psychiatric disorders. Curr Neurovasc Res 3:149–157

Mednick SA, Machon RA, Huttunen MO, Bonett D (1988) Adult schizophrenia following prenatal exposure to an influenza epidemic. Arch Gen Psychiatry 45:189–192

Metcalf SA, Jones PB, Nordstrom T et al (2017) Serum C-reactive protein in adolescence and risk of schizophrenia in adulthood: a prospective birth cohort study. Brain Behav Immun 59:253–259

Meyer U, Schwarz MJ, Müller N (2011a) Inflammatory processes in schizophrenia: a promising neuroimmunological target for the treatment of negative/cognitive symptoms and beyond. Pharmacol Ther 132:96–110

Meyer U, Feldon J, Dammann O (2011b) Schizophrenia and autism: both shared and disorder-specific pathogenesis via perinatal inflammation? Pediatr Res 69:26R–33R

Mors O, Mortensen PB, Ewald H (1999) A population-based register study of the association between schizophrenia and rheumatoid arthritis. Schizophr Res 40:67–74

Mortensen PB, Norgaard-Pedersen B, Waltoft BL et al (2007) Toxoplasma gondii as a risk factor for early-onset schizophrenia: analysis of filter paper blood samples obtained at birth. Biol Psychiatry 61:688–693

Muller NJ, Schwarz M (2010) The role of immune system in schizophrenia. Curr Immunol Rev 6:213–220

Niebuhr DW, Millikan AM, Yolken R, Li Y, Weber NS (2008) Results from a hypothesis generating case-control study: herpes family viruses and schizophrenia among military personnel. Schizophr Bull 34:1182–1188

Nielsen PR, Laursen TM, Mortensen PB (2013) Association between parental hospital-treated infection and the risk of schizophrenia in adolescence and early adulthood. Schizophr Bull 39:230–237

Nielsen PR, Benros ME, Mortensen PB (2014) Hospital contacts with infection and risk of schizophrenia: a population-based cohort study with linkage of Danish national registers. Schizophr Bull 40:1526–1532

Nikkilä HV, Müller K, Ahokas A, Rimón R, Andersson LC (2001) Increased frequency of activated lymphocytes in the cerebrospinal fluid of patients with acute schizophrenia. Schizophr Res 49:99–105

Noll R (2007) Kraepelin's 'lost biological psychiatry'? Autointoxication, organotherapy and surgery for dementia praecox. Hist Psychiatry 18:301–320

Ozawa K, Hashimoto K, Kishimoto T, Shimizu E, Ishikura H, Iyo M (2006) Immune activation during pregnancy in mice leads to dopaminergic hyperfunction and cognitive impairment in the offspring: a neurodevelopmental animal model of schizophrenia. Biol Psychiatry 59:546–554

Patterson PH (2009) Immune involvement in schizophrenia and autism: etiology, pathology and animal models. Behav Brain Res 204:313–321

Pedersen A, Zachariae R, Bovbjerg DH (2010) Influence of psychological stress on upper respiratory infection – a meta-analysis of prospective studies. Psychosom Med 72:823–832

Pedersen MG, Stevens H, Pedersen CB, Norgaard-Pedersen B, Mortensen PB (2011) Toxoplasma infection and later development of schizophrenia in mothers. Am J Psychiatry 168:814–821

Pedersen MS, Benros ME, Agerbo E, Børglum AD, Mortensen PB (2012) Schizophrenia in patients with atopic disorders with particular emphasis on asthma: a Danish population-based study. Schizophr Res 138:58–62

Potvin S, Stip E, Sepehry AA, Gendron A, Bah R, Kouassi E (2008) Inflammatory cytokine alterations in schizophrenia: a systematic quantitative review. Biol Psychiatry 63:801–808

Pouget JG, Gonçalves VF, Spain SL et al (2016) Genome-wide association studies suggest limited immune gene enrichment in schizophrenia compared to 5 autoimmune diseases. Schizophr Bull 42:1176–1184

Prins BP, Abbasi A, Wong A et al (2016) Investigating the causal relationship of C-reactive protein with 32 complex somatic and psychiatric outcomes: a large-scale cross-consortium Mendelian randomization study. PLoS Med 13:1–29

Ripke S, Neale BM, Corvin A et al (2014) Biological insights from 108 schizophrenia-associated genetic loci. Nature 511:421–427

Rose NR (1998) The role of infection in the pathogenesis of autoimmune disease. Semin Immunol 10:5–13

Saetre P, Emilsson L, Axelsson E, Kreuger J, Lindholm E, Jazin E (2007) Inflammation-related genes up-regulated in schizophrenia brains. BMC Psychiatry 7:46

Sainz J, Mata I, Barrera J, Varela I, Arranz MJ, Rodriguez MC (2012) Inflammatory and immune response genes have significantly altered expression in schizophrenia. Mol Psychiatry 18:1056–1057

Samaroo D, Dickerson F, Kasarda DD et al (2010) Novel immune response to gluten in individuals with schizophrenia. Schizophr Res 118:248–255
Sanders AR, Drigalenko EI, Duan J et al (2017) Transcriptome sequencing study implicates immune-related genes differentially expressed in schizophrenia: new data and a meta-analysis. Transl Psychiatry 7:e1093
Sekar A, Bialas AR, de Rivera H et al (2016) Schizophrenia risk from complex variation of complement component 4. Nature 530:177–183
Sellgren C, Frisell T, Lichtenstein P, Landen M, Askling J (2014) The association between schizophrenia and rheumatoid arthritis: a nationwide population-based Swedish study on intraindividual and familial risks. Schizophr Bull 40:1552–1559
Shankar V, Kao M, Hamir AN, Sheng H, Koprowski H, Dietzschold B (1992) Kinetics of virus spread and changes in levels of several cytokine mRNAs in the brain after intranasal infection of rats with Borna disease virus. J Virol 66:992–998
Sopori M (2002) Effects of cigarette smoke on the immune system. Nat Rev Immunol 2:372–377
Sperner-Unterweger B (2005) Immunological aetiology of major psychiatric disorders: evidence and therapeutic implications. Drugs 65:1493–1520
Stringer S, Kahn RS, de Witte LD, Ophoff RA, Derks EM (2014) Genetic liability for schizophrenia predicts risk of immune disorders. Schizophr Res 159:347–352
Sullivan PF, Daly MJ, O'Donovan M (2012) Genetic architectures of psychiatric disorders: the emerging picture and its implications. Nat Rev Genet 13:537–551
Suvisaari J, Mautemps N, Haukka J, Hovi T, Lönnqvist J (2003) Childhood central nervous system viral infections and adult schizophrenia. Am J Psychiatry 160:1183–1185
Torrey EF, Yolken RH (2001) The schizophrenia-rheumatoid arthritis connection: infectious, immune, or both? Brain Behav Immun 15:401–410
Torrey E, Yolken R, Winfrey J (1982) Cytomegalovirus antibody in cerebrospinal fluid of schizophrenic patients detected by enzyme immunoassay. Science 216:892–894
Torrey EF, Leweke MF, Schwarz MJ et al (2006) Cytomegalovirus and schizophrenia. CNS Drugs 20:879–885
Torrey EF, Bartko JJ, Lun ZR, Yolken RH (2007) Antibodies to Toxoplasma gondii in patients with schizophrenia: a meta-analysis. Schizophr Bull 33:729–736
Torrey EF, Bartko JJ, Yolken RH (2012) Toxoplasma gondii and other risk factors for schizophrenia: an update. Schizophr Bull 38:642–647
Trevathan R, Tatum JC (1953) Rarity of concurrence of psychosis and rheumatoid arthritis in individual patients. J Nerv Ment Dis 120:83–84
Wang WC, Lu ML, Chen VC et al (2017) Asthma, corticosteroid use and schizophrenia: a nationwide population-based study in Taiwan. PLoS One 12:e0173063
Weiser M, Werbeloff N, Levine A et al (2010) CNS infection in childhood does not confer risk for later schizophrenia: a case-control study. Schizophr Res 124:231–235
Wilkinson J, Radkowski M, Eschbacher JM, Laskus T (2010) Activation of brain macrophages/ microglia cells in hepatitis C infection. Gut 59:1394–1400
Wium-Andersen MK, Ørsted DD, Nordestgaard BG (2014) Elevated C-reactive protein associated with late-and very-late-onset schizophrenia in the general population: a prospective study. Schizophr Bull 40:1117–1127
Wright P, Sham PC, Gilvarry CM et al (1996) Autoimmune diseases in the pedigrees of schizophrenic and control subjects. Schizophr Res 20:261–267
Xu J, Sun J, Chen J et al (2012) RNA-Seq analysis implicates dysregulation of the immune system in schizophrenia. BMC Genomics 13:S2
Yolken RH, Torrey EF (2008) Are some cases of psychosis caused by microbial agents? A review of the evidence. Mol Psychiatry 13:470–479

Microglial Activation and Psychotic Disorders: Evidence from Pre-clinical and Clinical Studies

Tatiana Barichello, Lutiana R. Simoes, Joao Quevedo, and Xiang Y. Zhang

Contents

T. Barichello (✉)
Translational Psychiatry Program, Department of Psychiatry and Behavioral Sciences, McGovern Medical School, The University of Texas Health Science Center at Houston (UTHealth), Houston, TX, USA

Experimental Pathophysiology Laboratory, Graduate Program in Health Sciences, University of Southern Santa Catarina (UNESC), Criciúma, SC, Brazil

Neuroscience Graduate Program, The University of Texas MD Anderson Cancer Center, UTHealth Graduate School of Biomedical Sciences, Houston, TX, USA
e-mail: Tatiana.Barichello@uth.tmc.edu

L. R. Simoes
Experimental Pathophysiology Laboratory, Graduate Program in Health Sciences, University of Southern Santa Catarina (UNESC), Criciúma, SC, Brazil

J. Quevedo
Translational Psychiatry Program, Department of Psychiatry and Behavioral Sciences, McGovern Medical School, The University of Texas Health Science Center at Houston (UTHealth), Houston, TX, USA

Translational Psychiatry Laboratory, Graduate Program in Health Sciences, University of Southern Santa Catarina (UNESC), Criciúma, SC, Brazil

Center of Excellence on Mood Disorders, Department of Psychiatry and Behavioral Sciences, McGovern Medical School, The University of Texas Health Science Center at Houston (UTHealth), Houston, TX, USA

Neuroscience Graduate Program, The University of Texas MD Anderson Cancer Center, UTHealth Graduate School of Biomedical Sciences, Houston, TX, USA

X. Y. Zhang
Psychosis Research Program, Department of Psychiatry and Behavioral Sciences, McGovern Medical School, The University of Texas Health Science Center at Houston (UTHealth), Houston, TX, USA

Curr Topics Behav Neurosci (2020) 44: 161–206
DOI 10.1007/7854_2018_81
Published Online: 4 March 2019

Abstract Clinical and pre-clinical studies have demonstrated an important role of neuroinflammation in the etiology of schizophrenia. While the underlying mechanisms remain poorly understood, there are some studies demonstrating an association between maternal immune activation and behavioral changes in adult offspring and identifying early life infection as a trigger for schizophrenia; in addition, inflammatory markers were found to be increased in the schizophrenic post-mortem brain. During maternal immune activation, pro-inflammatory mediators such as cytokines, chemokines, antibodies, and acute-phase proteins are released in the maternal bloodstream, thus increasing the permeability of the placental barrier and the fetal blood-brain barrier, allowing the inflammatory mediators to enter the fetal brain. In the central nervous system (CNS), these pro-inflammatory mediators are able to activate microglial cells that can release pro-inflammatory cytokines such as tumor necrosis factor alpha (TNF-α), interleukin (IL)-1β, and IL-6. As a consequence, circulating immune cells may infiltrate the brain, increasing cytokine levels and releasing antibodies that aggravate the neuroinflammation. Neuroinflammation may affect processes that are pivotal for normal brain maturation such as myelination, synaptic pruning, and neuronal remodeling. Microglial cell activation and pro-inflammatory mediators have been extensively studied in schizophrenic post-mortem brain samples. Some results of these investigations demonstrated an increase in microglial activation markers, cytokines, and chemokines in post-mortem brain samples from individuals with schizophrenia. In contrast, there are studies that have demonstrated low levels of microglial activation makers in the schizophrenic post-mortem brain. Thus, based on the important role of neuroinflammation as a trigger in the development of schizophrenia, this chapter aims (1) to enumerate evidence of neuroinflammation and microglial activation from pre-clinical schizophrenia models, (2) to show links between schizophrenia and neuroinflammation in clinical studies, and (3) to identify mechanisms by which microglial activation may influence in the development of schizophrenia.

Keywords Microglia · Neuroinflammation · Psychosis · Schizophrenia · Schizophrenia-like behavior

Abbreviations

ATP	Adenosine 5-triphosphate
CCL3	Chemokine (C-C motif) ligand 3
CLR	C-type lectin receptors
CNS	Central nervous system
COX-2	Cyclooxygenase-2
CRP	C-reactive protein
DAMPS	Damage-associated molecular patterns
DM	Damaged processes
DNA	Deoxyribonucleic acid
G-CSF	Granulocyte colony-stimulating factor
GM-CSF	Granulocyte/monocyte colony-stimulating factor
HLA-DR	Human leukocyte antigen-antigen D related
HMGB-1	High mobility group box-1 protein
HSPs	Heat shock proteins
Iba1	Ionized calcium-binding adaptor molecule
IFITM	Interferon-induced transmembrane protein
IFN	Interferon
IkB	Inhibitors of NF-B
IL	Interleukin
IL-13RA1	IL-13 receptor alpha-1
IL-1RA	IL-1 receptor antagonist
iNOS	Inducible nitric oxide synthase
IRF	Interferon regulatory factor
KC	Keratinocyte chemoattractant
LIX	Lipopolysaccharide-induced CXC chemokine
Mal	MyD88 adapter-like
MAPKs	Mitogen-activate protein kinases
MCP-1	Monocyte chemoattractant protein-1
MD-2	Myeloid differentiation protein-2
MDA	Malondialdehyde
MHC	Major histocompatibility complex
MIP-1	Macrophage inflammatory protein-1
mRNA	Messenger ribonucleic acid
MYD88	Myeloid differentiation factor 88
NF-κB	Nuclear factor kappa B
NLR	NOD-like receptors
PAMPs	Pathogen-associated molecular patterns
pIRF3	Phosphorylated-IRF3
PK 11195	1-(2-chlorophenyl)-*N*-methyl-*N*-(1-methylpropyl)-3-isoquinoline carboxamide
PND	Postnatal day
Poly I:C	Polyinosinic-polycytidylic acid

PRRs	Pattern-recognition receptors
RAGE	Receptors for advanced glycation end products
RANTES	Regulated upon activation normal T-cell expressed and secreted
RIG-1	Retinoic acid-inducible gene-1
RLR	RIG-1-like receptors
RM	Ramification
RNA	Ribonucleic acid
SERPINA-3	Serpin family A member-3
ssRNA	Double-stranded ribonucleic acid
STAT-1	Signal transducer and activator of transcription-1
TIR	Toll/IL-1 receptor
TIRAP	Domain-containing adaptor protein
TLR	Toll-like receptor
TNFR1	TNF receptor 1
TNF-α	Tumor necrosis factor alpha
TRAF	Tumor necrosis factor receptor-associated factor
TRIF	Toll/IL-1 receptor domain-containing adaptor-inducing interferon-β
TSPO	Translocator protein

1 Introduction

Clinical and pre-clinical studies have demonstrated an important role of neuro-inflammation in the etiology of schizophrenia. While the underlying mechanisms remain poorly understood, there are some studies showing evidence of microglial activation and increased levels of cytokines and chemokines in post-mortem schizophrenic brain samples, as well as in fetal and adult brains of offspring subjected to maternal immune activation during fetal life. In 1999, the first evidence of microglial and macrophage activation in the brains of patients with psychiatric disorders was reported. In the study, 3 of the 14 samples of post-mortem brains from patients with schizophrenia presented immunoreactivity to human leukocyte antigen-antigen D related (HLA-DR) protein in the frontal cortex and the hippocampus (Bayer et al. 1999). After that, a number of studies showed an increase in microglial markers in post-mortem schizophrenic brains, whereas few studies found no effect or a decrease in microglial markers. The studies that followed demonstrated an important role of maternal immune activation in releasing cytokines, chemokines, antibodies, and C-reactive protein (CRP) as an inductor of schizophrenia rather than the pathogen involved in maternal infection (Feigenson et al. 2014; Khandaker et al. 2014a, b). Pre-clinical studies have shown that during maternal immune activation, cytokines, chemokines, antibodies, and acute-phase proteins are released into the maternal bloodstream, thus increasing the permeability of the placental barrier and the fetal blood-brain barrier and allowing the inflammatory mediators to reach the fetal brain. In the central nervous system (CNS), these pro-inflammatory mediators are able to activate microglia that can release pro-inflammatory cytokines such as tumor necrosis factor alpha (TNF-α), interleukin (IL)-1β, and IL-6. In addition, circulating immune cells may infiltrate the brain, increasing the cytokine levels and releasing antibodies that aggravate the neuroinflammation (Garay et al. 2013; Feigenson et al. 2014; van den Eynde et al. 2014; Reus et al. 2017). Thus, based on the important role of neuroinflammation as a trigger for the development of schizophrenia, this chapter aims (1) to enumerate evidence of neuroinflammation and microglial activation in pre-clinical schizophrenia models, (2) to highlight links between schizophrenia and neuroinflammation in clinical studies, and (3) to identify mechanisms by which microglial activation may influence the development of schizophrenia.

2 Evidence of Neuroinflammation and Microglial Activation from Pre-clinical and Clinical Schizophrenia Studies

2.1 Microglia Overview

Microglia comprise approximately 10–15% of all glial cells and are tissue-resident macrophages that present important functions in the CNS, including in supporting newborn neurons, cell death and clearance, homeostasis, and regulation of neuronal and synaptic plasticity (Salter and Stevens 2017). Microglia are derived from primitive myeloid progenitors emanating from the embryonic yolk sac during development and then populate the CNS (embryonic day 8.5 in mice) prior to its blood vessel formation (Ginhoux et al. 2010). Resting microglia have a small cell body and possess long branching; after being activated, the cells replace their ramified branches with highly amoeboid, motile protrusions (Stence et al. 2001). A modern transcriptome profiling of microglia in mice showed that the response phenotypes fail to conform to M1 or M2 patterns, though the functional significance and ontogeny of microglia had not yet been characterized (Ransohoff 2016; Salter and Stevens 2017). Microglia present class I (HLA-A, HLA-B, HLA-C) and class II (HLA-DR, HLA-DP, HLA-DQ) major histocompatibility complex (MHC) molecules. The MHC class II is found only on antigen-presenting cells, such as microglia, dendritic cells, mononuclear phagocytes, and B cells, because these cells are essential in initiating an immune response. Microglia are an important component of the innate immune system, and during their resting states, they are active with extremely motile processes and protrusions; thus, this cell type is referred to as a "housekeeper" in the adult brain (Nimmerjahn et al. 2005) (see Fig. 1).

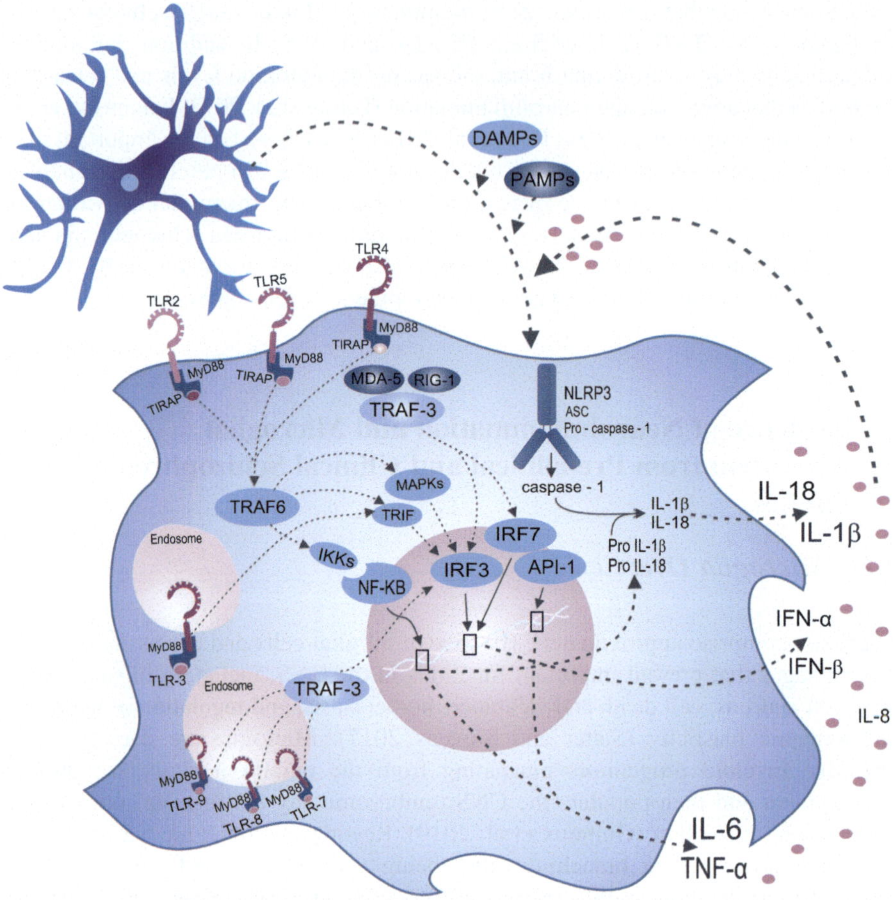

Fig. 1 Microglial activation. Resting microglial cell with ramified shape is activated by DAMPs, PAMPs, or pro-inflammatory mediators. After microglial activation, these cells present with highly amoeboid motile protrusions and release cytokines and chemokines. *API-1* apoptosis inhibitor gene-1; *ASC* caspase-recruitment domain; *DAMPS* damage-associated molecular patterns; *IFN-α, IFN-β* interferon-α, interferon-β; *IKKs* IκB kinase complex; *IL-1β, IL-6, IL-8, IL-18* interleukin-1β, interleukin-6, interleukin-8, and interleukin-18; *IRF-3, IRF-7* interferon regulatory factor-3, interferon regulatory factor-7; *MAPKs* mitogen-activated protein kinases; *MDA-5* melanoma differentiation-associated gene 5; *MyD88* myeloid differentiation factor 88; *NF-κB* nuclear factor kappa B; *NLRP-3* NLR family pyrin domain containing-3; *PAMPs* pathogen-associated molecular patterns; *RIG-1* retinoic acid-inducible gene-1; *TIRAP* domain-containing adaptor protein; *TLR* Toll-like receptor; *TNF-α* tumor necrosis factor alpha; *TRAF-3, TRAF-6* TNF receptor-associated factor-3, TNF receptor-associated factor-6; *TRIF* Toll/IL-1 receptor domain-containing adaptor-inducing interferon-β

2.2 *Evidence of Neuroinflammation and Microglial Activation in Pre-clinical Schizophrenia Models*

Several pre-clinical studies have demonstrated and supported evidence for the role of neuroinflammation in the development of schizophrenia. Among the different pre-clinical models that aim at recapitulating the development of schizophrenia, a subset of these is based on gestational exposure to maternal immune activation, a clinically relevant risk factor for schizophrenia. The experimental maternal immune activation induced by polyinosinic-polycytidylic acid (Poly I:C) mimics a viral infection because this chemical compound is a synthetic analogue of double-stranded ribonucleic acid (ssRNA). A multitude of studies that implement this model have observed long-lasting alterations of microglial markers, suggesting persistent microglial activation in adult animals exposed to gestational Poly I:C. For example, a study evaluated ionized calcium-binding adapter molecule-1 (Iba1), a microglia- and macrophage-specific calcium-binding protein that has actin-bundling activity and participates in membrane ruffling and phagocytosis in activated microglia. On gestation day 15, pregnant dams were given a single i.v. injection to the tail vein of Poly I:C or saline. The number of Iba1-positive cells was increased in the Poly I:C offspring's hippocampus and nucleus accumbens but was unchanged in the prefrontal cortex. In addition, MHC class II expression in microglia increased in the Poly I:C prefrontal cortex, but not in the hippocampus of adult male offspring at 18 weeks of life (Hadar et al. 2017). Similarly, Mattei et al. observed an increase in Iba1 immunoreactivity in the proximity of the hippocampal dentate gyrus of adult mice on PND 60 that were subjected to maternal immune activation by Poly I:C at embryonic day 15 compared with control offspring in adult life (Mattei et al. 2017). Further, using the same microglial marker, Iba1, the offspring of mice exposed to Poly I:C at embryonic day 9 were shown to have an elevated number of activated microglial cells in the hippocampus and striatum, but not in the frontal cortex, on PND 30 (Juckel et al. 2011). In another study, OX-42, an antibody designed to detect CD11b, was used as a marker of microglia in the brain. OX-42 immunoreactivity was detected on postnatal day (PND) 180 in adult Poly I:C offspring, showing an increase in the concentration of OX-42-positive staining and microglial density, and reduced microglia ramifications, indicating that the microglia were in the activated state in all brain regions. Additionally, there was a difference in the form of an overall significant increase in microglia score in the corpus callosum, hippocampus, and thalamus; however, this difference was not found in the pons, cortex, or striatum obtained from adult offspring of dams treated with Poly I:C on embryonic day 15 (Van den Eynde, Missault et al. 2014).

Activated microglial cells can increase the production and expression of pro-inflammatory cytokines, such as TNF-α and IL-1β, and neurotoxic substances, resulting in neuroinflammatory and neurodegenerative processes. Adult mice subjected to maternal immune activation by Poly I:C during the fetal stage presented high expression of proteins involved in the Toll-like receptor (TLR)-3 signaling pathway, such as signal transducer and activator of transcription-1 (STAT-1),

Toll/IL-1 receptor (TIR)-domain-containing adapter-inducing interferon-β (TRIF), nuclear factor kappa-light-chain-enhancer of activated B cells (NF-κB), and phosphorylated-IRF3 (pIRF3), in the frontal cortex. Increased oxidative and nitrosative stress, as evidenced by increased malondialdehyde (MDA) and inducible nitric oxide synthase (iNOS), and increased levels of TNF-α, interferon (IFN)-α, and IFN-β in the frontal cortex were also observed (MacDowell et al. 2017a). In another study, mice were subjected to maternal immune activation on embryonic day 12.5. On PNDs 0, 7, 14, 30, and 60, the offspring brains were removed, and the frontal cortex, cingulate cortex, and hippocampus were used to evaluate the presence of cytokines and chemokines. On PND 0, the frontal cortex showed increased levels of IL-1β, IL-10, IL-12, and granulocyte/monocyte colony-stimulating factor (GM-CSF). On PND 7, the levels of granulocyte colony-stimulating factor (G-CSF) were increased, and on PND 60, the levels of IL-1α, IL-6, IL-9, and IL-10 were also increased. On PND 0, the cingulate cortex showed increased levels of IFN-γ, IL-12, and monocyte chemoattractant protein-1 (MCP-1). On PND 7, the levels of IL-17 increased, and on PND 60, the levels of IL-10 and IFN-γ increased. On PND 0, the hippocampus showed an increase in the level of IL-6. On PND 7, increased levels of IL-9, keratinocyte chemoattractant (KC), and macrophage inflammatory protein-1 alpha (MIP-1α) were observed, and on PND 14, increased levels of IL-1α and IL-6 were found (Garay et al. 2013).

Pratt et al. injected pregnant mice on embryonic day 12.5 with Poly I:C, and fetal brains were collected at embryonic day 16.5 to evaluate the inflammatory profile of microglial cells, which included cytokine and chemokine expression. Fetal microglia expressed high levels of cytokines and chemokines such as IL-1α, IL-4, IL-6, IL-9, GM-CSF, and M-CSF, which were regulated upon activation by normal T-cell expressed and secreted (RANTES), lipopolysaccharide-induced CXC chemokine (LIX), exotoxin, and MIP-1β (Pratt et al. 2013). Using another approach, Arad et al. injected dams with Poly I:C on day 4 after birth, and the offspring were breastfed. Two hours after Poly I:C injection, the milk of the dams presented elevated levels of IL-1β, IL-6, and corticosterone. At 6 and 24 h after the dams received the Poly I:C injection, the male offspring presented high levels of IL-6 and IFN-γ in the hippocampus. Twenty-four hours after the dams received the Poly I:C injection, both male and female offspring presented high levels of TNF-α in the hippocampus. In addition, lactational Poly I:C exposure triggered behavioral abnormalities in the adult offspring (PND 90 to 120), with male, but not female, offspring exhibiting attentional and executive function abnormalities (manifested in persistent latent inhibition and slow reversal) and female, but not male, offspring exhibiting despair and anhedonia (Arad et al. 2017).

A subset of studies aimed at characterizing the role of single cytokines. For example, Smith et al. demonstrated the important role of IL-6 in schizophrenia-like behavior. Specifically, an intraperitoneal injection of IL-6 on embryonic day 12.5 in pregnant mice triggered prepulse inhibition and latent inhibition deficits in the adult offspring, but IFN-γ maternal injection did not affect the schizophrenia-like behavior of adult offspring (Smith et al. 2007). The section above highlights several studies demonstrating that TLR-3 activation and pro-inflammatory cytokines could influence the development of schizophrenia-like behavior in adult offspring. In contrast to other studies, on PND 90 to 104, adult offspring did not present any significant difference in the level of microglial activation compared to the control adult offspring (Missault et al. 2014). In this study, despite the confirmation of systemic inflammation in the pregnant mice, there was no difference in fetal microglial cell density or in the activation level on embryonic days 11.5–17.5 between the control and Poly I:C group (Smolders et al. 2015); see Table 1.

The Gunn rat is another animal model of schizophrenia (Gunn 1944). Gunn rats present behavioral abnormalities, deficits in prepulse inhibition, and neuropathological changes that are similar to the characteristics of schizophrenia-like behavior (Liaury et al. 2012). CD11b immunoreactivity is increased in microglial cells of the hippocampal dentate gyrus of Gunn rats (Liaury et al. 2012, 2014). Gunn rats showed a prepulse inhibition deficit compared to Wistar rats. The amount of CD11b microglial cell marker increased in the hippocampus of Gunn rats compared to the same brain structure of the Wistar rats (Limoa et al. 2016).

3 Evidence of Neuroinflammation from Schizophrenic Patients

3.1 Microglia Evaluation in Post-mortem Schizophrenic Brain

Brain samples from 3 of 14 patients with schizophrenia exhibited HLA-DR-positive tests in the frontal cortex and hippocampus (Bayer et al. 1999). HLA-DR is an MHC class II cell surface receptor that interacts with antigen-presenting cells such as microglia, mononuclear phagocytes, dendritic cells, and B cells. The microglial marker HLA-DR was increased in paranoid schizophrenic hippocampal samples compared with residual schizophrenic and matched control samples. In the same study, higher expression levels of CD3+ and CD20+ lymphocytes were found in the hippocampus of residual schizophrenics compared with paranoid schizophrenics and matched controls (Busse et al. 2012). The density of HLA-DR cells that were morphologically similar to microglia was increased in the dorsolateral prefrontal cortex of individuals with schizophrenia (Fillman et al. 2013). In a previous study, the frontal and temporal lobes of chronic schizophrenics presented greater microglial cell activation compared with control brains. However, the first layer of the cerebral cortex presented the same amounts of well-developed ramifications (RM), degenerative traits, and damaged processes (DM), and the number of DM cells in the remaining regions was higher than that of the RM cells (Wierzba-Bobrowicz et al. 2005). Another study evaluated 12 brains of female chronic schizophrenics. The schizophrenic frontal and temporal lobe samples presented ramified microglial cells with expression of MHC class II. Most cells presented with cytoplasm shrinkage, thinning, shortening and fragmentation of their processes, and apoptotic changes.

Table 1 Microglial cell and inflammatory markers in pre-clinical schizophrenia models

Pre-clinical model	Species	Animal age	Sex	Anatomical area	Biomarkers evaluated	Results	Reference
Poly I:C (5 mg/kg) on GD 9	Mutant hDISC1 and C57BL6/J mice	ED 9 at 6 h after Poly I:C	No	Whole brain	IL-1β, IL-2, IL-4, IL-5, IL-6, IL-10, and TNF-α	IL-1β levels increased in the brains of C57BL6/J and mutant hDISC1 mice. IL-4 and IL-5 levels increased only in C57BL6/J mice	Abazyan et al. (2010)
Poly I:C (4 mg/kg)	Wistar rats	On PND 4 at 6 h and 24 h following lactation with saline or Poly I:C exposure	Female and male	Whole brain	IL-6, IFN-γ, and TNF-α	IL-6 and IFN-γ were elevated at both time points only in male Poly I:C offspring compared to their controls [IL-6: main effect of sex, F (1.55) = 5.74, $p < 0.05$; main effect of immune activation, F (1.55) = 4.63, $p < 0.05$; immune activation vs. sex interaction, F (1.55) = 7.76, $p < 0.01$, and a significant difference in post hoc comparisons, $p < 0.01$; IFN-γ: sex vs. treatment interaction, F (1.56) = 4.43, $p < 0.05$, and a significant difference in post hoc comparisons, $p < 0.01$]. Both male and female	Arad et al. (2017)

						Poly I:C offspring had a lower hippocampal level of TNF-α at 24 h, but not at 6 h, following exposure to Poly I:C compared to the saline offspring [treatment x time interaction, F (1.56) = 6.78, $p < 0.05$, and significant differences in post hoc comparisons, $p < 0.005$]	
Poly I:C (20 mg/kg) on GD 12.5	C57BL/6 J mice	PND 0, PND 7, PND 14, PND 30, and PND 60	No	Frontal cortex, cingulate cortex, and hippocampus	Iba1, IL-1α, IL-1β, IL-2, IL-3, IL-4, IL-5, IL-6, IL-9, IL-10, IL-12 (p40), IL-12 (p70), IL-13, eotaxin, G-CSF, GM-CSF, IFN-γ, KC, MCP-1, MIP-1α, MIP-1β, RANTES, and TNF-α	IL-1β levels in the hippocampus and cingulate cortex were similar with age but quite different in the frontal cortex where they were highest on PND 0 and PND 7 and then dramatically decreased at PND 14. Levels of IL-9 were steadily high with age in the cingulate cortex and hippocampus and were also relatively high in the frontal cortex during the period of rapid synaptogenesis (PND 0–PND 14) but decreased in the frontal cortex with maturity. Conversely,	Garay et al. (2013)

(continued)

Table 1 (continued)

Pre-clinical model	Species	Animal age	Sex	Anatomical area	Biomarkers evaluated	Results	Reference
						IL-6 decreased at PND 60 in the hippocampus and cingulate cortex but not in the frontal cortex, where it remains high in adulthood. A second point of interest is that some cytokines, including IL-4, IL-2, and IL-17, are higher in mid-postnatal life but lower at birth and in the adult. Third, several cytokines dip in concentration specifically at PND 14, a period of intense synaptogenesis; these include IL-3, IL-13, IL-12 (p40), eotaxin, MIP-1α, and KC in addition to IL-2 and IL-5 specifically in the cingulate cortex and hippocampus. Fourth, another set of cytokines increased in concentration with age, including IL-6, IL-10, IFN-γ, GM-CSF, IL-12 (p70), IL-17, and MIP-1β	

Poly I:C (4 mg/kg) on GD 15	Wistar rats	PND 33–34	Female and male	Hippocampus, nucleus accumbens, and medial prefrontal cortex	IL-1β, TNF-α, IL-6, Iba1, and MHC II	Pro-inflammatory cytokine mRNA levels were unchanged in all brain areas. The density of Iba1-positive cells (microglia) was increased in the Poly I:C hippocampus and nucleus accumbens but unchanged in the medial prefrontal cortex. Using FACS, detected an increase in MHC II expression in microglia derived from the Poly I:C medial prefrontal cortex, but not the hippocampus	Hadar et al. (2017)
Poly I:C (20 mg/kg) on GD 9	BALB/c mice	PND 30	Females and males	Hippocampus, frontal cortex, striatum, and as a control region the occipital cortex	CD11b/Iba1	Poly I:C treatment caused a significant increase of microglial markers in the hippocampus ($p = 0.028$) and a significant increase in the striatum ($p = 0.028$). Poly I:C offspring from LPS mothers exhibited significantly less branches and processes in microglial cells compared to the control mice ($p = 0.002$), showing a	Juckel et al. (2011)

(continued)

Table 1 (continued)

Pre-clinical model	Species	Animal age	Sex	Anatomical area	Biomarkers evaluated	Results	Reference
						reduced surface of processes in Poly I:C mice, suggesting that Poly I:C treatment of mothers caused a higher activation status in the offspring generation	
Gunn rats	Gunn and Wistar rats (8 weeks old)	0 h	Male	Hippocampal dentate gyrus	CD11b	There was no significant difference between cell numbers in the Gunn rats and controls. However, there was a significant increase in CD11b expression in the hippocampal dentate gyrus in GUNN rats	Liaury et al. (2012)
Gunn rats Minocycline hydrochloride (40 mg/kg)	Gunn and Wistar rats (6 weeks old)	14 days	Male	Whole brain	CD11b	Immunohistochemistry analysis revealed that microglial cells in the minocycline-treated Gunn rat group showed less expression of CD11b compared to vehicle-treated Gunn and Wistar groups	Liaury et al. (2014)
Poly I:C (5 mg/kg) on GD 9.5	C57BL/6 J mice	PND 60	Female and male	Frontal cortical areas	IL-1β, IL-6, IL-10, IFN-α, IFN-β,	MIA by Poly I:C induced an increase in the main	MacDowell et al. (2017a)

					CX3CL1, STAT1, TGF-β, and TNF-α	pro-inflammatory cytokines TNF-α and IL-6 mRNA levels, but no changes were seen in the IL-1β mRNA levels	
Poly I:C (4.0 mg/kg) on GD 15	Wistar rats	PND 128 (Poly I:C/minocycline; Poly I:C/H_2O; NaCl/minocycline; NaCl/H_2O)	Male	Ventral striatum, cingulate gyrus, medial prefrontal cortex, nucleus accumbens core, dentate gyrus of the hippocampus, and cerebellum	Iba1, IL-1β and TNF-α	In the dentate gyrus, a significant decrease in microglia Iba1 reactivity in the Poly I:C/H_2O group compared to control NaCl/H_2O was detected. In the hippocampus a significant increase of IL-1β mRNA in the Poly I:C/H_2O compared to NaCl/H_2O was found. In addition, they detected a significant effect of minocycline on IL-1β mRNA levels in the Poly I:C/minocycline group compared to Poly I:C/H_2O in the hippocampus. The increase in TNF-α mRNA in Poly I:C/H_2O in the hippocampus did not reach significance compared to NaCl/H_2O but was significantly higher compared to NaCl/minocycline	Mattei et al. (2014)

(continued)

Table 1 (continued)

Pre-clinical model	Species	Animal age	Sex	Anatomical area	Biomarkers evaluated	Results	Reference
Poly I:C (2, 4, or 8 mg/kg) on GD 9 and GD 15	Wistar-Hannover rats	PND 90 until PND 104	No	Maternal serum and fetal brains	IL-1β, IL-6, IL-10, and TNF-α	Surprisingly, not the highest dose tested, but the 4 mg/kg dose induced the largest increase in IL-1β mRNA in maternal blood, which was significant at GD 15 ($p < 0.05$). The highest increase in TNF-α mRNA expression in the blood of GD 15 mothers was also observed at the 4 mg/kg dose, while at GD 9 the strongest expression was observed using 8 mg/kg Poly I:C, an effect which was statistically significant ($p < 0.05$). The brains of fetuses exhibited a moderate increase in the IL-1β and TNF-α levels compared to controls. The largest increase in pro-inflammatory cytokines was observed in offspring belonging to the 4 mg/kg group.	Missault et al. (2014)

						While at GD 9 this rise in pro-inflammatory cytokines was balanced by a rise in anti-inflammatory IL-10, this was not the case at GD 15	
LPS (500 or 10 μg/kg) on GD 17	Mice	1.5 h after injected to LPS	Female and male	Maternal serum, amniotic fluid, fetal liver, and fetal brain	TNF-α	TNF-α increased in maternal serum and amniotic fluid in response to LPS. Although maternally administered LPS also increased the level of TNF-α protein in the fetal liver and brain, no significant difference in TNF-α mRNA level in fetal liver and brain was found. When the pregnant mice were pretreated with 10 μg/kg at 4, 12, 24, or 48 h before LPS 500 μg/kg, TNF-α in maternal serum and amniotic fluid was inhibited. Low doses of LPS pretreatment attenuated LPS-induced increases in TNF-α protein in the fetal liver and fetal brain.	Ning et al. (2008)

(continued)

Table 1 (continued)

Pre-clinical model	Species	Animal age	Sex	Anatomical area	Biomarkers evaluated	Results	Reference
						Perinatal exposure to low doses of LPS induced a reduced sensitivity to subsequent LPS challenge	
Poly I:C (20 mg/kg) On GD 12.5	C57BL/6 J mice	ED 16.5	Females and males	Fetal brains	IL-1α, IL-1β, IL-2, IL-3, IL-4, IL-5, IL-6, IL-7, IL-9, IL-10, IL-12 (p40), IL-12 (p70), IL-13, IL-15, IL-17, eotaxin, G-CSF, GM-CSF, IFN-γ, LIF, LIX, KC, MCP-1, MIP-1α, MIP-1β, MIP-2, M-CSF, VEGF, and RANTES	Although the mRNA for IL-6 had increased, the IL-6 protein levels failed to reach statistical significance in the Luminex assay, although the trend was upward. Of the classic pro-inflammatory cytokines, IL-1α was significantly elevated. Other cytokines showing significant elevations include G-CSF, GM-CSF, M-CSF, IL-4, and IL-9. Unexpectedly, a number of chemokines also showed statistically significant increases, including eotaxin, MIP-1β, LIX-CXC chemokine, and RANTES. $CD11b^+$ fractions, the $CD11b^-$ fractions also produced significant	Pratt et al. (2013)

						quantities of cytokines following maternal inflammation, often in larger quantities than the $CD11b^+$ fractions. These included IL-1β, IL-9, IL-10, and IL-13	
LPS (50 μg/kg) on PND 3 and keta-mine (5, 15, and 25 mg/kg) for 7 days during adulthood	Wistar rats	PND 60	Male	Prefrontal cortex hippocampus and striatum	IL-1β, IL-6, IL-10, and TNF-α	In the prefrontal cortex, hippocampus, and stria-tum, two-way ANOVA revealed an interaction for ketamine versus LPS in the levels of IL-1β. A decrease in the IL-1β levels in the prefrontal cortex for LPS plus ketamine 5 mg/kg ($p = 0.019$), a decrease in the hippocampus ($p = 0.008$) and striatum ($p = 0.018$) for LPS plus ketamine 15 mg/kg was shown. The two-way ANOVA dem-onstrated LPS effects on IL-1 in the prefrontal cortex ($p < 0.001$), in the striatum ($p < 0.001$). In the hippocampus no effects were found for LPS ($p = 0.056$). The levels of IL-1β were not	Reus et al. (2017)

(continued)

Table 1 (continued)

Pre-clinical model	Species	Animal age	Sex	Anatomical area	Biomarkers evaluated	Results	Reference
						altered in the ketamine group in the prefrontal cortex ($p = 0.306$), in the hippocampus ($p = 0.060$), and striatum ($p = 0.093$)	
Poly I:C (20 mg/kg) on GD 11.5, 12.5, 15.5, and 17.5	Transgenic CX3CR1-eGFP knock-in mice	3 and 5 h after injection	Female and male	Cortex and hippocampus	IL-1β and iNOS	Despite the presence of a systemic inflammation in the pregnant mice, there was no significant difference in fetal microglial cell density or immunohistochemically determined activation level between the control and inflammation group	Smolders et al. (2015)
LPS (0.5 or 2.5 mg/kg) at GD 16	Sprague-Dawley rats	2 or 8 h after injected to LPS	Female and male	Amniotic fluid, fetal brain, and placental	IL-1β, IL-6, and TNF-α	The low dose (0.5 mg/kg) of LPS increased the levels of cytokines in the placenta with significant increases of IL-1β ($p < 0.0001$), IL-6 ($p < 0.0001$), and TNF-α ($p = 0.0001$) over the 2 and 8 h time course. In the amniotic fluid, there was an increase of IL-6 levels ($p = 0.0006$). Two	Urakubo et al. (2001)

						hours after maternal administration of a high dose (2.5 mg/kg) of LPS, there were significant elevations of cytokines in placenta IL-6 ($p < 0.0001$), TNF-α ($p < 0.0001$), a significant increase of TNF-α in amniotic fluid ($p = 0.008$), and a small but significant decrease in TNF-α ($p = 0.035$) in the fetal brain	
Poly I:C (4 mg/kg) on GD 15	Sprague-Dawley rat	PND180	Female and male	Corpus callosum, hippocampus, thalamus, pons, cortex, and striatum	CD11b/OX-42 and CD68/ED-1	Significant differences over the different brain regions were observed ($p \leq 0.001$), with an overall significant increase being indicated in microglia scores in the corpus callosum ($p \leq 0.05$), hippocampus ($p \leq 0.05$), and thalamus ($p \leq 0.01$) but not in the pons, cortex, and striatum ($p = 0.1$, $p = 0.5$, and $p = 0.1$, respectively). In the corpus callosum, both microglia intensity ($p \leq 0.001$) and density	van den Eynde et al. (2014)

(continued)

Table 1 (continued)

Pre-clinical model	Species	Animal age	Sex	Anatomical area	Biomarkers evaluated	Results	Reference
						($p \leq 0.05$) significantly contributed to the overall significant increase in OX-42. The ED-1 staining on the other hand revealed very few reactive microglia. No difference in ED-1 immunoreactivity was found between Poly I:C and control offspring	

ANOVA analysis of variance, *CX3CL1* fractalkine, *CD11b* microglial marker, *ED* embryonic day, *FACS* fluorescence-activated cell sorting, *G-CSF* granulocyte colony-stimulating factor, *GD* gestation day, *GM-CSF* granulocyte/monocyte colony-stimulating factor, H_2O water, *Iba1* ionized calcium-binding adaptor molecule, *IL* interleukin, *IFN* interferon, *iNOS* inducible nitric oxide synthase, *KC* keratinocyte chemoattractant, *LIF* leukemia inhibitory factor, *LIX* lipopolysaccharide-induced CXC chemokine, *LPS* lipopolysaccharide, *MCP* monocyte chemoattractant protein, *M-CSF* monocyte colony-stimulating factor, *MIA* maternal immune activation, *MIP* macrophage inflammatory protein, *NaCl* sodium chloride, *OX-42* general macrophage-associated marker, *PND* postnatal day, *Poly I:C* polyinosinic-polycytidylic acid, *RANTES* regulated upon activation normal T-cell expressed and secreted, *STAT1* signal transducer and activator of transcription 1, *TGF-β* transforming growth factor-β, *TNF* tumor necrosis factor, *VEGF* vascular endothelial growth factor

Several microglial cells presented phagosomes and/or degenerated mitochondria (Wierzba-Bobrowicz et al. 2004). The dorsolateral prefrontal cortex, anterior cingulate cortex, hippocampus, and mediodorsal thalamus were evaluated in 16 schizophrenic brain samples. HLA-DR-positive cell expression was not different between the schizophrenia and control groups. The post-mortem interval correlated with the ramified cell numbers in the anterior cingulate cortex and the dorsolateral prefrontal cortex and with the amoeboid cell density in the hippocampus. Two schizophrenic patients who had committed suicide during acute psychosis presented highly elevated microglial cell numbers in the anterior cingulate cortex and the mediodorsal thalamus (Steiner et al. 2006). In another study from the same research group, microglial HLA-DR expression was evaluated in the dorsolateral prefrontal cortex, anterior cingulate cortex, mediodorsal thalamus, and hippocampus of 16 schizophrenic patients. Microglial HLA-DR expression did not presently affect the diagnosis of microglial density in the dorsolateral prefrontal cortex, anterior cingulate cortex, mediodorsal thalamus, and hippocampus. However, the study found microgliosis in the dorsolateral prefrontal cortex, cingulate cortex, and mediodorsal thalamus of the schizophrenic suicide patients (Steiner et al. 2008). In a study by Sinkus et al., the mRNA levels for the MHC class I antigen HLA-B was increased in schizophrenic nonsmokers, while the levels for smokers were indistinguishable from those of controls. HLA-A was expressed in a pattern where inflammatory illness was associated with increased expression in controls but not in subjects with schizophrenia (Sinkus et al. 2013). Radewicz et al. found an increase of HLA-DR expression in the dorsolateral prefrontal cortex in eight schizophrenics compared with ten controls. Regarding the superior temporal gyrus, there was an increase in microglia in seven schizophrenics compared with ten controls. In the anterior cingulate gyrus, the results did not reach significance (Radewicz et al. 2000). Calprotectin is a calcium-binding protein of the S100 family and is a nonspecific inflammatory marker. Samples of post-mortem brain tissue from Brodmann area 9 were obtained from the prefrontal cortices of subjects with schizophrenia and of controls. Calprotectin presented higher levels in the schizophrenic brains (Brodmann area 9 from prefrontal cortex) compared to the controls, and this protein was found to localize in microglial cells (Foster et al. 2006).

Through investigation of the microglial activation using Iba1 antibody marker, which is expressed in macrophages and microglia and is upregulated during the activation of these cells, the brain samples presented unaltered immunoreactivity in the cingulate white matter (Connor et al. 2009) and the dorsolateral prefrontal cortex in the post-mortem schizophrenic brain (Hercher et al. 2014). Moreover, in another study, the regional differences in the ependymal and subventricular zone cytoarchitecture were unchanged in schizophrenic brain samples (Comte et al. 2012). CD68 was evaluated for resting and active microglia in the caudate nucleus and the mediodorsal nucleus of the thalamus in a post-mortem study of 11 elderly people with schizophrenia. No differences were found between the schizophrenic and control subjects (Falke et al. 2000). HLA-DRA did not present any differences in the dorsolateral prefrontal cortex or the parietal cortex samples between the schizophrenic and control groups (Nakatani et al. 2006). Messenger RNA expression of

HLA-A did not present any differences in the frontal cortex of schizophrenic subjects compared to control subjects (Saetre et al. 2007). The temporal cortex of the schizophrenic brain samples did not present differences in HLA-DRB3 and HLA-DPA1 expression compared with control brain samples (Schmitt et al. 2011). Another study evaluated the MHC class I and complement protein C3 expression in two frontal cortical regions of post-mortem brains of schizophrenic patients. MHC class I protein expression was decreased in the dorsolateral prefrontal cortex, but the protein expression did not present any change in the orbitofrontal cortex of nonsmoking schizophrenic patients, and this study did not find any association between schizophrenia and changes in C3 mRNA expression (Kano et al. 2011). A subsequent study presented a reduction in microglial immunoreactivity for the endogenous NMDA receptor agonist, quinolinic acid, in the hippocampus of schizophrenic patients and presented no difference in HLA-DR expression between schizophrenic and the control group brain samples (Gos et al. 2014). The MHC class II receptors HLA-DR and HLA-DRBA were downregulated in the temporal lobe of schizophrenic post-mortem brain samples (Durrenberger et al. 2015). No differences were found in the CD40 and HLA-DP/DQ/DR markers in four brain samples of schizophrenic patients (Togo et al. 2000). CD68 for resting and active microglia was evaluated in the entorhinal cortex, the subiculum and CA1 of the hippocampus, midfrontal cortex, orbitofrontal cortex, and calcarine cortex in schizophrenic brain samples. There were no differences between the schizophrenic and the control brain samples in the densities of any of the markers (Arnold et al. 1998). Kurumaji et al. evaluated [3H] PK 11195 as a ligand for the translocator protein (TSPO) receptor in the cerebral cortex, thalamus, and extrapyramidal system of the post-mortem brains of 13 chronic schizophrenics and 10 control subjects. The [^{3}H] PK 11195-specific binding was decreased in the superior parietal cortex, primary visual area, and putamen of schizophrenics, although there were no changes in this binding in the other brain areas (Kurumaji et al. 1997); see Table 2.

3.2 Cytokine and Chemokine Evaluation in Post-mortem Schizophrenic Brain Samples

In a clinical study, the mRNA expression of IL-6, IL-8, and SERPINA-3 presented higher levels in the dorsolateral prefrontal cortex of individuals with schizophrenia compared with their controls (Fillman et al. 2013). IL-6, IL-1β, IL-8, and SERPINA-3 mRNA levels were quantified in the contralateral fresh frozen orbitofrontal cortex. The volumes of the cortical gray matter and the superior frontal gyrus had a significant negative correlation with IL-1β, IL-6, and SERPINA-3 mRNA levels in the schizophrenic group. Thus, cortical gray matter volume reduction in schizophrenic patients was associated with neuroinflammation, and the researchers also found that the expression of inflammatory mRNA in the orbitofrontal cortex was correlated with those found by Fillman et al. (2013), in the dorsolateral prefrontal

Table 2 Microglial and neuroinflammatory markers in schizophrenic post-mortem brain samples

Number of patients and controls	Sample	Biomarkers evaluated	Results	Reference
Schizophrenia ($n = 23$) and control ($n = 14$)	Ventromedial temporal and frontal lobe and the calcarine	CD68	No statistically significant differences were found between the patients with schizophrenia and the control patients without neuropsychiatric disease for the densities of any of the markers	Arnold et al. (1998)
Schizophrenia (residual $n = 9$ and paranoid $n = 9$) and control ($n = 22$)	Dorsal raphe nucleus	HLA-DR and AgNOR	There was no change in the density of HLA-DR-positive microglial reaction in schizophrenic patients (residual and paranoid) compared to controls. Thus, a positive correlation existed between microglial densities evaluated by the AgNOR silver staining parameter in a residual subgroup of schizophrenic patients, which revealed a significant increase in this subgroup	Brisch et al. (2017)
Schizophrenia ($n = 15$) and control ($n = 15$)	Zona subventricular	MHC II	There were no differences between schizophrenic patient groups and controls in the width of the hypocellular gap or in the density of cells in the hypocellular gap. Because ventricular enlargement in schizophrenia may disrupt ependymal cells, we quantified them but observed no difference between diagnostic groups and controls	Comte et al. (2012)

(continued)

Table 2 (continued)

Number of patients and controls	Sample	Biomarkers evaluated	Results	Reference
Schizophrenia ($n = 22$) and control ($n = 45$)	Cingulate cortex	Iba1	No significant difference in Iba1 immunoreactivity between groups. It was not associated with NeuN+ density in white matter	Connor et al. (2009)
Schizophrenia ($n = 10$) and control ($n = 10$)	Temporal lobe	IL-13RA1, MHC II, HLA-DRA, and HLA-DPA1	The MHC II receptors, HLA-DRA, and HLA-DPA1 were significantly upregulated in neurodegenerative disorders and downregulated in schizophrenia. IL-13RA1 was significantly downregulated in schizophrenia	Durrenberger et al. (2015)
Schizophrenia ($n = 12$) and control ($n = 11$)	Caudate nucleus and mediodorsal nucleus	CD68	No statistically significant differences were found between schizophrenic and control subjects for the densities of any markers. There was no evidence that abnormal neurodegeneration occurs in these two important subcortical structures	Falke et al. (2000)
Schizophrenia ($n = 37$) and control ($n = 37$)	Dorsolateral prefrontal cortex	IL-1β, IL-6, IL-6ST, IL-8, and SERPINA-3	The individuals presented increased levels of IL-1β, IL-6, IL-8, and SERPINA-3 mRNA expression ($p < 0.001$). Other characteristics of this group included high mRNA expression of IL-6ST ($p < 0.033$)	Fillman et al. (2013)

(continued)

Table 2 (continued)

Number of patients and controls	Sample	Biomarkers evaluated	Results	Reference
Schizophrenia ($n = 35$) and control ($n = 35$)	Dorsolateral pre-frontal cortex	IL-1β, IL-1RL1, IL-6, IL-8, IL-6ST, PTGS2, IL-18, SERPINA-3, and TNF	The SERPINA-3 mRNA was specifically increased in schizophrenic brain samples ($p < 0.05$) compared with both controls; IL-8 mRNA showed a significant diagnostic effect ($p < 0.05$), but surprisingly, with a decreased expression in individuals with schizophrenia compared with controls. IL-1β, IL-18, TNF, and PTGS2 mRNAs showed no significant diagnostic effects overall and no consistent pattern of expression according to diagnosis. IL-1RL1 and IL-6 mRNAs were not significantly changed	Fillman et al. (2014)
Schizophrenia ($n = 15$) and control ($n = 15$)	Dorsolateral pre-frontal cortex	Calprotectin	Calprotectin was detectable in all samples, and mean levels were noted to be highest in schizophrenic brains ($p < 0.05$) and lowest in controls	Foster et al. (2006)
Schizophrenia ($n = 30$) and control ($n = 30$)	Prefrontal cortex	TLR-4, MyD88, mRNA, NF-κB, p65, κBα, RNA, IL-1β, IL-6, iNOS, COX-2, MDA, and NO_2	TLR-4, MyD88, and NF-κB expression increased in the prefrontal cortex of patients with schizophrenia. These alterations seem to depend on the presence/absence of antipsychotic treatment at death	Garcia-Bueno et al. (2016)

(continued)

Table 2 (continued)

Number of patients and controls	Sample	Biomarkers evaluated	Results	Reference
Schizophrenia ($n = 13$) and control ($n = 12$)	CA1, CA2/CA3, and dentate gyrus hippocampal	HLA-DR	Fewer quinolinic acid-immunoreactive microglial cells were observed in the CA1 hippocampal subregion of schizophrenic patients compared to controls (left $p = 0.028$, right $p = 0.018$). No significant diagnosis-dependent changes were observed in the CA2/CA3 and dentate gyrus regions	Gos et al. (2014)
Schizophrenia ($n = 35$) and control ($n = 33$)	Frontal cortex	IFN-γ	IFN-γ demonstrated high levels significantly different between schizophrenia and control samples ($p = 0.043$)	Harris et al. (2012)
Schizophrenia ($n = 20$) and control ($n = 20$)	Dorsolateral prefrontal cortex	Iba1	The density of Iba1-stained microglia did not differ among the groups; however, a qualitative assessment of microglial morphology found numerous activated microglial cells in three schizophrenic samples, but not in the controls	Hercher et al. (2014)
Schizophrenia ($n = 35$) and control ($n = 35$)	Dorsolateral prefrontal cortex	MHC I	The MHC I protein expression was reduced in the dorsolateral prefrontal cortex but not in the orbitofrontal cortex of schizophrenic patients	Kano et al. (2011)

(continued)

Table 2 (continued)

Number of patients and controls	Sample	Biomarkers evaluated	Results	Reference
Schizophrenia ($n = 13$) and control ($n = 10$)	Cerebral cortex, thalamus, and extrapyramidal system	[3H] PK 11195	The specific [3H] PK 11195 binding was significantly decreased in three brain areas (superior parietal cortex, primary visual area, and putamen) of schizophrenics, although there were no changes in the binding in the other brain areas	Kurumaji et al. (1997)
Schizophrenia ($n = 16$) and control ($n = 14$)	Prefrontal cortex and cerebellum	TLR-4, MyD88, κBα, iNOS, MDA, NF-κB, and COX-2	In the prefrontal cortex, TLR-4, MyD88, and IκBα protein levels were lower in schizophrenic patients, while NF-κB activity, COX-2 expression, and the MDA appeared to be increased. In the cerebellum it occurred opposite, except for COX-2 expression that remained augmented and MDA levels unaltered	MacDowell et al. (2017b)
Schizophrenia ($n = 7$) and control ($n = 7$)	Dorsolateral prefrontal cortex	HLA-DRA, HLA-DRB4, and CCL3	The expression of CCL3 was downregulated in schizophrenia. The expression of the HLA-DRA and HLA-DRB4 genes was not altered in schizophrenia	Nakatani et al. (2006)
Schizophrenia ($n = 8$) and control ($n = 10$)	Dorsolateral prefrontal cortex, the superior temporal gyrus, and the anterior cingulate gyrus	GFAP and HLA-DR	There was an increase of HLA-DR expression in the dorsolateral prefrontal cortex in eight schizophrenics	Radewicz et al. (2000)

(continued)

Table 2 (continued)

Number of patients and controls	Sample	Biomarkers evaluated	Results	Reference
			compared with ten controls. For the superior temporal gyrus, there was an increase in microglia in seven schizophrenics compared with ten controls. In the anterior cingulate gyrus, the results did not find significance	
Schizophrenia ($n = 55$) and control ($n = 55$)	Frontal cortex	IFITM-2, IFITM-3, SERPINA-3, GBP1, SCD, MAG, and TF	IFITM-2, IFITM-3, SERPINA-3, and GBP1 showed increased mRNA levels in schizophrenic brain samples ($p \leq 0.01$)	Saetre et al. (2007)
Schizophrenia ($n = 10$) and control ($n = 10$)	Temporal cortex	IL-1α, IL-1β, IL8, IL-1RAP, CCL2, HLA-DPA1, and HLA-DRB3	A microarray analysis, followed by qPCR validation, found a decrease in IL-8 and IL-1α mRNA expression in the temporal cortex of schizophrenic patients compared with healthy control patients. However, increases detected in the microarray were not reproduced by qPCR for cytokines and chemokines such as IL-1β and CCL2	Schmitt et al. (2011)
Schizophrenic smokers ($n = 28$), schizophrenic non-smokers ($n = 14$), control smokers ($n = 23$), and	Hippocampus	HLA-A and HLA-B	Messenger RNA levels for the class I major histocompatibility complex antigen HLA-B were increased in schizophrenic non-smokers, while levels for smokers	Sinkus et al. (2013)

(continued)

Table 2 (continued)

Number of patients and controls	Sample	Biomarkers evaluated	Results	Reference
control non-smokers ($n = 24$)			were indistinguishable from those of controls. β2-macroglobulin, HLA-A, and Notch4 were all expressed in a pattern where inflammatory illness was associated with increased expression in controls but not in subjects with schizophrenia	
Schizophrenia ($n = 17$) and control ($n = 11$)	Hippocampus	CD3, CD20, and HLA-DR	Higher densities of CD3 and CD20 lymphocytes were observed in residual versus paranoid schizophrenia. In contrast, HLA-DR microglia was increased in paranoid schizophrenia versus residual schizophrenia	Steiner et al. (2006)
Schizophrenia ($n = 16$) and control ($n = 16$)	Dorsolateral prefrontal cortex, anterior cingulate cortex, hippocampus and mediodorsal thalamus	HLA-DR	Immunostaining was found in all brain regions and was not restricted to macrophage-like amoeboid cells but also appeared in ramified cells. Region-specific HLA-DR-positive cell density was not significantly different between cases with schizophrenia and controls. However, amoeboid microglial cells were lateralized toward the right hemisphere in healthy subjects but not in the schizophrenia group	Steiner et al. (2006)

(continued)

Table 2 (continued)

Number of patients and controls	Sample	Biomarkers evaluated	Results	Reference
			($p = 0.01$). Post-mortem interval correlated with ramified cell numbers in the anterior cingulate cortex ($p = 0.01$), the dorsolateral prefrontal cortex ($p = 0.04$), and amoeboid cell density in hippocampus ($p = 0.03$)	
Schizophrenia ($n = 9$) and control ($n = 6$)	Frontal lobes and gyrus temporal inferior	HLA-DP, DQ, and DR	Frontal and temporal lobes of chronic schizophrenic patients presented higher microglial cell activation compared with control brains. However, the first layer of the cerebral cortex presented the same amounts of well-developed ramification and degenerative traits and damaged processes; the number of DM cells in the remaining regions was higher than that of RM cells	Steiner et al. (2008)
Schizophrenia ($n = 16$) and control ($n = 10$)	Dorsolateral prefrontal cortex, anterior cingulate cortex, hippocampus and mediodorsal thalamus	HLA-DR	The results revealed no effect of diagnosis on microglial density (dorsolateral prefrontal cortex ($p = 0.469$), anterior cingulate cortex ($p = 0.349$), mediodorsal thalamus ($p = 0.569$), and hippocampus ($p = 0.497$)). However, significant microgliosis was observed in the	Steiner et al. (2008)

(continued)

Table 2 (continued)

Number of patients and controls	Sample	Biomarkers evaluated	Results	Reference
			dorsolateral prefrontal cortex ($p = 0.004$), anterior cingulate cortex ($p = 0.012$), and mediodorsal thalamus ($p = 0.004$) of suicide patients. A similar trend was seen in the hippocampus ($p = 0.057$)	
Schizophrenia ($n = 4$)	Hippocampus and temporal exocortex	CD40	Vascular expression of CD40 was enhanced in the lesions of schizophrenia disease	Togo et al. (2000)
Schizophrenia ($n = 22$) and control ($n = 14$)	Prefrontal cortex	IL-1β and IL-1RA	Both protein and mRNA levels of IL-1RA were specifically decreased in the prefrontal cortex of schizophrenic patients, whereas IL-1β levels were not significantly altered in all the regions examined. The IL-1RA decrease was not correlated with the dose of antipsychotics given to patients. There was no influence of this illness on protein levels for IL-1β receptor type 1 in the prefrontal cortex	Toyooka et al. (2003)
Schizophrenia ($n = 62$) and control ($n = 62$)	Frontal cortex	IL-1β, IL-6, IL-8, IFN-β, NF-κB, and Schnurri-2	Schizophrenic subjects had markedly higher mRNA levels for IL-1β, IL-6, and IFN-β, which induce IFITM expression; lower mRNA levels	Volk et al. (2015)

(continued)

Table 2 (continued)

Number of patients and controls	Sample	Biomarkers evaluated	Results	Reference
			for Schnurri-2, a transcriptional inhibitor that lowers IFITM expression; and higher mRNA levels for NF-κB. IL-8 did not quite reach statistical significance	
Schizophrenia ($n = 14$) and control ($n = 13$)	Frontal cortex and hippocampus	HLA-DR	Schizophrenic patients revealed subjects with abundant microglial immunostaining in both gray and white matter. This finding provides evidence for distinct neuropathological changes in brains of patients with schizophrenia	Wierzba-Bobrowicz et al. (2004)
Schizophrenia ($n = 12$)	Frontal and temporal cortex	MHC II	Most cells showed degenerative traits (cytoplasm shrinkage, thinning, shortening, and fragmentation of their processes) up to apoptotic changes. Perivascular microglia displayed the lowest intensity of degenerative changes. Ultrastructurally, some damaged microglial cells contained phagosomes and/or degenerated mitochondria. Most abnormal microglia showed morphological signs of the former normal function of immunocompetent and phagocytosing cells	Wierzba-Bobrowicz et al. (2004)

(continued)

Table 2 (continued)

Number of patients and controls	Sample	Biomarkers evaluated	Results	Reference
Schizophrenia ($n = 47$) and control ($n = 45$)	Orbitofrontal cortex	IL-1β, IL-6, IL-8, and SERPINA-3	The volumes of cortical gray matter and superior frontal gyrus were significantly negatively correlated with IL-1β, IL-6, and SERPINA-3 mRNAs levels in the schizophrenia group. Thus, cortical gray matter volume reduction in schizophrenic patients was associated with neuroinflammation. The expression of inflammatory mRNAs in the orbitofrontal cortex was significantly correlated with those found in studies in the dorsolateral prefrontal cortex	Zhang et al. (2016)

[3H] PK 11195 1-(2-chlorophenyl)-*N*-methyl-*N*-(1-methylpropyl)-3-isoquinoline carboxamide, *AgNOR* argyrophilic nucleolar organizing region, *CCL2* chemokine (C-C motif) ligand 2, *CCL3* chemokine (C-C motif) ligand 3, *CD11b* microglial marker, *COX-2* cyclooxygenase-2, *GBP1* guanylate-binding protein 1, *GFAP* glial fibrillary acidic protein, *GM-CSF* granulocyte/monocyte colony-stimulating factor, *HLA* human leukocyte antigen class I (A, B, C), *HLA* human leukocyte antigen class II (DR [A, alpha; B, beta], DQ, DM, and DP), *Iba1* ionized calcium-binding adaptor molecule, *IFITM* interferon-induced transmembrane, *IFN* interferon, *IL* interleukin, *IL-13RA1* interleukin 13 receptor alpha 1, *IL-1RA* interleukin 1 receptor accessory, *IL-1RAP* interleukin 1 receptor accessory protein, *IL-1RL1* interleukin 1 receptor-like 1, *IL-2R* interleukin-2 receptor, *IL-6ST* interleukin 6 signal transducer, *iNOS* inducible nitric oxide synthase, *κBα* inhibitory protein, *MAG* myelin-associated glycoprotein, *MDA* malondialdehyde, *MHC I and II* major histocompatibility complex class I and II, *mRNA* relative messenger RNA, *MyD88* myeloid differentiation factor 88, *NeuN* neuron-specific nuclear protein, *NF-κB* nuclear factor kappa B, *NO_2* nitrite levels, *PTGS2* prostaglandin-endoperoxide synthase 2, *qPCR* quantitative polymerase chain reaction, *RNA* ribonucleic acid, *SCD* stearoyl-CoA desaturase, *Schnurri-2* inhibits NF-κB function, *TF* transferrin, *TGF-β* transforming growth factor-β, *TLR-4* Toll-like receptors 4, *TNF* tumor necrosis factor

cortex, except for IL-8 (Zhang et al. 2016). SERPINA-3 mRNA was also present at high levels in the dorsolateral prefrontal cortex of individuals with schizophrenia (Fillman et al. 2014). IFN-γ, as evaluated by the ELISA technique, was elevated in the BA10 brain region of schizophrenic patients (Harris et al. 2012). Schizophrenic subjects presented markedly higher mRNA levels of IL-6, IFN-β, and NF-κB transcription factor in the prefrontal cortex compared with the control group (Volk et al. 2015). TNF receptor 1 (TNFR1) mRNA was significantly increased in both Brodmann areas 24 (BA24) and 46 (BA46) in patients with schizophrenia (Dean et al. 2013). In contrast, IL-1β mRNA levels were not changed in post-mortem brain tissues of the prefrontal or parietal cortices, putamen, or the hypothalamus. In addition, endogenous IL-1 receptor antagonist (IL-1RA) decreased in the prefrontal cortex of schizophrenic patients (Toyooka et al. 2003). IL-13 receptor alpha-1 (IL-13RA1) was downregulated in the temporal lobe of schizophrenic patients (Durrenberger et al. 2015). Chemokine (C-C motif) ligand 3 (CCL3) gene expression was also downregulated in the dorsolateral prefrontal cortex and parietal cortex (Nakatani et al. 2006), and in the temporal cortex, the expression of IL-1α and IL-8 was downregulated in schizophrenic brain samples (Schmitt et al. 2011); see Table 2.

4 Mechanisms by Which Neuroinflammation Could Influence the Development of Schizophrenia

During pregnancy or early life infection, replication of microorganisms and the release of their immunogenic compounds can occur. These immunogenic compounds derived from microorganisms are denominated as pathogen-associated molecular patterns (PAMPs), and they are recognized by the immune system through equipped receptors denominated as pattern-recognition receptors (PRRs) (Barichello et al. 2015; Morris et al. 2018). These receptors, such as Toll-like receptors (TLR), nucleotide-binding oligomerization domain (NOD)-like receptors (NLR), C-type lectin receptors (CLR), retinoic acid-inducible gene-1 (RIG-1)-like receptors (RLR), receptors for advanced glycation end products (RAGE), and intracytosolic deoxyribonucleic acid (DNA) sensors, are crucial components in the activation of the innate immune system (Keestra-Gounder and Tsolis 2017; Zhou et al. 2017). The PRRs can also recognize a broader array of endogenous danger signals such as adenosine 5-triphosphate (ATP), heat shock proteins (HSPs), and high mobility group box-1 proteins (HMGB-1) that are denominated as damage-associated molecular patterns (DAMPs) (Nakahira et al. 2015; Wilkins et al. 2017); see Figs. 1 and 2.

TLR receptors are divided into two groups, of which one is expressed on the cell membrane for ligand recognition (TLR-1, TLR-2, TLR-4, TLR-5, TLR-6, and TLR-10) and the other is localized in the intracellular endosomal space for the recognition of pathogen nucleic acids: TLR-3, TLR-7, TLR-8, and TLR-9 (Kigerl et al. 2014). TLR-3 can signal through a TRIF-dependent pathway that recruits the TNF receptor-associated factor-3 (TRAF-3), thus resulting in the activation of interferon regulatory factor-3 (IRF-3) and IRF-7. This pathway triggers the

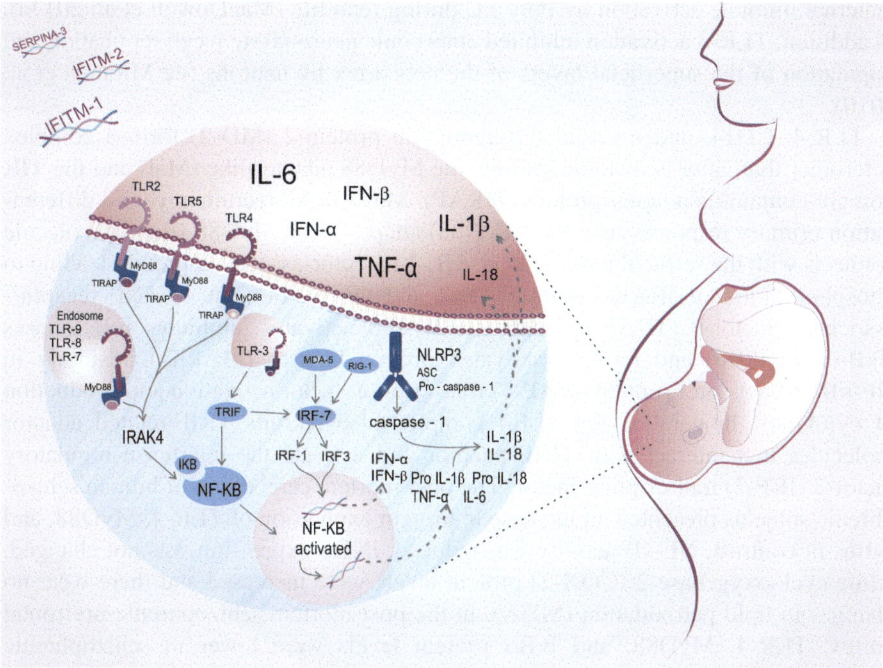

Fig. 2 Maternal immune activation and a possible mechanism in which neuroinflammation could influence the development of schizophrenia. TLRs, MDA-5, and RIG-1 are innate immune sensors involved in the detection of microorganisms. TLR-3, RIG-1, and MDA-5 promote the expression of type I and type III IFNs and the NF-kappa B-dependent expression of pro-inflammatory cytokines. Maternal immune activation increases the levels of cytokines such as IL-6, TNF-α, and IL-1β in the serum, as well as in the amniotic fluid, placenta, and fetal brain. *ASC* caspase-recruitment domain; *IFITM-1, IFITM-2* interferon-induced transmembrane protein-1, interferon-induced transmembrane protein-2; *IFN-α, IFN-β* interferon-α, interferon-β; *IκB* inhibitors of NF-κB; *IL-1β, IL-6, IL-18* interleukin-1β, interleukin-6, interleukin-18; *IRAK-4* interleukin receptor-associated kinase-4; *IRF-1, IRF-3, IRF-7* interferon regulatory factor-1, interferon regulatory factor-3, interferon regulatory factor-7; *MDA-5* melanoma differentiation-associated gene 5; *MyD88* myeloid differentiation factor 88; *NF-κB* nuclear factor kappa B; *NLRP-3* NLR family pyrin domain containing-3; *RIG-1* retinoic acid-inducible gene-1; *SERPINA-3* serpin family A member-3; *TIRAP* domain-containing adaptor protein; *TLR* Toll-like receptor; *TNF-α* tumor necrosis factor alpha; *TRAF-3, TRAF-6* TNF receptor-associated factor-3, TNF receptor-associated factor-6; *TRIF* Toll/IL-1 receptor domain-containing adaptor-inducing interferon-β

production of type I interferons, such as IFN-α or IFN-β. In another pathway, TLR-3 activates TRIF, AP1, and NF-κB, inducing the expression of pro-inflammatory cytokine genes. TLR-3 serves as a sensor of dsRNA produced during the replication of single-stranded RNA (ssRNA) and is also activated by a synthetic chemical compound analogue of dsRNA, Poly I:C (Verma and Bharti 2017). TLR-3 is an essential sensor of the host's immune responses to protect it against viral infections. A pre-clinical model of schizophrenia demonstrated high expression of TLR-3 signaling, IFN-α, and IFN-β in the frontal cortex of adult offspring subjected to maternal immune activation by Poly I:C during fetal life (MacDowell et al. 2017a). In addition, TLR-3 activation inhibited embryonic neuronal stem cell replication and population of the superficial layers of the neocortex by neurons (de Miranda et al. 2010).

TLR-4, CD14, and myeloid differentiation protein-2 (MD-2) form a complex heteromer that, after activation, recruits the MyD88 adapter-like (Mal) and the TIR domain-containing adaptor protein (TIRAP). Mal/TIRAP recruits myeloid differentiation primary response gene 88 (MYD88) adaptor. The MyD88 adaptor molecule connects with the serine/threonine kinase IL-1 receptor-associated protein leading to phosphorylation of IRAK-1 and IRAK-2 and the recruitment of TNF receptor-associated factor-6 (TRAF-6) adaptor. TRAF-6 activates inhibitory IκB kinases (IkBα and IkBβ) and mitogen-activated protein kinases (MAPKs), resulting in NF-κB and activator protein-1 (AP-1) transcription factor activation and production of cytokines. In parallel, the TLR4 complex also recruits TRIF-related adaptor molecules that interact with TRIF adaptor and activate the interferon regulatory factor-3 (IRF-3) transcription factor. The post-mortem cerebellum of human schizophrenic subjects presented an increase in protein expression of TLR-4, MyD88, and IκBα. In contrast, NF-κB activity was reduced, iNOS expression was not changed, while cyclooxygenase-2 (COX-2) protein levels were increased and there were no changes in lipid peroxidation (MDA). In the post-mortem schizophrenic prefrontal cortex, TLR-4, MyD88, and IκBα protein levels were lower in schizophrenic patients, while nuclear transcription NF-κB activity, COX-2 expression, and malondialdehyde (MDA) were increased (MacDowell et al. 2017b). Another study found evidence of alterations in the expression of the TLR-4 signaling and MyD88 and NF-κB in the prefrontal cortex of patients with schizophrenia. However, there were no changes in the IκBα protein levels, IL-1β, and IL-6 mRNA levels in the prefrontal cortex. An additional study evaluated the effect of antipsychotic treatment on schizophrenic post-mortem brain samples. The antipsychotic treatment schizophrenic group presented higher levels of TLR-4, MyD88 protein, and MyD88 mRNA compared to control samples. An MDA decrease was observed in the antipsychotic-free group compared to the control and antipsychotic treatment groups, but the antipsychotic-free group presented high levels of NF-κB protein compared with controls. This study demonstrated that it is necessary to pay special attention to the potentially confounding factor of antipsychotic treatment, because these alterations seem to depend on the presence or absence of antipsychotic treatment at death (Garcia-Bueno et al. 2016).

A number of studies have shown an increase in expression of SERPINA-3 and its gene in schizophrenic brain samples (Arion et al. 2007; Fillman et al. 2013, 2014; Zhang et al. 2016). The transcriptome signature of altered genes related to immune function may be a consequence of high levels of pro-inflammatory cytokines such as IL-6, TNF-α, IL-1β, or IFNs, during the stages of prenatal development or during early life. These pro-inflammatory mediators could not only alter brain development but also be responsible for these immune-/inflammation-related genes such as SERPINA-3, interferon-induced transmembrane protein (IFITM)-1, IFITM-2, and IFITM-3 that are found in the schizophrenic adult brain (Arion et al. 2007; Saetre et al. 2007; Hwang et al. 2013; Volk et al. 2015); see Tables 2 and 3.

Table 3 Neuroinflammatory markers in the CSF of schizophrenic patients

Number of patients and controls	Sample	Biomarkers evaluated	Results	Reference
Schizophrenia ($n = 16$) and control ($n = 10$)	CSF	IL-lβ, IL-2, IL-2R, IL-6, and TNF-α	No significant differences were found in levels of TNF-α and IL-2 or IL-6 in CSF fluid. IL-1β and IL-2R were significantly decreased in patients' CSF compared to controls	Barak et al. (1995)
Schizophrenia ($n = 14$) and control ($n = 16$)	CSF	IL-1β, IFN-γ, IL-10, IL-6, and TNF-α	Schizophrenia patients showed a significant increase in IL-6 in CSF ($p = 0.02$). IL-1β, IFN-γ, IL-10, and TNF-α were often below the levels of detection of this assay	Coughlin et al. (2016)
Schizophrenia ($n = 16$) and control ($n = 11$)	CSF	IL-lα and IL-2	IL-1α levels were found below the detection limits of the assay in both controls and the schizophrenic groups, and there were no statistically significant differences of IL-1α and IL-2 between the schizophrenic and control groups	el-Mallakh et al. (1993)
Schizophrenia ($n = 31$) and control ($n = 14$)	CSF	IL-6	In the CSF, IL-6 was found to be significantly higher in the subtypes of schizophrenics "delayed responder" than the "poor responders" ($p = 0.017$) and the controls ($p = 0.013$)	Garver et al. (2003)
Schizophrenia ($n = 14$) and control ($n = 9$)	CSF	IL-1β and IL-6	IL-1β and IL-6 CSF levels did not present a significant difference between medicated schizophrenic patients and controls	Katila et al. (1994)
Schizophrenia ($n = 10$) and control ($n = 10$)	CSF	IL-1α and IL-2	IL-2 levels in the CSF were found higher in neuroleptic-free schizophrenic patients compared to control group. The levels of IL-1α did not present significant difference	Licinio et al. (1993)
Schizophrenia ($n = 60$) and control ($n = 21$)	CSF	IL-1α and IL-2	There were no differences between schizophrenic patients and normal volunteers in measures of CSF IL-lα and IL-2	Rapaport et al. (1997)

(continued)

Table 3 (continued)

Number of patients and controls	Sample	Biomarkers evaluated	Results	Reference
Schizophrenia ($n = 32$) and control ($n = 35$)	CSF	IL-6	Patients with schizophrenia had significantly higher CSF IL-6 levels compared to the controls ($p = 0.0027$)	Sasayama et al. (2013)
Schizophrenia ($n = 23$) and control ($n = 37$)	CSF	IL-1β, IL-2, IL-4, IL-6, IL-8, 1L-10, IL-18, TNF-α, IFN-α-2a, and IFN-γ	Patients with schizophrenia had increased CSF levels of IL-6 compared with healthy volunteers	Schwieler et al. (2015)
Schizophrenia ($n = 26$) and control ($n = 30$)	CSF	IL-1β, IL-6, IL-8, IL-2, IL-4, IL-5, IL-10, GM-CSF, IFN-γ, and TNF-α	IL-1β, IL-6, and IL-8 were reliably detectable in CSF of both patients and controls. IL-2, IL-4, IL-5, IL-10, GM-CSF, IFN-γ, and TNF-α were found in low concentrations or were undetectable in both patients and controls. In patients, IL-1β concentrations were markedly elevated compared with controls	Soderlund et al. (2009)
Schizophrenia ($n = 85$) and control ($n = 51$)	CSF	TGF-β1 and TGF-β2	TGF-β1 and TGF-β2 did not present any differences in the CSF of chronic schizophrenic patients and the control group	Vawter et al. (1997)

CSF cerebrospinal fluid, *GM-CSF* granulocyte/monocyte colony-stimulating factor, *IFN* interferon, *IL* interleukin, *IL-2R* interleukin-2 receptor, *TGF-β* transforming growth factor-β, *TNF* tumor necrosis factor

5 Conclusion

There are a significant number of studies showing an increase in microglial markers and pro-inflammatory gene expression in the post-mortem brains of schizophrenic patients compared with controls. The transcriptome signature of altered genes related to immune function may be a consequence of high levels of pro-inflammatory cytokines during the stages of prenatal development or during early life. These pro-inflammatory mediators could not only alter brain development but also be responsible for these immune/inflammation-related genes found in the schizophrenic adult brain.

Acknowledgments The Translational Psychiatry Program (USA) is funded by the Department of Psychiatry and Behavioral Sciences, McGovern Medical School, The University of Texas Health Science Center at Houston (UTHealth). The Laboratory of Neurosciences (Brazil) is one of the centers of the National Institute for Molecular Medicine (INCT-MM) and one of the members of the Center of Excellence in Applied Neurosciences of Santa Catarina (NENASC). This research is supported by grants from CNPq (TB, JQ), FAPESC (TB, JQ), Instituto Cérebro e Mente (JQ), and UNESC (TB and JQ).

Conflicts of Interest The authors declare that they have no conflicts of interest.

References

Abazyan B, Nomura J, Kannan G, Ishizuka K, Tamashiro KL, Nucifora F, Pogorelov V, Ladenheim B, Yang C, Krasnova IN, Cadet JL, Pardo C, Mori S, Kamiya A, Vogel MW, Sawa A, Ross CA, Pletnikov MV (2010) Prenatal interaction of mutant DISC1 and immune activation produces adult psychopathology. Biol Psychiatry 68(12):1172–1181

Arad M, Piontkewitz Y, Albelda N, Shaashua L, Weiner I (2017) Immune activation in lactating dams alters sucklings' brain cytokines and produces non-overlapping behavioral deficits in adult female and male offspring: a novel neurodevelopmental model of sex-specific psychopathology. Brain Behav Immun 63:35–49

Arion D, Unger T, Lewis DA, Levitt P, Mirnics K (2007) Molecular evidence for increased expression of genes related to immune and chaperone function in the prefrontal cortex in schizophrenia. Biol Psychiatry 62(7):711–721

Arnold SE, Trojanowski JQ, Gur RE, Blackwell P, Han LY, Choi C (1998) Absence of neurodegeneration and neural injury in the cerebral cortex in a sample of elderly patients with schizophrenia. Arch Gen Psychiatry 55(3):225–232

Barak V, Barak Y, Levine J, Nisman B, Roisman I (1995) Changes in interleukin-1 beta and soluble interleukin-2 receptor levels in CSF and serum of schizophrenic patients. J Basic Clin Physiol Pharmacol 6(1):61–69

Barichello T, Generoso JS, Goularte JA, Collodel A, Pitcher MR, Simoes LR, Quevedo J, Dal-Pizzol F (2015) Does infection-induced immune activation contribute to dementia? Aging Dis 6(5):342–348

Bayer TA, Buslei R, Havas L, Falkai P (1999) Evidence for activation of microglia in patients with psychiatric illnesses. Neurosci Lett 271(2):126–128

Brisch R, Steiner J, Mawrin C, Krzyzanowska M, Jankowski Z, Gos T (2017) Microglia in the dorsal raphe nucleus plays a potential role in both suicide facilitation and prevention in affective disorders. Eur Arch Psychiatry Clin Neurosci 267(5):403–415

Busse S, Busse M, Schiltz K, Bielau H, Gos T, Brisch R, Mawrin C, Schmitt A, Jordan W, Muller UJ, Bernstein HG, Bogerts B, Steiner J (2012) Different distribution patterns of lymphocytes and microglia in the hippocampus of patients with residual versus paranoid schizophrenia: further evidence for disease course-related immune alterations? Brain Behav Immun 26 (8):1273–1279

Comte I, Kotagiri P, Szele FG (2012) Regional differences in human ependymal and subventricular zone cytoarchitecture are unchanged in neuropsychiatric disease. Dev Neurosci 34(4):299–309

Connor CM, Guo Y, Akbarian S (2009) Cingulate white matter neurons in schizophrenia and bipolar disorder. Biol Psychiatry 66(5):486–493

Coughlin JM, Wang Y, Ambinder EB, Ward RE, Minn I, Vranesic M, Kim PK, Ford CN, Higgs C, Hayes LN, Schretlen DJ, Dannals RF, Kassiou M, Sawa A, Pomper MG (2016) In vivo markers of inflammatory response in recent-onset schizophrenia: a combined study using [(11)C]DPA-713 PET and analysis of CSF and plasma. Transl Psychiatry 12(6):40

de Miranda J, Yaddanapudi K, Hornig M, Villar G, Serge R, Lipkin WI (2010) Induction of Toll-like receptor 3-mediated immunity during gestation inhibits cortical neurogenesis and causes behavioral disturbances. MBio 1(4):e00176–e00110

Dean B, Gibbons AS, Tawadros N, Brooks L, Everall IP, Scarr E (2013) Different changes in cortical tumor necrosis factor-alpha-related pathways in schizophrenia and mood disorders. Mol Psychiatry 18(7):767–773

Durrenberger PF, Fernando FS, Kashefi SN, Bonnert TP, Seilhean D, Nait-Oumesmar B, Schmitt A, Gebicke-Haerter PJ, Falkai P, Grunblatt E, Palkovits M, Arzberger T,

Kretzschmar H, Dexter DT, Reynolds R (2015) Common mechanisms in neurodegeneration and neuroinflammation: a BrainNet Europe gene expression microarray study. J Neural Transm (Vienna) 122(7):1055–1068
el-Mallakh RS, Suddath RL, Wyatt RJ (1993) Interleukin-1 alpha and interleukin-2 in cerebrospinal fluid of schizophrenic subjects. Prog Neuro-Psychopharmacol Biol Psychiatry 17(3):383–391
Falke E, Han LY, Arnold SE (2000) Absence of neurodegeneration in the thalamus and caudate of elderly patients with schizophrenia. Psychiatry Res 93(2):103–110
Feigenson KA, Kusnecov AW, Silverstein SM (2014) Inflammation and the two-hit hypothesis of schizophrenia. Neurosci Biobehav Rev 38:72–93
Fillman SG, Cloonan N, Catts VS, Miller LC, Wong J, McCrossin T, Cairns M, Weickert CS (2013) Increased inflammatory markers identified in the dorsolateral prefrontal cortex of individuals with schizophrenia. Mol Psychiatry 18(2):206–214
Fillman SG, Sinclair D, Fung SJ, Webster MJ, Shannon Weickert C (2014) Markers of inflammation and stress distinguish subsets of individuals with schizophrenia and bipolar disorder. Transl Psychiatry 4:e365
Foster R, Kandanearatchi A, Beasley C, Williams B, Khan N, Fagerhol MK, Everall IP (2006) Calprotectin in microglia from frontal cortex is up-regulated in schizophrenia: evidence for an inflammatory process? Eur J Neurosci 24(12):3561–3566
Garay PA, Hsiao EY, Patterson PH, McAllister AK (2013) Maternal immune activation causes age- and region-specific changes in brain cytokines in offspring throughout development. Brain Behav Immun 31:54–68
Garcia-Bueno B, Gasso P, MacDowell KS, Callado LF, Mas S, Bernardo M, Lafuente A, Meana JJ, Leza JC (2016) Evidence of activation of the Toll-like receptor-4 proinflammatory pathway in patients with schizophrenia. J Psychiatry Neurosci 41(3):E46–E55
Garver DL, Tamas RL, Holcomb JA (2003) Elevated interleukin-6 in the cerebrospinal fluid of a previously delineated schizophrenia subtype. Neuropsychopharmacology 28(8):1515–1520
Ginhoux F, Greter M, Leboeuf M, Nandi S, See P, Gokhan S, Mehler MF, Conway SJ, Ng LG, Stanley ER, Samokhvalov IM, Merad M (2010) Fate mapping analysis reveals that adult microglia derive from primitive macrophages. Science 330(6005):841–845
Gos T, Myint AM, Schiltz K, Meyer-Lotz G, Dobrowolny H, Busse S, Muller UJ, Mawrin C, Bernstein HG, Bogerts B, Steiner J (2014) Reduced microglial immunoreactivity for endogenous NMDA receptor agonist quinolinic acid in the hippocampus of schizophrenia patients. Brain Behav Immun 41:59–64
Gunn CK (1944) Hereditary acholuric jaundice in the rat. Can Med Assoc J 50(3):230–237
Hadar R, Dong L, Del-Valle-Anton L, Guneykaya D, Voget M, Edemann-Callesen H, Schweibold R, Djodari-Irani A, Goetz T, Ewing S, Kettenmann H, Wolf SA, Winter C (2017) Deep brain stimulation during early adolescence prevents microglial alterations in a model of maternal immune activation. Brain Behav Immun 63:71–80
Harris LW, Pietsch S, Cheng TM, Schwarz E, Guest PC, Bahn S (2012) Comparison of peripheral and central schizophrenia biomarker profiles. PLoS One 7(10):30
Hercher C, Chopra V, Beasley CL (2014) Evidence for morphological alterations in prefrontal white matter glia in schizophrenia and bipolar disorder. J Psychiatry Neurosci 39(6):376–385
Hwang Y, Kim J, Shin JY, Kim JI, Seo JS, Webster MJ, Lee D, Kim S (2013) Gene expression profiling by mRNA sequencing reveals increased expression of immune/inflammation-related genes in the hippocampus of individuals with schizophrenia. Transl Psychiatry 3:e321
Juckel G, Manitz MP, Brune M, Friebe A, Heneka MT, Wolf RJ (2011) Microglial activation in a neuroinflammational animal model of schizophrenia – a pilot study. Schizophr Res 131 (1–3):96–100
Kano S, Nwulia E, Niwa M, Chen Y, Sawa A, Cascella N (2011) Altered MHC class I expression in dorsolateral prefrontal cortex of nonsmoker patients with schizophrenia. Neurosci Res 71 (3):289–293

Katila H, Hurme M, Wahlbeck K, Appelberg B, Rimon R (1994) Plasma and cerebrospinal fluid interleukin-1 beta and interleukin-6 in hospitalized schizophrenic patients. Neuropsychobiology 30(1):20–23

Keestra-Gounder AM, Tsolis RM (2017) NOD1 and NOD2: beyond peptidoglycan sensing. Trends Immunol 38(10):758–767

Khandaker GM, Pearson RM, Zammit S, Lewis G, Jones PB (2014a) Association of serum interleukin 6 and C-reactive protein in childhood with depression and psychosis in young adult life: a population-based longitudinal study. JAMA Psychiat 71(10):1121–1128

Khandaker GM, Zammit S, Lewis G, Jones PB (2014b) A population-based study of atopic disorders and inflammatory markers in childhood before psychotic experiences in adolescence. Schizophr Res 152(1):139–145

Kigerl KA, de Rivero Vaccari JP, Dietrich WD, Popovich PG, Keane RW (2014) Pattern recognition receptors and central nervous system repair. Exp Neurol 258:5–16

Kurumaji A, Wakai T, Toru M (1997) Decreases in peripheral-type benzodiazepine receptors in postmortem brains of chronic schizophrenics. J Neural Transm 104(11–12):1361–1370

Liaury K, Miyaoka T, Tsumori T, Furuya M, Wake R, Ieda M, Tsuchie K, Taki M, Ishihara K, Tanra AJ, Horiguchi J (2012) Morphological features of microglial cells in the hippocampal dentate gyrus of Gunn rat: a possible schizophrenia animal model. J Neuroinflammation 9:56

Liaury K, Miyaoka T, Tsumori T, Furuya M, Hashioka S, Wake R, Tsuchie K, Fukushima M, Limoa E, Tanra AJ, Horiguchi J (2014) Minocycline improves recognition memory and attenuates microglial activation in Gunn rat: a possible hyperbilirubinemia-induced animal model of schizophrenia. Prog Neuro-Psychopharmacol Biol Psychiatry 50:184–190

Licinio J, Seibyl JP, Altemus M, Charney DS, Krystal JH (1993) Elevated CSF levels of interleukin-2 in neuroleptic-free schizophrenic patients. Am J Psychiatry 150(9):1408–1410

Limoa E, Hashioka S, Miyaoka T, Tsuchie K, Arauchi R, Azis IA, Wake R, Hayashida M, Araki T, Furuya M, Liaury K, Tanra AJ, Horiguchi J (2016) Electroconvulsive shock attenuated microgliosis and astrogliosis in the hippocampus and ameliorated schizophrenia-like behavior of Gunn rat. J Neuroinflammation 13(1):230

MacDowell KS, Munarriz-Cuezva E, Caso JR, Madrigal JL, Zabala A, Meana JJ, Garcia-Bueno B, Leza JC (2017a) Paliperidone reverts Toll-like receptor 3 signaling pathway activation and cognitive deficits in a maternal immune activation mouse model of schizophrenia. Neuropharmacology 116:196–207

MacDowell KS, Pinacho R, Leza JC, Costa J, Ramos B, Garcia-Bueno B (2017b) Differential regulation of the TLR4 signalling pathway in post-mortem prefrontal cortex and cerebellum in chronic schizophrenia: relationship with SP transcription factors. Prog Neuropsychopharmacol Biol Psychiatry 79(Pt B):481–492

Mattei D, Djodari-Irani A, Hadar R, Pelz A, de Cossio LF, Goetz T, Matyash M, Kettenmann H, Winter C, Wolf SA (2014) Minocycline rescues decrease in neurogenesis, increase in microglia cytokines and deficits in sensorimotor gating in an animal model of schizophrenia. Brain Behav Immun 38:175–184

Mattei D, Ivanov A, Ferrai C, Jordan P, Guneykaya D, Buonfiglioli A, Schaafsma W, Przanowski P, Deuther-Conrad W, Brust P, Hesse S, Patt M, Sabri O, Ross TL, Eggen BJL, Boddeke E, Kaminska B, Beule D, Pombo A, Kettenmann H, Wolf SA (2017) Maternal immune activation results in complex microglial transcriptome signature in the adult offspring that is reversed by minocycline treatment. Transl Psychiatry 7(5):e1120

Missault S, van den Eynde K, van den Berghe W, Fransen E, Weeren A, Timmermans JP, Kumar-Singh S, Dedeurwaerdere S (2014) The risk for behavioural deficits is determined by the maternal immune response to prenatal immune challenge in a neurodevelopmental model. Brain Behav Immun 42:138–146

Morris G, Barichello T, Stubbs B, Kohler CA, Carvalho AF, Maes M (2018) Zika virus as an emerging neuropathogen: mechanisms of neurovirulence and neuro-immune interactions. Mol Neurobiol 55:4160–4184

Nakahira K, Hisata S, Choi AM (2015) The roles of mitochondrial damage-associated molecular patterns in diseases. Antioxid Redox Signal 23(17):1329–1350

Nakatani N, Hattori E, Ohnishi T, Dean B, Iwayama Y, Matsumoto I, Kato T, Osumi N, Higuchi T, Niwa S, Yoshikawa T (2006) Genome-wide expression analysis detects eight genes with robust alterations specific to bipolar I disorder: relevance to neuronal network perturbation. Hum Mol Genet 15(12):1949–1962

Nimmerjahn A, Kirchhoff F, Helmchen F (2005) Resting microglial cells are highly dynamic surveillants of brain parenchyma in vivo. Science 308(5726):1314–1318

Ning H, Wang H, Zhao L, Zhang C, Li XY, Chen YH, Xu DX (2008) Maternally-administered lipopolysaccharide (LPS) increases tumor necrosis factor alpha in fetal liver and fetal brain: its suppression by low-dose LPS pretreatment. Toxicol Lett 176(1):13–19

Pratt L, Ni L, Ponzio NM, Jonakait GM (2013) Maternal inflammation promotes fetal microglial activation and increased cholinergic expression in the fetal basal forebrain: role of interleukin-6. Pediatr Res 74(4):393–401

Radewicz K, Garey LJ, Gentleman SM, Reynolds R (2000) Increase in HLA-DR immunoreactive microglia in frontal and temporal cortex of chronic schizophrenics. J Neuropathol Exp Neurol 59(2):137–150

Ransohoff RM (2016) A polarizing question: do M1 and M2 microglia exist? Nat Neurosci 19 (8):987–991

Rapaport MH, McAllister CG, Pickar D, Tamarkin L, Kirch DG, Paul SM (1997) CSF IL-1 and IL-2 in medicated schizophrenic patients and normal volunteers. Schizophr Res 25(2):123–129

Reus GZ, Simoes LR, Colpo GD, Scaini G, Oses JP, Generoso JS, Prossin AR, Kaddurah-Daouk R, Quevedo J, Barichello T (2017) Ketamine potentiates oxidative stress and influences behavior and inflammation in response to lipolysaccharide (LPS) exposure in early life. Neuroscience 353:17–25

Saetre P, Emilsson L, Axelsson E, Kreuger J, Lindholm E, Jazin E (2007) Inflammation-related genes up-regulated in schizophrenia brains. BMC Psychiatry 7:46

Salter MW, Stevens B (2017) Microglia emerge as central players in brain disease. Nat Med 23 (9):1018–1027

Sasayama D, Hattori K, Wakabayashi C, Teraishi T, Hori H, Ota M, Yoshida S, Arima K, Higuchi T, Amano N, Kunugi H (2013) Increased cerebrospinal fluid interleukin-6 levels in patients with schizophrenia and those with major depressive disorder. J Psychiatr Res 47 (3):401–406

Schmitt A, Leonardi-Essmann F, Durrenberger PF, Parlapani E, Schneider-Axmann T, Spanagel R, Arzberger T, Kretzschmar H, Herrera-Marschitz M, Gruber O, Reynolds R, Falkai P, Gebicke-Haerter PJ (2011) Regulation of immune-modulatory genes in left superior temporal cortex of schizophrenia patients: a genome-wide microarray study. World J Biol Psychiatry 12 (3):201–215

Schwieler L, Larsson MK, Skogh E, Kegel ME, Orhan F, Abdelmoaty S, Finn A, Bhat M, Samuelsson M, Lundberg K, Dahl ML, Sellgren C, Schuppe-Koistinen I, Svensson C, Erhardt S, Engberg G (2015) Increased levels of IL-6 in the cerebrospinal fluid of patients with chronic schizophrenia – significance for activation of the kynurenine pathway. J Psychiatry Neurosci 40(2):126–133

Sinkus ML, Adams CE, Logel J, Freedman R, Leonard S (2013) Expression of immune genes on chromosome 6p21.3-22.1 in schizophrenia. Brain Behav Immun 32:51–62

Smith SE, Li J, Garbett K, Mirnics K, Patterson PH (2007) Maternal immune activation alters fetal brain development through interleukin-6. J Neurosci 27(40):10695–10702

Smolders S, Smolders SM, Swinnen N, Gartner A, Rigo JM, Legendre P, Brone B (2015) Maternal immune activation evoked by polyinosinic:polycytidylic acid does not evoke microglial cell activation in the embryo. Front Cell Neurosci 9:301

Soderlund J, Schroder J, Nordin C, Samuelsson M, Walther-Jallow L, Karlsson H, Erhardt S, Engberg G (2009) Activation of brain interleukin-1beta in schizophrenia. Mol Psychiatry 14 (12):1069–1071. https://doi.org/10.1038/mp.2009.52

Steiner J, Mawrin C, Ziegeler A, Bielau H, Ullrich O, Bernstein HG, Bogerts B (2006) Distribution of HLA-DR-positive microglia in schizophrenia reflects impaired cerebral lateralization. Acta Neuropathol 112(3):305–316

Steiner J, Bielau H, Brisch R, Danos P, Ullrich O, Mawrin C, Bernstein HG, Bogerts B (2008) Immunological aspects in the neurobiology of suicide: elevated microglial density in schizophrenia and depression is associated with suicide. J Psychiatr Res 42(2):151–157

Stence N, Waite M, Dailey ME (2001) Dynamics of microglial activation: a confocal time-lapse analysis in hippocampal slices. Glia 33(3):256–266

Togo T, Akiyama H, Kondo H, Ikeda K, Kato M, Iseki E, Kosaka K (2000) Expression of CD40 in the brain of Alzheimer's disease and other neurological diseases. Brain Res 885(1):117–121

Toyooka K, Watanabe Y, Iritani S, Shimizu E, Iyo M, Nakamura R, Asama K, Makifuchi T, Kakita A, Takahashi H, Someya T, Nawa H (2003) A decrease in interleukin-1 receptor antagonist expression in the prefrontal cortex of schizophrenic patients. Neurosci Res 46 (3):299–307

Urakubo A, Jarskog LF, Lieberman JA, Gilmore JH (2001) Prenatal exposure to maternal infection alters cytokine expression in the placenta, amniotic fluid, and fetal brain. Schizophr Res 47 (1):27–36

van den Eynde K, Missault S, Fransen E, Raeymaekers L, Willems R, Drinkenburg W, Timmermans JP, Kumar-Singh S, Dedeurwaerdere S (2014) Hypolocomotive behaviour associated with increased microglia in a prenatal immune activation model with relevance to schizophrenia. Behav Brain Res 258:179–186

Vawter MP, Dillon-Carter O, Issa F, Wyatt RJ, Freed WJ (1997) Transforming growth factors beta 1 and beta 2 in the cerebrospinal fluid of chronic schizophrenic patients. Neuropsychopharmacology 16(1):83–87

Verma R, Bharti K (2017) Toll like receptor 3 and viral infections of nervous system. J Neurol Sci 372:40–48

Volk DW, Chitrapu A, Edelson JR, Roman KM, Moroco AE, Lewis DA (2015) Molecular mechanisms and timing of cortical immune activation in schizophrenia. Am J Psychiatry 172 (11):1112–1121

Wierzba-Bobrowicz T, Lewandowska E, Kosno-Kruszewska E, Lechowicz W, Pasennik E, Schmidt-Sidor B (2004) Degeneration of microglial cells in frontal and temporal lobes of chronic schizophrenics. Folia Neuropathol 42(3):157–165

Wierzba-Bobrowicz T, Lewandowska E, Lechowicz W, Stepien T, Pasennik E (2005) Quantitative analysis of activated microglia, ramified and damage of processes in the frontal and temporal lobes of chronic schizophrenics. Folia Neuropathol 43(2):81–89

Wilkins HM, Weidling IW, Ji Y, Swerdlow RH (2017) Mitochondria-derived damage-associated molecular patterns in neurodegeneration. Front Immunol 8:508

Zhang Y, Catts VS, Sheedy D, McCrossin T, Kril JJ, Shannon Weickert C (2016) Cortical grey matter volume reduction in people with schizophrenia is associated with neuro-inflammation. Transl Psychiatry 6(12):e982

Zhou Y, He C, Wang L, Ge B (2017) Post-translational regulation of antiviral innate signaling. Eur J Immunol 47(9):1414–1426

Early-Life Adversity, Systemic Inflammation and Comorbid Physical and Psychiatric Illnesses of Adult Life

Maria Antonietta Nettis, Carmine M. Pariante, and Valeria Mondelli

Contents

M. A. Nettis and C. M. Pariante
King's College London, Institute of Psychiatry, Psychology and Neuroscience, Department of Psychological Medicine, London, UK

National Institute for Health Research (NIHR) Mental Health Biomedical Research Centre at South London and Maudsley NHS Foundation Trust and King's College London, London, UK

V. Mondelli (✉)
King's College London, Institute of Psychiatry, Psychology and Neuroscience, Department of Psychological Medicine, London, UK

National Institute for Health Research (NIHR) Mental Health Biomedical Research Centre at South London and Maudsley NHS Foundation Trust and King's College London, London, UK

Maurice Wohl Clinical Neuroscience Institute, London, UK
e-mail: valeria.mondelli@kcl.ac.uk

Curr Topics Behav Neurosci (2020) 44: 207–226
DOI 10.1007/7854_2019_89
Published Online: 21 March 2019

Abstract Recently, the evidence of increased immune activation in patients with schizophrenia has suggested a role for the immune system in the development of psychosis. However, what is causing this increased immune activation and how this leads to the development of psychopathology remain still unclear. In this chapter we discuss the evidence about the role of childhood trauma as possible underlying cause of the increased immune activation in patients with schizophrenia. According to preclinical and clinical models, early adverse events can disrupt the homeostatic control of immune responses and lead to enduring inflammatory dysregulation at a peripheral and central level. In fact, persisting systemic inflammation may facilitate peripheral tissues damage and breach the blood-brain barrier, leading to microglia activation and to neuroinflammation.

Such chronic immune dysregulation also appear to partially explain the frequent comorbidity between psychosis and metabolic abnormalities, which have previously mainly considered as side effect of antipsychotic treatment.

Overall, this evidence suggests that early stress may contribute to development of schizophrenia spectrum disorders through a modulation of the peripheral and central immune system and support the immune pathways as possible future therapeutic approach for psychosis.

Keywords Childhood maltreatment · Childhood trauma · Comorbidities · Immune activation · Inflammation · Metabolic abnormalities · Psychosis · Schizophrenia

1 Introduction

In recent decades, research on schizophrenia has increasingly focused on the role of the immune system in the development of psychosis. In particular, the evidence of comorbidity between psychosis and metabolic abnormalities, together contributing to a reduced life expectancy, suggests that specific biological factors, such as an immune dysregulation, may link the aetiology of psychosis and physical health problems. A chronic low-grade immune activation has been consistently shown by studies in patients with schizophrenia spectrum disorders. These immune abnormalities have been detected not only in peripheral tissues (such as blood) but also in the brain, and they have been suggested to contribute to the development of other medical comorbidities, including metabolic and cardiovascular disorders (Leonard et al. 2012). However, what is causing this increased immune activation and how this leads to the development of psychopathology remain still unclear. Different causes have been considered, including perinatal infection, disruption of microbiota and psychosocial stress. In this chapter we are going to first discuss the evidence about the role of psychosocial stress, and more specifically childhood trauma, as possible underlying cause of the increased immune activation in patients with schizophrenia. In the second part of this chapter, we will analyse the evidence supporting the underlying role of childhood trauma and related immune dysregulation in the link between psychosis and metabolic disorders.

Childhood trauma is defined as harm, potential of harm or threat of a harm resulting from commission or omission by child's caregiver (Sideli et al. 2012). This definition includes a number of adverse experiences, such as physical, sexual and emotional abuse, neglect, parental death and bullying, but the most common forms of trauma, reported by both men and women, are physical abuse, physical neglect and emotional abuse, all of which are likely to co-occur (Scherrer et al. 2004). Neglect, the failure to provide for all aspects of the child's well-being, is considered the most frequent form with 78.5% of children exposed in the general population; physical abuse is reported in 17.6%, and it is defined as the use of physical force that harms the child's health, survival, development or dignity (Dubowitz et al. 2004). Approximately 8% of males and 20% of females universally experienced childhood sexual abuse (the involvement in sexual activity that a child is unable to give consent to or is not developmentally prepared for), with the highest prevalence in Africa (34.4%), followed by Asia (in particular China and India), America and Europe, with 23.9%, 10.1% and 9.2%, respectively (Verdolini et al. 2015). Finally, emotional abuse is the failure to provide children with a supportive environment. Actually, it seems to have higher prevalence than sexual and physical abuse, but it is more difficult to measure and quantify (Holmes and Slap 1998).

2 Association Between Childhood Trauma and Psychosis

Childhood exposure to adverse experiences has been associated with an increased risk of psychosis, mood disorders and other medical disorders in general, like cardiovascular disorders or diabetes (Varese et al. 2012; Coelho et al. 2014). Several studies have outlined the high prevalence of a history of childhood adversities among patients with psychosis. For example, in a comprehensive review in 2005, Read and colleagues reviewed 39 studies on female in- and outpatients with schizophrenia and demonstrated that the majority of female and male patients reported either childhood sexual abuse (CSA) or childhood physical abuse (CPA) (Read et al. 2005). A more recent study by Varese et al. (2012) found that childhood trauma increases the risk of psychosis with an OR of 2.8. Moreover, they showed that if the adversities considered as risk factors were entirely removed from the population, the number of people with psychosis would be reduced by 33%. It has also been showed that, with the exception of parental death, all types of adversities were related to an increased risk of psychosis. This suggests that psychosis risk is increased by the exposure to adverse experiences in general, rather than to a specific type of trauma.

When looking at studies investigating subjects at "ultrahigh risk" of developing psychosis, trauma has been repeatedly found to predict transition to psychosis in this population (Mayo et al. 2017). Sexual abuse has been shown to be the most common form of childhood trauma associated with later psychosis conversion, followed by physical abuse (Bechdolf et al. 2005; Conus 2010). Finally, similarly to findings on sexual abuse history, increased severity and duration of individuals' bullying history has also been linked to the emergence of psychotic symptoms (Arseneault et al. 2011).

The most consistent finding about the association between childhood maltreatment and psychosis is that a history of childhood trauma increases specifically positive symptoms. Already in 1994, Ross and colleagues found that in-patients with schizophrenia with a history of sexual or physical abuse in their childhood had significantly more positive schizophrenia symptoms, especially hallucinations, and slightly fewer negative symptoms than those not abused (Ross et al. 1994). This evidence has been replicated in more recent studies (Ajnakina et al. 2016). Similarly, Bendall et al. (2013) reported that first-episode psychosis (FEP) patients exposed to childhood sexual abuse had more severe hallucinations and delusions (Bendall et al. 2013). According to their findings, childhood sexual abuse, especially rape, was associated with auditory verbal hallucinations, whereas victimization (physical abuse and bullying) predicted paranoia as well as auditory verbal hallucinations. Separation experiences (placements in foster care or institutions) were also associated with paranoia.

Trauma has also been suggested to influence the content of hallucinations, with patients experiencing hallucinations with themes similar to their trauma (Hardy et al. 2005). Furthermore, childhood trauma has a negative effect on cognitive functions in healthy individuals as well as in patients with psychosis and their high-risk offspring, in particular in relation to general cognitive abilities, memory and executive functions (Aas et al. 2011; Berthelot et al. 2015; Bucker et al. 2012). By contrast, several studies (Read and Ross 2003; Resnick et al. 2003; Lysaker et al. 2001) have found no differences in negative symptoms prevalence between abused and non-abused in-patients, whereas two adult in-patient studies found slightly fewer negative symptoms in abused subjects (McCormick and Goff 1991; Ross et al. 1994; Ajnakina et al. 2016).

3 Childhood Trauma and Immune Activation

Childhood adversities can be considered as environmental insults happening in a critical developmental phase of life. Interestingly, they have been suggested to alter the immune system function and to finally result in a chronic immune activation. Supporting this hypothesis, childhood adversities are strongly associated with acute and persistent inflammatory dysregulation, namely, increased blood levels of C reactive protein (CRP), interleukin-6 (IL-6) and tumour necrosis factor alpha (TNF-α), and such inflammation persists throughout adulthood (Coelho et al. 2014) offering a potential molecular pathway by which early trauma may increase susceptibility to psychiatric disorders (Baumeister et al. 2016). The most supported hypothesis is that early stress causes a pro-inflammatory status that may influence the response to inflammatory challenges during adult life, further contributing to behavioural and cognitive alterations. As a result, childhood adversities may become an important preventable cause of psychiatric disorders, including psychosis.

As far as evidence of peripheral inflammation is concerned, we have recently conducted a meta-analysis of all studies investigating history of childhood maltreatment and inflammatory markers in adulthood (Baumeister et al. 2016). The meta-analysis included studies on both general and clinical populations (with psychiatric and physical disorders) and focused on CRP, IL-6 and TNF-α as these are the most examined inflammatory markers in psychiatric research. Results showed a significant association between childhood trauma and inflammatory markers in adulthood, with greatest effect size for TNF-α ($z = 0.20$, 95% CI = 0.10–0.29) followed by IL-6 ($z = 0.09$, 95% CI = 0.04–0.15) and then CRP ($z = 0.08$, 95% CI = 0.04–0.11). Another important finding from this study was that different types of trauma exposure impacted differentially on the inflammatory markers: physical and sexual abuse was associated with significant increased TNF-α and IL-6, but not CRP. Conversely, CRP was primarily related to parental absence during early development. How different types of trauma may impact different aspects of immune activity remains unknown; however, it has been suggested that factors such as context, time and duration of stress exposure may interact with individual trauma type in modulating immune response.

3.1 *Preclinical Evidence of the Link Between Early Stress and Increased Inflammation*

The first evidence of an association between early stress and immune system dysregulation was derived by preclinical studies. After Ader and colleagues found that rats handled before weaning showed slower development of a transplanted tumour (Ader and Friedman 1965), a number of authors started investigating the association between early-life stress and later immune functioning in rodents and non-human primates (Hennessy et al. 2010; Shanks and Lightman 2001). Stress exposure in early life (i.e., across postnatal days 1–20 in rodents) has been re-created in experimental models including neonatal handling, maternal separation, maternal deprivation, nursery rearing, early weaning and dexamethasone treatment. For example, maternal separation has been associated with an increased stress reactivity. Measures of immune functioning have included pro-inflammatory cytokines in the plasma, antigen-induced immunoproliferation in the spleen, expression levels of pro-inflammatory genes in the brain and intestinal microflora. Maternal separation in rats has been associated with elevated pro-inflammatory cytokines in the plasma (Reus et al. 2013; Wieck et al. 2013).

In studies with non-human primates, maternal separation led to an increase in macrophage activity (Coe et al. 1988) and long-term up-regulation in pro-inflammatory gene transcription in monocytes (Cole et al. 2012).

3.2 The Emerging Evidence of Neuroinflammation

The second interesting line of evidence coming from preclinical models is the link between early-life stress and markers of immune function in the central nervous system (CNS) (Reus et al. 2013) in adulthood. For example, in adult rats, early maternal separation is associated with greater synaptic levels of the receptor for the pro-inflammatory cytokine interleukin-1 (IL-1) (Viviani et al. 2014), greater number and motility of cortical microglial processes (Takatsuru et al. 2015) and greater microglia activation (Ganguly and Brenhouse 2015; Mondelli et al. 2017). Microglia are myeloid cells which provide the main form of adaptive immune response in the central nervous system (CNS) and play a crucial role in neuroinflammation. Indeed, neuroinflammation may lead to cognitive and behavioural dysfunction and psychopathology. A model of stress-induced central inflammation has already been created in animals by using stressful physical stimuli, such as restraint and foot shock, which could be applied to parent animals to induce maternal or perinatal stress (Mondelli et al. 2017).

The immediate consequences of early-life stress on animals' brain were investigated by Roque and colleagues in 2016 (Roque et al. 2016). The authors analysed the effect of maternal separation on astrocyte and microglial cell morphology in the hippocampus and hypothalamus of male rat pups and found that maternal separation (MS) caused microglia activation and decrease in astrocyte density in both areas. Moreover, compared with controls, MS rats showed increased IL-1β in the hippocampus and increased TNF-α and IL-6 levels in the hypothalamus. Similarly, Diz-Chaves et al. (2012) showed that prenatal stress induces a basal pro-inflammatory status in the hippocampal formation during adulthood that results in an enhanced activation of microglia and astrocytes in response to pro-inflammatory insults. These findings suggest that early stress may contribute to decreased hippocampal neurogenesis and alter the neuroendocrine axis, leading to both psychopathological manifestation and metabolic abnormalities.

3.3 From Animal to Human Models

Preclinical evidence has led to a new insight about the link between stress and inflammation in humans. As far as peripheral inflammation is concerned, it is clear that the immune activation resulting from stress is able to protect the body from possible external challenges. Once released, inflammatory mediators are able to coordinate a variety of cell functions that stimulate and enhance immune activation. In particular, IL-1, IL-6 and TNF-α promote the differentiation of lymphocytes called *cytotoxic T cells*, which kill pathogens that are introduced into the body during physical wounding. These cytokines also promote increased vascular permeability and cellular adhesion, which allows immune cells to leave the blood vessels and migrate to tissues (Dhabhar et al. 2012). However, when such immune activation persists chronically, it may damage vessels and peripheral tissues and lead to cardiovascular diseases.

Moreover, the hypothesis that stress may also reach the brain has progressively been investigated in humans. On one hand, stress-induced peripheral inflammation has been showed to affect the brain and to alter neural activity through different pathways: by active transport of cytokines, by involving macrophage-like cells residing in circumventricular organs or by the release of second messengers, which in turn stimulate local cytokines. These, in turn, can cross the blood-brain barrier and determine microglia activation (Cattaneo et al. 2015). They influence cell proliferation and survival depending on their inflammatory state (Mondelli et al. 2017). In response to harmful stimuli, microglial cells undergo a number of changes (Walker et al. 2014) including production of pro-inflammatory cytokines and the expression of several cell surface antigens that promote oxidative stress in the brain.

In humans, central inflammation (in terms of microglial activation) can be investigated with positron emission tomography (PET) using radio ligands for the 18-kDa translocator protein (TSPO), that is, a five-membrane domain protein localized mainly in the outer mitochondrial membrane of steroid-synthesizing cells, including those in the central nervous system. TSPO is involved in the transport of cholesterol into mitochondria. Peripheral lipopolysaccharide injection, used as immune challenge in primates, has been shown to increase TSPO expression in a uniform manner across the brain. This, in turn, seems to lead to microglial activation and to the expression of histological markers of brain activation in human postmortem tissue (Mondelli et al. 2017).

4 An Inflammatory Pathway Linking Childhood Trauma and Psychosis?

As a result of the association of early adversities with both elevated inflammation and with psychosis risk, it is not surprising that patients with schizophrenia who have history of childhood trauma tend to show increased levels of some pro-inflammatory markers, including IL-6 and TNF-α (Dennison et al. 2012). In agreement with these findings, we have also previously shown that FEP patients with history of childhood sexual abuse have higher body mass index (BMI) and increased CRP levels in comparison with controls and patients without a history of sexual abuse (Hepgul et al. 2012). In addition, in another study we have also reported that FEP patients with childhood trauma have significantly higher serum levels of TNF-α and monocyte chemo-attractant protein-1 compared with patients without childhood trauma (di Nicola et al. 2013). The biological pathways linking stress to psychosis remain partly unclear, and although peripheral immune activation has been identified as an important link between the two, other biological systems and mechanisms appear to play a role. In particular, the hypothalamic-pituitary-adrenal (HPA) axis and the effects of neuroinflammation have been both explored in the link between stress and psychosis and offer further understanding behind the presence of an increased peripheral inflammation in psychotic patients with experience of childhood trauma.

4.1 The Allostatic Systems: The Role of the Nervous System and of the Hypothalamic-Pituitary-Adrenal Axis

A suggested mechanism involved in childhood trauma embedding and then leading to chronic inflammation is the association between childhood adversities and enduring changes in the nervous, the endocrine and the immune systems that are highly integrated. Thus, the activation of one of these systems commonly triggers responses in the others (Danese and McEwen 2012).

4.1.1 The Nervous System

A neurobiological network including cerebral areas such as the thalamus, the sensory cortex and the amygdala is involved in detecting environmental threats. In response to psychosocial stressors, such as early adversities, the amygdala triggers firing in the locus coeruleus, which increases alertness and attention to the environment and induces a bodily response through the activation of the sympathetic nervous system (SNS) (the "fight or flight response").

The SNS, in turn, affects the immune system and regulates pro-inflammatory cytokine production by releasing the neurotransmitter norepinephrine into peripheral tissues. In fact, by binding to β-adrenergic and α-adrenergic receptors, norepinephrine regulates the transcription of pro-inflammatory cytokine genes interleukin (IL)-1 and tumor necrosis factor (TNF)-α, leading to systemic immune activation (Danese and McEwen 2012). Then, a chronic or repeated stimulation of the sympathetic nervous system leads to chronically increased levels of inflammation. In the same time, key anti-inflammatory pathways, such as the *hypothalamic-pituitary-adrenal* (*HPA*) axis, are progressively downregulated under stressful conditions.

4.1.2 The Endocrine System: The Hypothalamic-Pituitary-Adrenal Axis

The principal endocrine effectors of the stress response are localized in the paraventricular nucleus (PVN) of the hypothalamus, the anterior lobe of the pituitary gland and the adrenal gland. Together, these structures are known as the hypothalamic-pituitary-adrenal axis.

Neurons localized in the medial parvocellular subdivision of the PVN secrete corticotropin-releasing factor (CRF), the principle regulator of the HPA axis. Stress (in this case, early adversities) triggers the release of CRF into hypophysial portal vessels that access the anterior pituitary gland. In turn, CRF induces the release of adrenocorticotropic hormone (ACTH) from corticotropic cells into the systemic circulation. Then circulating ACTH targets the adrenal cortex, where it stimulates glucocorticoid synthesis and secretion from the zona fasciculata (Smith and Vale 2006).

Glucocorticoids, such as cortisol, exert their activity by binding to glucocorticoid receptors (GR). However, chronic stress can cause persistent elevated basal glucocorticoids levels and impair glucocorticoids ability to exert the negative feedback on the HPA axis itself. Moreover, glucocorticoids exert a natural anti-inflammatory action, binding GR in the body, but chronic stress leads to an impaired GR sensitivity (glucocorticoid resistance) and then to a prolonged/increased immune activation (Miller et al. 1999). Both HPA axis hyperactivity and increased immune activation are present in maltreated children (Danese and McEwen 2012) and persist in adulthood with several detrimental effects. As a consequence, subsequent responses to stressful situations may be abnormal. Second, the elevation of glucocorticoids levels due to glucocorticoid resistance may endanger the hippocampus and make it vulnerable to potential injury as well as stimulate striatal dopamine release that make patients more prone to psychosis (Read et al. 2001). Finally, this chronic alteration of the endocrine and immune system may progressively promote metabolic abnormalities and explain their common comorbidity with psychotic disorders. In fact, high cortisol levels can increase glucose levels and fat deposition (Dinan et al. 2004); clinical evidence has also shown that stress can lead to an increased intake of high-calories food (Dallman et al. 2005) and precipitate binge eating (Freeman and Gil 2004).

4.2 The Role of Neuroinflammation

Another hypothesized mechanism linking early trauma to psychosis is neuroinflammation. Preclinical studies suggest that microglia could be an important mediator of the association between psychosocial stress, in particular early stress, and psychiatric disorders, including psychosis. In fact, microglia activation may in turn be associated with impairment of neurogenesis and lead to structural and functional changes in the brain that predispose individuals to mental illnesses. In humans, ten studies have been published that investigated microglial activity using PET brain imaging in patients with psychosis, but results are inconsistent, probably because of the use of different radiotracers and different ways of analysing the data. Indeed, four of these studies found an increase in TSPO binding, one found an increase in medicated patients but not in drug-naive patients, and the other five did not find significant differences between patients and matched controls (Mondelli et al. 2017).

Findings suggesting the presence of neuroinflammation have already led researchers to support the development of new treatments. In particular, drugs that inhibit or reduce microglial activation are being tested in a number of studies (Mondelli et al. 2017). A preclinical study from Giovanoli et al. (2016)

have explored whether an early anti-inflammatory intervention with minocycline during peripubertal stress exposure might prevent the subsequent emergence of adult behavioural pathology (Giovanoli et al. 2016). They used an environmental two-hit model in mice, in which prenatal maternal administration of the viral mimetic poly(I:C) served as the first hit and exposure to subchronic unpredictable stress during peripubertal maturation as the second hit. Using this model, they examined the effectiveness of the tetracycline antibiotic minocycline, a broad-spectrum tetracyclic antibiotic displaying neuroprotective properties, given during stress exposure, to block stress-induced inflammatory responses and to prevent subsequent behavioural abnormalities. They found that combined exposure to prenatal immune activation and peripubertal stress caused significant behavioural dysfunctions, which minocycline treatment during stress exposure was able to prevent. In addition, the same pharmacological intervention blocked hippocampal and prefrontal microglia activation and interleukin-1β expression in offspring exposed to prenatal infection and peripubertal stress. Minocycline has been suggested as a new potential therapy for negative symptoms in clinical studies of patients with schizophrenia. In two clinical trials comparing minocycline versus placebo, both added to the standard care, patients receiving minocycline showed a greater reduction in negative symptoms (Levkovitz et al. 2010; Chaudhry et al. 2012).

4.3 Possible Role of the Interaction of Childhood Trauma with Other Environmental Factors

A number of environmental insults have been associated with a risk of psychosis, including socioeconomic status, cannabis use and urbanicity. Epidemiological studies have focused their interest on the interaction between childhood adversities and other environmental risk factors in order to provide broader insights into inflammatory trajectories leading to psychosis onset. In particular, there are studies showing that childhood traumatic events are associated with increased cannabis use in adulthood (Harley et al. 2010; Houston et al. 2011; Konings et al. 2012). Some preclinical data demonstrate that Δ^9-THC in adolescent mice triggers immune dysfunctions that last long after the end of abuse, switching the murine immune system to pro-inflammatory status in adulthood.

Such a pro-inflammatory state they may play a key role in the pathogenesis of neuropsychiatric disorders by modulating neurotransmitter and neuropeptide systems (Kronfol and Remick 2000; Muller and Ackenheil 1998) such as central monoamine activity (DeLisi 1992; Zalcman et al. 1994). This may partly explain why the onset of cannabis consumption at an earlier age has been identified by several authors as a factor contributing to poor prognosis in schizophrenia (Veen et al. 2004; Busse et al. 2012). A modulation of microglia function by cannabinoids, both endogenous and synthetic ones, has also been suggested. This dysregulation might cause an alteration in the neuronal architecture or the neurotransmitter flow (Busse et al. 2012; Leweke and Koethe 2008; Bernstein et al. 2009) meaning that cannabinoid action may not be limited to direct changes at the neurotransmitter level; it may alter the "immune atmosphere" of the brain (Skaper et al. 2013).

5 Inflammation and Physical Health

Another interesting aspect of high levels of inflammation is that they are generally associated with physical health comorbidities, in particular cardiovascular diseases, in both the general population and in patients with psychosis (Russell et al. 2015).

This is because the functions of the metabolic and immune system are highly interdependent. An example is the overlapping function of macrophages and adipocytes. In obesity, these two kinds of cells contribute together to the production of inflammatory mediators, which, together with fatty acids, are able to inhibit the downstream signalling of insulin receptors, eventually leading to insulin resistance. Consistently, metabolic-related conditions as obesity, type 2 diabetes and insulin resistance are usually associated with chronic inflammation (Wellen and Hotamisligil 2005), in particular with high levels of IL-6 and CRP (Capuzzi and Freeman 2007). Inflammation is also involved in all stages of atherothrombosis, the underlying cause of approximately 80% of all sudden cardiac death (SCD). In fact, inflammatory cells such as macrophages and T lymphocytes eventually contribute to the formation of the atheromatous lesion, which consists of a lipid pool protected by a fibrous cap. Moreover, the activation of these cells leads to the release of additional mediators, including adhesion molecules, cytokines, chemokines and growth factors, all of which play important roles in atherogenesis and in the development of cardiovascular events (Willerson and Ridker 2004). In patients with psychosis, cardiovascular issues are even more present than in the general population. For example, as early as 1960 a survey in the USA showed that 40% of hospitalized patients with schizophrenia were overweight, in contrast to 20% of the general population (Gordon et al. 1960), and in 2004, a cross-sectional study by Kato and colleagues found that the prevalence of metabolic syndrome in patients with schizophrenia was 63% and that they had threefold greater risk to develop this syndrome than the general population (Kato et al. 2004).

5.1 *Inflammation and Metabolic Complications in Psychosis*

Metabolic abnormalities and weight gain in patients with psychosis are widely associated to antipsychotic treatment (Allison and Casey 2001; Mondelli et al. 2013). However, studies have shown impaired glucose tolerance, increased visceral fat and increased obesity and hypertension also in drug-naïve patients (Correll et al. 2014), suggesting an impaired metabolism regardless of medication. For instance, some authors found that prevalence of diabetes in schizophrenia exceeds that in the general population well before the widespread use of the new (atypical) antipsychotic drugs (Dixon et al. 2000). Therefore, impaired glucose metabolism may be associated with schizophrenia rather than be only a side effect of antipsychotic treatment. To this regard, Perry and colleagues have conducted an interesting meta-analysis on prediabetic markers in antipsychotic naïve FEP. They included 12 studies (a total of 1,137 participants), and they found first-episode psychosis to be related to insulin resistance and impaired glucose tolerance (but not fasting plasma glucose) in the absence of medication (Perry et al. 2016).

On the other hand, in diabetic patients, untreated hyperglycaemia has been associated with alterations of mood state and acute psychosis manifestations (Sahoo et al. 2016).

Again, stress and inflammation are possible common pathways explaining the link between psychosis and metabolic abnormalities. In particular, childhood trauma may have a key role in mediating both.

Interestingly, inflammation has already been suggested to link cardiovascular abnormalities and depression. CRP, IL-1 and IL-6 have been associated with atherosclerosis and depression alike, both in healthy subjects and in cardiac patients. Previous studies have suggested both that inflammation increases risk of depression and that depression causes inflammation in patients with cardiovascular disorders (Elderon and Whooley 2013). Thus, although it remains difficult to disentangle whether such inflammatory mediators serve as triggers of both depression and CVD, act on the causal pathway between them or result from both conditions, at least it is now clear that inflammatory pathways may be the shared biological mechanisms between mental and physical diseases.

5.2 Childhood Trauma and Metabolic Abnormalities

Again, animal models, like those involving non-human primates, have been a useful source of evidence suggesting that adverse childhood experiences may influence physical health, especially obesity risk.

Kaufman et al. (2007) showed that compared with normally reared monkeys, those exposed to early stress exhibit greater weight, BMI, abdominal circumference, glucagon-like peptide-1 and decreased glucose disposal rates during hyperinsulinaemic-euglycaemic clamps. In their model of early-life stress (variable foraging demand [VFD]), food insecurity is imposed on monkey mothers for 16 weeks beginning when their nursing offspring are 3–5 months of age. VFD resulted in a range of neurobiological abnormalities, including dysregulation of the HPA axis, manifested in abnormal cerebrospinal fluid cortisol and corticotropin-releasing factor levels. These data suggest that early-life stress during a critical period of neurodevelopment can result in the prepubertal emergence of obesity and insulin resistance.

Evidence of the link between early trauma and metabolic abnormalities is also quite clear from more recent human studies. A meta-analysis conducted by Danese and Tan (2014) on 41 studies in humans suggests that childhood maltreatment predicts obesity, independently from the measures and the definitions used and other potential confounding variables. Stressful psychosocial experiences in childhood might thus be conceptualized as potentially modifiable risk factors for obesity. Thus, prevention or effective treatment of severe cases of childhood maltreatment could avoid development of obesity in adulthood. However, it is still unclear if and how the effect of maltreatment on obesity could be modified through intervention.

Li et al. (2017) showed that childhood maltreatment is an independent risk factor for developing prediabetes. They recruited 121 participants from the general population, either with ($n = 69$) or without history of childhood maltreatment ($n = 52$). The authors found a 15% higher glucose area under the OGTT curve in the maltreated group, together with impaired insulin sensitivity. This group also showed higher CRP and TNF-α levels, both positively correlating with severity of childhood trauma ($r = 0.21, 0.23$, respectively, both $p < 0.05$). These data suggest an important relationship between childhood maltreatment and increased risk for prediabetic state due to glucose intolerance and impaired insulin sensitivity and beta cells function. Finally, Rich-Edwards et al. (2012) also showed that severe child abuse is a risk factor for early adult cardiovascular disorders, while Suglia et al. (2014) found that women who experienced sexual abuse in early childhood had a higher prevalence of hypertension (prevalence ratio (PR) 1.43 95% CI 1.00, 2.05) compared with women who did not experience such maltreatment.

5.3 *Childhood Trauma and Metabolic Abnormalities in Patients with Psychosis*

When looking at schizophrenia spectrum patients, only three studies have addressed the contribution of childhood trauma to cardio-metabolic risk.

In our study by Hepgul et al. (2012), we found that FEP patients with a history of childhood sexual abuse had higher BMI and CRP levels when compared with healthy controls and patients without childhood sexual abuse. CRP has been recently considered a marker of increased risk of diabetes and other metabolic dysfunction (Bassuk et al. 2004) and has also been similarly linked to chronic psychosocial stress (Miller 2008). Another study (Misiak et al. 2015) revealed that a history of childhood adversities, especially sexual and emotional abuse, is associated with higher systolic and diastolic blood pressure as well as greater levels of low-density lipoproteins (LDL) in FEP patients. Moreover, a study by Rajkumar (2015) found an association between childhood trauma and higher BMI and systolic blood pressure in patients with schizophrenia. In particular, physical abuse was linked to elevated systolic blood pressure, while emotional abuse and neglect in women were linked to being overweight.

Finally, whether psychotropic drugs may play a role in leading to inflammation, by interacting with childhood adversities, is still unclear. The association between antipsychotic medication and an impaired metabolism is well known (Martin Otano et al. 2013), so that treated patients often show and overall altered metabolic-inflammatory status. Given the already mentioned association between sexual abuse and severity of hallucinations and delusions, it could be hypothesized that higher doses of antipsychotics are prescribed to patients with a history of childhood maltreatment. As a consequence, side effects of medication would be another mechanism linking childhood maltreatment and adulthood inflammation. However, studies investigating this hypothesis are currently lacking, so this should be the aim of future works.

6 Conclusion

In conclusion, susceptibility to psychosis is likely to be influenced by personal history of exposure to early traumatic events. In this model, priming adverse events can disrupt the homeostatic control of immune responses leading to enduring inflammatory dysregulation at a peripheral and central level. In particular, persisting systemic low-grade chronic inflammation may facilitate peripheral tissues damage and breach the blood-brain barrier with activation of microglia and development of neuroinflammation. The immune dysregulation also explains the development of multiaxial comorbidity including metabolic abnormalities and cardiovascular disorders, which have previously mainly considered as side effect of antipsychotic treatment (Fig. 1).

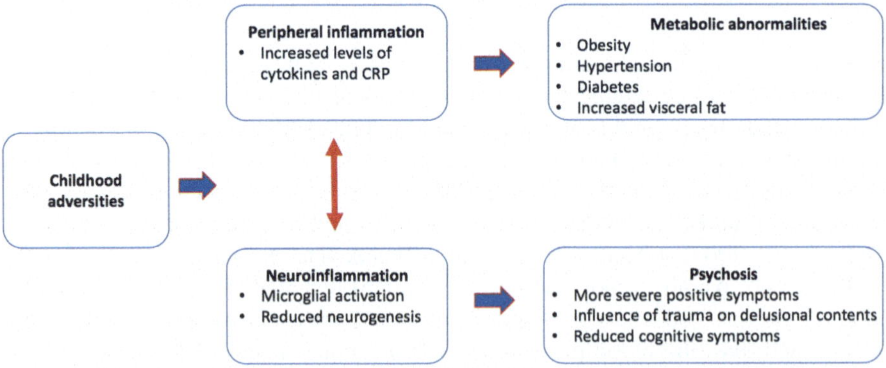

Fig. 1 Proposed model through which childhood adversities contribute to development of psychosis and comorbid metabolic abnormalities via activation of peripheral and central immune system

Two important considerations should be drawn from this work: First, research in the field of immunopsychiatry may lead to the development of new treatments targeting immune pathways in future therapeutic approaches to psychosis. Second, it is clear that early stress might be an important factor to early detect, prevent and better manage the development and treatment of schizophrenia spectrum disorders and any related physical comorbidities.

Acknowledgements This work has been supported by the South London and Maudsley (SLaM) NHS Foundation Trust & Institute of Psychiatry NIHR Biomedical Research Centre (BRC) for Mental Health; a strategic award from the Wellcome Trust (Consortium of Neuroimmunology of Mood Disorders and Alzheimer's Disease); and the grants 'Persistent Fatigue Induced by Interferon-alpha: A New Immunological Model for Chronic Fatigue Syndrome' (MR/J002739/1), and 'Immuno-psychiatry: a consortium to test the opportunity for immunotherapeutics in psychiatry' (MR/L014815/1), from the Medical Research Council (UK).

References

Aas M, Dazzan P, Fisher HL, Morgan C, Morgan K, Reichenberg A, Zanelli J, Fearon P, Jones PB, Murray RM, Pariante CM (2011) Childhood trauma and cognitive function in first-episode affective and non-affective psychosis. Schizophr Res 129(1):12–19

Ader R, Friedman SB (1965) Differential early experiences and susceptibility to transplanted tumor in the rat. J Comp Physiol Psychol 59:361–364

Ajnakina O, Trotta A, Oakley-Hannibal E, di Forti M, Stilo SA, Kolliakou A, Gardner-Sood P, Gaughran F, David AS, Dazzan P, Pariante C, Mondelli V, Morgan C, Vassos E, Murray RM, Fisher HL (2016) Impact of childhood adversities on specific symptom dimensions in first-episode psychosis. Psychol Med 46(2):317–326

Allison DB, Casey DE (2001) Antipsychotic-induced weight gain: a review of the literature. J Clin Psychiatry 62(Suppl 7):22–31

Arseneault L, Cannon M, Fisher HL, Polanczyk G, Moffitt TE, Caspi A (2011) Childhood trauma and children's emerging psychotic symptoms: a genetically sensitive longitudinal cohort study. Am J Psychiatry 168(1):65–72

Bassuk SS, Rifai N, Ridker PM (2004) High-sensitivity C-reactive protein: clinical importance. Curr Probl Cardiol 29(8):439–493

Baumeister D, Akhtar R, Ciufolini S, Pariante CM, Mondelli V (2016) Childhood trauma and adulthood inflammation: a meta-analysis of peripheral C-reactive protein, interleukin-6 and tumour necrosis factor-alpha. Mol Psychiatry 21(5):642–649

Bechdolf A, Pohlmann B, Geyer C, Ferber C, Klosterkotter J, Gouzoulis-Mayfrank E (2005) Motivational interviewing for patients with comorbid schizophrenia and substance abuse disorders: a review. Fortschr Neurol Psychiatr 73(12):728–735

Bendall S, Hulbert CA, Alvarez-Jimenez M, Allott K, McGorry PD, Jackson HJ (2013) Testing a model of the relationship between childhood sexual abuse and psychosis in a first-episode psychosis group: the role of hallucinations and delusions, posttraumatic intrusions, and selective attention. J Nerv Ment Dis 201(11):941–947

Bernstein HG, Steiner J, Bogerts B (2009) Glial cells in schizophrenia: pathophysiological significance and possible consequences for therapy. Expert Rev Neurother 9(7):1059–1071

Berthelot N, Paccalet T, Gilbert E, Moreau I, Merette C, Gingras N, Rouleau N, Maziade M (2015) Childhood abuse and neglect may induce deficits in cognitive precursors of psychosis in high-risk children. J Psychiatry Neurosci 40(5):336–343

Bucker J, Kapczinski F, Post R, Cereser KM, Szobot C, Yatham LN, Kapczinski NS, Kauer-Sant'Anna M (2012) Cognitive impairment in school-aged children with early trauma. Compr Psychiatry 53(6):758–764

Busse S, Busse M, Schiltz K, Bielau H, Gos T, Brisch R, Mawrin C, Schmitt A, Jordan W, Muller UJ, Bernstein HG, Bogerts B, Steiner J (2012) Different distribution patterns of lymphocytes and microglia in the hippocampus of patients with residual versus paranoid schizophrenia: further evidence for disease course-related immune alterations? Brain Behav Immun 26 (8):1273–1279

Capuzzi D, Freeman J (2007) C-reactive protein and cardiovascular risk in the metabolic syndrome and type 2 diabetes: controversy and challenge. Clin Diabetes 25(1):16–22

Cattaneo A, Macchi F, Plazzotta G, Veronica B, Bocchio-Chiavetto L, Riva MA, Pariante CM (2015) Inflammation and neuronal plasticity: a link between childhood trauma and depression pathogenesis. Front Cell Neurosci 9:40

Chaudhry IB, Hallak J, Husain N, Minhas F, Stirling J, Richardson P, Dursun S, Dunn G, Deakin B (2012) Minocycline benefits negative symptoms in early schizophrenia: a randomised double-blind placebo-controlled clinical trial in patients on standard treatment. J Psychopharmacol 26 (9):1185–1193

Coe CL, Rosenberg LT, Levine S (1988) Effect of maternal separation on the complement system and antibody responses in infant primates. Int J Neurosci 40(3–4):289–302

Coelho R, Viola TW, Walss-Bass C, Brietzke E, Grassi-Oliveira R (2014) Childhood maltreatment and inflammatory markers: a systematic review. Acta Psychiatr Scand 129(3):180–192

Cole SW, Conti G, Arevalo JM, Ruggiero AM, Heckman JJ, Suomi SJ (2012) Transcriptional modulation of the developing immune system by early life social adversity. Proc Natl Acad Sci U S A 109(50):20578–20583

Conus P (2010) First episode of mood disorders: an opportunity for early intervention in bipolar disorders. Encéphale 36(Suppl 3):S71–S76

Correll CU, Robinson DG, Schooler NR, Brunette MF, Mueser KT, Rosenheck RA, Marcy P, Addington J, Estroff SE, Robinson J, Penn DL, Azrin S, Goldstein A, Severe J, Heinssen R, Kane JM (2014) Cardiometabolic risk in patients with first-episode schizophrenia spectrum disorders: baseline results from the RAISE-ETP study. JAMA Psychiat 71(12):1350–1363

Dallman MF, Pecoraro NC, la Fleur SE (2005) Chronic stress and comfort foods: self-medication and abdominal obesity. Brain Behav Immun 19(4):275–280

Danese A, McEwen BS (2012) Adverse childhood experiences, allostasis, allostatic load, and age-related disease. Physiol Behav 106(1):29–39

Danese A, Tan M (2014) Childhood maltreatment and obesity: systematic review and meta-analysis. Mol Psychiatry 19(5):544–554

DeLisi LE (1992) The significance of age of onset for schizophrenia. Schizophr Bull 18(2):209–215

Dennison U, McKernan D, Cryan J, Dinan T (2012) Schizophrenia patients with a history of childhood trauma have a pro-inflammatory phenotype. Psychol Med 42(9):1865–1871

Dhabhar FS, Malarkey WB, Neri E, McEwen BS (2012) Stress-induced redistribution of immune cells – from barracks to boulevards to battlefields: a tale of three hormones – Curt Richter Award winner. Psychoneuroendocrinology 37(9):1345–1368

di Nicola M, Cattaneo A, Hepgul N, Di Forti M, Aitchison KJ, Janiri L, Murray RM, Dazzan P, Pariante CM, Mondelli V (2013) Serum and gene expression profile of cytokines in first-episode psychosis. Brain Behav Immun 31:90–95

Dinan T, Peveler R, Holt R (2004) Understanding schizophrenia and diabetes. Hosp Med 65 (8):485–488

Dixon L, Weiden P, Delahanty J, Goldberg R, Postrado L, Lucksted A, Lehman A (2000) Prevalence and correlates of diabetes in national schizophrenia samples. Schizophr Bull 26 (4):903–912

Diz-Chaves Y, Pernia O, Carrero P, Garcia-Segura LM (2012) Prenatal stress causes alterations in the morphology of microglia and the inflammatory response of the hippocampus of adult female mice. J Neuroinflammation 9:71

Dubowitz H, Pitts SC, Black MM (2004) Measurement of three major subtypes of child neglect. Child Maltreat 9(4):344–356

Elderon L, Whooley MA (2013) Depression and cardiovascular disease. Prog Cardiovasc Dis 55 (6):511–523

Freeman LM, Gil KM (2004) Daily stress, coping, and dietary restraint in binge eating. Int J Eat Disord 36(2):204–212

Ganguly P, Brenhouse HC (2015) Broken or maladaptive? Altered trajectories in neuroinflammation and behavior after early life adversity. Dev Cogn Neurosci 11:18–30

Giovanoli S, Engler H, Engler A, Richetto J, Feldon J, Riva MA, Schedlowski M, Meyer U (2016) Preventive effects of minocycline in a neurodevelopmental two-hit model with relevance to schizophrenia. Transl Psychiatry 5(6):38

Gordon HL, Law A, Hohman KE, Groth C (1960) The problem of overweight in hospitalized psychotic patients. Psychiatry Q 34:69–82

Hardy A, Fowler D, Freeman D, Smith B, Steel C, Evans J, Garety P, Kuipers E, Bebbington P, Dunn G (2005) Trauma and hallucinatory experience in psychosis. J Nerv Ment Dis 193 (8):501–507

Harley M, Kelleher I, Clarke M, Lynch F, Arseneault L, Connor D, Fitzpatrick C, Cannon M (2010) Cannabis use and childhood trauma interact additively to increase the risk of psychotic symptoms in adolescence. Psychol Med 40(10):1627–1634

Hennessy MB, Deak T, Schiml-Webb PA (2010) Early attachment-figure separation and increased risk for later depression: potential mediation by proinflammatory processes. Neurosci Biobehav Rev 34(6):782–790

Hepgul N, Pariante CM, Dipasquale S, DiForti M, Taylor H, Marques TR, Morgan C, Dazzan P, Murray RM, Mondelli V (2012) Childhood maltreatment is associated with increased body mass index and increased C-reactive protein levels in first-episode psychosis patients. Psychol Med 42(9):1893–1901

Holmes WC, Slap GB (1998) Sexual abuse of boys: definition, prevalence, correlates, sequelae, and management. JAMA 280(21):1855–1862

Houston JE, Murphy J, Shevlin M, Adamson G (2011) Cannabis use and psychosis: re-visiting the role of childhood trauma. Psychol Med 41(11):2339–2348

Kato MM, Currier MB, Gomez CM, Hall L, Gonzalez-Blanco M (2004) Prevalence of metabolic syndrome in Hispanic and non-Hispanic patients with schizophrenia. Prim Care Companion J Clin Psychiatry 6(2):74–77

Kaufman D, Banerji MA, Shorman I, Smith EL, Coplan JD, Rosenblum LA, Kral JG (2007) Early-life stress and the development of obesity and insulin resistance in juvenile bonnet macaques. Diabetes 56(5):1382–1386

Konings M, Stefanis N, Kuepper R, de Graaf R, ten Have M, van Os J, Bakoula C, Henquet C (2012) Replication in two independent population-based samples that childhood maltreatment and cannabis use synergistically impact on psychosis risk. Psychol Med 42(1):149–159

Kronfol Z, Remick DG (2000) Cytokines and the brain: implications for clinical psychiatry. Am J Psychiatry 157(5):683–694

Leonard BE, Schwarz M, Myint AM (2012) The metabolic syndrome in schizophrenia: is inflammation a contributing cause? J Psychopharmacol 26(5 Suppl):33–41

Levkovitz Y, Mendlovich S, Riwkes S, Braw Y, Levkovitch-Verbin H, Gal G, Fennig S, Treves I, Kron S (2010) A double-blind, randomized study of minocycline for the treatment of negative and cognitive symptoms in early-phase schizophrenia. J Clin Psychiatry 71(2):138–149

Leweke FM, Koethe D (2008) Cannabis and psychiatric disorders: it is not only addiction. Addict Biol 13(2):264–275

Li L, Garvey WT, Gower BA (2017) Childhood maltreatment is an independent risk factor for prediabetic disturbances in glucose regulation. Front Endocrinol (Lausanne) 8:151

Lysaker PH, Meyer PS, Evans JD, Clements CA, Marks KA (2001) Childhood sexual trauma and psychosocial functioning in adults with schizophrenia. Psychiatr Serv 52(11):1485–1488

Martin Otano L, Barbadillo Izquierdo L, Galdeano Mondragon A, Alonso Pinedo M, Querejeta Ayerdi I (2013) After six months of anti-psychotic treatment: is the improvement in mental health at the expense of physical health? Rev Psiquiatr Salud Ment 6(1):26–32

Mayo D, Corey S, Kelly LH, Yohannes S, Youngquist AL, Stuart BK, Niendam TA, Loewy RL (2017) The role of trauma and stressful life events among individuals at clinical high risk for psychosis: a review. Front Psych 8:55

McCormick S, Goff DC (1991) Apparent decrease in schizophrenia. Br J Psychiatry 159:725

Miller AH (2008) Inflammation versus glucocorticoids as purveyors of pathology during stress: have we reached the tipping point? Biol Psychiatry 64(4):263–265

Miller AH, Pariante CM, Pearce BD (1999) Effects of cytokines on glucocorticoid receptor expression and function. Glucocorticoid resistance and relevance to depression. Adv Exp Med Biol 461:107–116

Misiak B, Kiejna A, Frydecka D (2015) The history of childhood trauma is associated with lipid disturbances and blood pressure in adult first-episode schizophrenia patients. Gen Hosp Psychiatry 37(4):365–367

Mondelli V, Anacker C, Vernon AC, Cattaneo A, Natesan S, Modo M, Dazzan P, Kapur S, Pariante CM (2013) Haloperidol and olanzapine mediate metabolic abnormalities through different molecular pathways. Transl Psychiatry 3:e208

Mondelli V, Vernon AC, Turkheimer F, Dazzan P, Pariante CM (2017) Brain microglia in psychiatric disorders. Lancet Psychiatry 4(7):563–572

Muller N, Ackenheil M (1998) Psychoneuroimmunology and the cytokine action in the CNS: implications for psychiatric disorders. Prog Neuro-Psychopharmacol Biol Psychiatry 22 (1):1–33

Perry BI, McIntosh G, Weich S, Singh S, Rees K (2016) The association between first-episode psychosis and abnormal glycaemic control: systematic review and meta-analysis. Lancet Psychiatry 3(11):1049–1058

Rajkumar RP (2015) The impact of childhood adversity on the clinical features of schizophrenia. Schizophr Res Treat 2015:532082

Read J, Ross CA (2003) Psychological trauma and psychosis: another reason why people diagnosed schizophrenic must be offered psychological therapies. J Am Acad Psychoanal Dyn Psychiatry 31(1):247–268

Read J, Perry BD, Moskowitz A, Connolly J (2001) The contribution of early traumatic events to schizophrenia in some patients: a traumagenic neurodevelopmental model. Psychiatry 64 (4):319–345

Read J, van Os J, Morrison AP, Ross CA (2005) Childhood trauma, psychosis and schizophrenia: a literature review with theoretical and clinical implications. Acta Psychiatr Scand 112:330–350

Resnick SG, Bond GR, Mueser KT (2003) Trauma and posttraumatic stress disorder in people with schizophrenia. J Abnorm Psychol 112(3):415–423

Reus GZ, Dos Santos MA, Abelaira HM, Ribeiro KF, Petronilho F, Vuolo F, Colpo GD, Pfaffenseller B, Kapczinski F, Dal-Pizzol F, Quevedo J (2013) Imipramine reverses alterations in cytokines and BDNF levels induced by maternal deprivation in adult rats. Behav Brain Res 242:40–46

Rich-Edwards JW, Mason S, Rexrode K, Spiegelman D, Hibert E, Kawachi I, Jun HJ, Wright RJ (2012) Physical and sexual abuse in childhood as predictors of early-onset cardiovascular events in women. Circulation 126(8):920–927

Roque A, Ochoa-Zarzosa A, Torner L (2016) Maternal separation activates microglial cells and induces an inflammatory response in the hippocampus of male rat pups, independently of hypothalamic and peripheral cytokine levels. Brain Behav Immun 55:39–48

Ross CA, Anderson G, Clark P (1994) Childhood abuse and the positive symptoms of schizophrenia. Hosp Community Psychiatry 45(5):489–491

Russell A, Ciufolini S, Gardner-Sood P, Bonaccorso S, Gaughran F, Dazzan P, Pariante CM, Mondelli V (2015) Inflammation and metabolic changes in first episode psychosis: preliminary results from a longitudinal study. Brain Behav Immun 49:25–29

Sahoo S, Mehra A, Grover S (2016) Acute hyperglycemia associated with psychotic symptoms in a patient with type 1 diabetes mellitus: a case report. Innov Clin Neurosci 13(11–12):25–27

Scherrer JF, Waterman BM, Heath AC, Bucholz KK, True WR, Jacob T (2004) Are substance use, abuse and dependence associated with study participation? Predictors of offspring nonparticipation in a twin-family study. J Stud Alcohol 65(1):140–144

Shanks N, Lightman SL (2001) The maternal-neonatal neuro-immune interface: are there long-term implications for inflammatory or stress-related disease? J Clin Invest 108(11):1567–1573

Sideli L, Mule A, La Barbera D, Murray RM (2012) Do child abuse and maltreatment increase risk of schizophrenia? Psychiatry Investig 9(2):87–99

Skaper SD, Facci L, Giusti P (2013) Glia and mast cells as targets for palmitoylethanolamide, an anti-inflammatory and neuroprotective lipid mediator. Mol Neurobiol 48(2):340–352

Smith SM, Vale WW (2006) The role of the hypothalamic-pituitary-adrenal axis in neuroendocrine responses to stress. Dialogues Clin Neurosci 8(4):383–395

Suglia SF, Clark CJ, Boynton-Jarrett R, Kressin NR, Koenen KC (2014) Child maltreatment and hypertension in young adulthood. BMC Public Health 14:1149

Takatsuru Y, Nabekura J, Ishikawa T, Kohsaka S, Koibuchi N (2015) Early-life stress increases the motility of microglia in adulthood. J Physiol Sci 65(2):187–194

Varese F, Smeets F, Drukker M, Lieverse R, Lataster T, Viechtbauer W, Read J, van Os J, Bentall RP (2012) Childhood adversities increase the risk of psychosis: a meta-analysis of patient-control, prospective- and cross-sectional cohort studies. Schizophr Bull 38(4):661–671

Veen ND, Selten JP, van der Tweel I, Feller WG, Hoek HW, Kahn RS (2004) Cannabis use and age at onset of schizophrenia. Am J Psychiatry 161(3):501–506

Verdolini N, Attademo L, Agius M, Ferranti L, Moretti P, Quartesan R (2015) Traumatic events in childhood and their association with psychiatric illness in the adult. Psychiatr Danub 27(Suppl 1): S60–S70

Viviani B, Boraso M, Valero M, Gardoni F, Marco EM, Llorente R, Corsini E, Galli CL, di Luca M, Marinovich M, Lopez-Gallardo M, Viveros MP (2014) Early maternal deprivation immunologically primes hippocampal synapses by redistributing interleukin-1 receptor type I in a sex dependent manner. Brain Behav Immun 35:135–143

Walker FR, Beynon SB, Jones KA, Zhao Z, Kongsui R, Cairns M, Nilsson M (2014) Dynamic structural remodelling of microglia in health and disease: a review of the models, the signals and the mechanisms. Brain Behav Immun 37:1–14

Wellen KE, Hotamisligil GS (2005) Inflammation, stress, and diabetes. J Clin Invest 115(5):1111–1119

Wieck A, Andersen SL, Brenhouse HC (2013) Evidence for a neuroinflammatory mechanism in delayed effects of early life adversity in rats: relationship to cortical NMDA receptor expression. Brain Behav Immun 28:218–226

Willerson JT, Ridker PM (2004) Inflammation as a cardiovascular risk factor. Circulation 109:II2–II10

Zalcman S, Green-Johnson JM, Murray L, Nance DM, Dyck D, Anisman H, Greenberg AH (1994) Cytokine-specific central monoamine alterations induced by interleukin-1, -2 and -6. Brain Res 643(1–2):40–49

Inflammation, Antipsychotic Drugs, and Evidence for Effectiveness of Anti-inflammatory Agents in Schizophrenia

Ananda K. Pandurangi and Peter F. Buckley

Contents

Abstract In recent years, there is a new optimism in schizophrenia therapeutics with the emergence of immunomodulation as a potential treatment approach. Current evidence points to various immunological abnormalities in schizophrenia, including cell-mediated processes, acute phase proteins, cytokines, and intracellular mediators. Trait- and state-related immune dysfunction appears to exist, and a strong case can therefore be made for immunomodulation therapies in the prevention, treatment, and/or moderating the course of schizophrenia.

A. K. Pandurangi (✉)
Virginia Commonwealth University, Richmond, VA, USA
e-mail: apandura@vcu.edu

P. F. Buckley
School of Medicine, Virginia Commonwealth University, Richmond, VA, USA

Medical Affairs, Virginia Commonwealth University, Richmond, VA, USA

Curr Topics Behav Neurosci (2020) 44: 227–244
DOI 10.1007/7854_2019_91
Published Online: 17 April 2019

Immunomodulation approaches include use of nonsteroidal anti-inflammatory agents to stop or moderate an over-activated inflammatory process, anti-oxidants, nutrients, vitamins, herbal products, and other neuroprotection agents that inhibit pro-inflammatory processes, optimal use of antipsychotic drugs (APDs) that may have anti-inflammatory actions or in certain cases such as clozapine may enhance blunted inflammatory responses, and biological agents to antagonize specific immune mediators such as the cytokines. A combination of two or more of the above approaches is also worthy of consideration.

In this chapter, the available data for each of the above approaches is reviewed and discussed. Strengths and limitations of current studies are identified, and suggestions are made for future studies. For example, identifying patients with high levels of specific biomarkers such as C-Reactive Protein, IL-6, IFN-γ, TNF-α, and genetic polymorphisms of cytokines, and match them with clinical subgroups such as prodromal, first episode psychosis, chronic psychosis, and negative symptoms with the aim of developing targeted treatment approaches and more personalized medicine. Meanwhile, since the science and trial data are not advanced enough to make definitive recommendations, clinicians should stay up to date with the literature, obtain detailed immunological histories, and review the risk-benefit ratio of adding available immune modulating agents to standard therapies, to provide optimal and state-of-the-art care to patients.

Keywords Anti-inflammatory therapies · Antioxidants · Antipsychotic drugs · Biological agents · Nonsteroidal agents · Nutrients and neuroprotectors

1 Introduction

The last decade has created a new optimism in schizophrenia therapeutics with the emergence of immunomodulation as a potential treatment approach. Current evidence points to various immunological abnormalities in schizophrenia, including cell-mediated processes, acute phase proteins, cytokines, and intracellular mediators. Broadly, a blunting of the Type 1 immune response and overactivation of the Type 2 response have been suggested. The Type 1 immune response is a largely cell-mediated phagocytic response involving Type 1 helper lymphocytes and mediated by cytokines such as IL-2, IFN-γ, and lymphotoxin-α, and Type 2 immune response is mediated by Type 2 helper cells with cytokines such as IL-4, IL-10, and IL-13 and exhibits high antibody titers (Müller 2014). Trait- and state-related immune dysfunction appears to exist. Dysregulation by genetic loci that control immune processes has been identified in schizophrenia, and certain infections have also been implicated as possible causes of impaired immune programming in schizophrenia (Müller 2014; Cox et al. 2015; Maes et al. 1997; Arias et al. 2012; Miller et al. 2011; Girgis et al. 2014; Müller et al. 2013). Although as yet there isn't any approved anti-inflammatory agent for treating schizophrenia, a strong case can be made for anti-inflammatory therapies in the prevention, treatment, and/or moderating the course of schizophrenia (Girgis et al. 2014; Müller et al. 2013; Bumb et al. 2015; Torrey and Davis 2012). Another driver for seeking new approaches to treat schizophrenia is the fact that despite 70 years of antipsychotic pharmacology research and trials, a humbling fact is that more than 30% of patients with schizophrenia are refractory to current medications and in those who do respond, negative and cognitive symptoms remain largely unimproved (Kane and Correll 2010).

2 Anti-inflammatory Therapies in Schizophrenia

There are several approaches to the use of immunological therapy in schizophrenia (Cox et al. 2015; Girgis et al. 2014; Müller et al. 2013; Bumb et al. 2015; Torrey and Davis 2012):

1. Anti-inflammatory agents to stop or moderate an over-activated inflammatory process in hopes of reducing the severity of symptoms and moderating the course of the illness
2. Agents that inhibit the pro-inflammatory processes in hopes of avoiding or delaying the onset of the pathological processes
3. Optimal use of antipsychotic drugs (APDs) with known or likely anti-inflammatory actions
4. Biological therapies to antagonize specific immune mediators such as the cytokines
5. Combination approach of two or more of the above

We present below brief summaries for each anti-inflammatory agent tried or recommended in schizophrenia. Although the actions significantly overlap, for convenience, individual agents used for therapy are categorized under the headings of (1) Nonsteroidal Anti-inflammatory Drugs (NSAIDs), (2) Antioxidants including Nutrients, (3) Antipsychotic Drugs, (4) Biologicals, and (5) Miscellaneous Agents. We caution the reader however that this field of therapeutics remains in its infancy and the trials reported so far are not comparable to the large multicenter randomized controlled trials (RCT) familiar to clinicians with APDs.

2.1 *Nonsteroidal Anti-inflammatory Drugs (NSAIDs)*

- *Acetyl Salicylic Acid (aspirin, ASA)*: ASA was synthesized in 1897 and is used for various medical conditions from mild analgesia to cardiovascular disease and neuroprotection. It irreversibly inhibits the inflammatory enzyme cyclooxygenase (COX-1 and COX-2) and decreases the synthesis of pro-inflammatory prostaglandins. It also inhibits thromboxane, an enzyme needed for platelet aggregation. To date, there are two placebo-controlled, double-blind (DB) RCTs of ASA 1000 mg given as an adjuvant to APD over 3–4 months to a total 270 patients.

ASA patients showed significantly lower positive and negative symptom scores, but not cognitive function scores. Subgroups such as inpatients (more acute/severe) and first-episode psychosis (FEP) appeared to have benefitted more. Also, symptom improvement was more pronounced in patients showing higher levels of inflammatory markers. Nitta et al. (2013) and Sommer et al. (2014) have reported results of meta-analyses of trials until 2013 and 2014, respectively. Effect size across the two ASA studies was 0.3 and significant (Nitta et al. 2013; Sommer et al. 2014).

- *Celecoxib* is an NSAID. It is a specific COX-2 inhibitor and inhibits the transformation of arachidonic acid to prostaglandins. It was the first selective COX inhibitor approved for clinical use and extensively prescribed in arthritic pain management. Its use decreased after the Food and Drug Administration (USA) issued a black box warning of adverse cardiac effects. Nevertheless, it may have a role in immune modulation therapy of psychosis. Oral celecoxib 400 mg was administered as an add-on agent to antipsychotic medications in 5 RCTs with a total sample size of 225, for periods of 5–11 weeks. One of the trials was restricted to first-episode patients (FEP). The findings were rather heterogeneous with three positive studies, including one with FEP, and two negative studies. Overall there was improvement in total symptom scores and negative symptoms but not positive symptoms. One study did not find any benefit in psychosis. Two trials reported associations between response to COX inhibition and an impaired Type 1 to Type 2 immune balance, which was normalized by celecoxib. The Sommer et al. (2014) meta-analysis showed a nonsignificant effect size of 0.15.
- *Minocycline* is a synthetic lipophilic broad-spectrum tetracycline antibiotic that easily crosses the blood-brain barrier. It acts as an antioxidant, anti-inflammatory, and anti-apoptotic agent. It is approved only as an antibiotic. Its anti-inflammatory and neuroprotective effects are likely through inhibition of nitric oxide (NO) synthase and 5-lipoxygenase, a pro-inflammatory enzyme. Minocycline 200 mg has been given over 4–12 months as an add-on agent in six RCTs in schizophrenia. Two of the studies were in FEP. Modest gains were seen in negative symptoms and executive function, but the effect size of 0.22 was not significant (Sommer et al. 2014). Most recently, Deakin et al. published the findings of their large study of 207 patients with recent onset psychoses who were given adjunct minocycline (Benemin) 200 mg by mouth in addition to a second-generation APD. Among other measures, clinical outcomes and IL-6 levels were measured. There was no discernible effect of minocycline on clinical outcomes, and IL levels were stable and unaffected by the drug (Deakin et al. 2018).

Despite a lack of robust and uniform improvement across studies, these trials are promising and can guide future research. They suggest using inflammatory markers to identify patients who might benefit and studying clinical subgroups of patients to define the optimal group for immunotherapy. Also, further investigation is needed to establish that it is the anti-inflammatory actions of the agents that results in benefit, as most agents tried so far have a broad spectrum of actions.

2.2 Antioxidants and Nutrients

Antioxidants inhibit the formation of free radicals that often have detrimental effects at the cellular level. They are not specific therapies for any single disease but enhance overall cellular integrity. N-Acetylcysteine, ascorbic acid (vitamin C), α-tocopherol (vitamin E), the polyunsaturated fatty acids (PUFA) eicosapentaenoic acid (EPA) and docosahexaenoic acid (DHA), melatonin, and L-theanine are some of the well-known antioxidants and free radical scavengers that have been tested in schizophrenia trials (Cox et al. 2015; Miller et al. 2011; Girgis et al. 2014; Müller et al. 2013; Bumb et al. 2015; Arroll et al. 2014; Joy et al. 2006).

- *N-Acetylcysteine (NAC)* is a precursor molecule of glutathione, an endogenous neuromodulator of glutamate receptors. It is an antioxidant, reduces hydroxyl radicals, and provides essential cysteine for the regeneration of endogenous glutathione. NAC also modulates the synthesis and degradation of anti- and pro-inflammatory cytokines. It is approved for use in overdose of acetaminophen where it binds to the metabolites of the drug. It is also used as a mucolytic in conditions with excessive pulmonary secretions. Two RCTs of NAC in schizophrenia have been published. Two grams of NAC was administered daily as an adjunct to APD in chronic schizophrenia patients for 8 and 24 weeks, respectively. There were modest improvement in negative symptoms and superiority over placebo with an effect size of 0.45 in the Sommer et al. meta-analysis (Sommer et al. 2014), although positive symptoms did not show improvement.
- *Polyunsaturated fatty acids (PUFAs)*

  Also known as omega-3 fatty acids, these naturally occurring nutrients have various biological roles including anti-inflammatory, anti-apoptotic, antioxidant, and free radical scavenging. In the central nervous system, they increase glutathione levels and modulate microglial activity in the expression of TNF-α, IL-6, NO synthase, and COX-2, resulting in increased glutathione, which may be reduced in schizophrenia patients. Combined with the findings of oxidative stress and increased generation of toxic free radicals in schizophrenia, a strong case has been made for using omega-3 fatty acids as an adjunct. There are eight trials to date, which used EPA 1.4–3 g over a period of 12–48 weeks (Arroll et al. 2014; Joy et al. 2006; McGorry et al. 2017). The results have been mixed with only two trials being clearly positive. The meta-analyses by Sommer et al. (2014) of six studies reported a very low and nonsignificant effect size of 0.09. Creating some hope however, PUFA supplementation had been shown to reduce conversion rates to first-episode psychosis in an ultrahigh-risk cohort, and prevention strategies rather than symptomatic treatment were considered more promising (Arroll et al. 2014; Joy et al. 2006). A recent large trial titled the Neurapro Trial by the same group (McGorry et al. 2017) has produced negative results leading to rethinking on this strategy. In this trial, 304 subjects at ultrahigh risk for conversion to psychosis were recruited. One hundred and fifty-two subjects received 1.4 g of daily omega-3 PUFA for 12 months, and the remainder received placebo. Both groups received cognitive behavior case management. There was no significant difference in the conversion rate to psychosis between the two groups at 6 or 12 months.

- *Melatonin* (N-acetyl-5-methoxytryptamine) is a naturally occurring compound that is well established to play a role in the sleep-wake cycle. It is also a free radical scavenger, stimulates antioxidant enzymes, and enhances intracellular glutathione to maintain cellular membrane integrity. Two RCTs of 19 and 40 schizophrenia patients each have reported improved sleep and mood but not core psychotic symptoms (Arroll et al. 2014).
- *Vitamins C and E*

  Vitamins C (ascorbic acid) and E (tocopherols and tocotrienols) are dietary antioxidants and helpful in reducing oxidative stress by breaking free radical chain formation. One RCT of 40 patients receiving 500 mg of daily vitamin C as an adjunct to APD reported improved schizophrenia symptoms compared to placebo (Arroll et al. 2014). Vitamin E had held promise as a treatment for tardive dyskinesia, an adverse effect especially of first-generation agent (FGA) APDs. However meta-analyses of the RCTs in this condition have not supported this. It is still possible that vitamin E slows down neurodegeneration and may have a value in neuropsychiatric conditions (Arroll et al. 2014).
- *L-Theanine* (gamma-glutamylethylamide), an amino acid with antioxidant activity through inhibition of peroxidation and found in tea especially green tea, has been investigated as a supplement in schizophrenia in 1 RCT with 40 patients. Only general symptoms such as anxiety seemed to benefit but not the specific symptoms of schizophrenia (Arroll et al. 2014).
- *Gluten-free diet:* Schizophrenia co-occurs at a 2–3 times higher rate in patients who suffer from celiac disease, an immune-mediated gastrointestinal disorder triggered by proteins such as gliadin and prolamines found in wheat gluten, barley, and rye. It is not clear whether there is a common pathophysiology between celiac disease and schizophrenia. Gluten-free diets are reported as benefitting psychiatric conditions including schizophrenia. It is speculated that this may be through intestinal protection against the formation of abnormal immune agents. Both positive and negative studies have been reported with this strategy warranting additional studies to include immune markers (Table 1).

2.3 Anti-inflammatory Effects of Antipsychotic Drugs (APDs)

2.3.1 Antipsychotic Drugs

We first provide a brief overview of antipsychotic drugs and then review their recently discovered inflammatory actions. Antipsychotic medications have been the mainstay of psychosis treatment for over 60 years. While the first-generation agents (FGA) worked predominantly by blocking the D2 receptor in the mesocortical and mesolimbic tracts, the second-generation antipsychotics (SGA) with the exception of aripiprazole/brexpiprazole work by both dopamine and serotonin antagonism. Aripiprazole and brexpiprazole are partial dopamine agonists and

Table 1 Anti-inflammatory therapies in Schizophrenia

Agent	Anti-inflammatory actions	Clinical evidence
Nonsteroidal anti-inflammatory agents (Nitta et al. 2013; Sommer et al. 2014)		
Aspirin	Irreversibly inhibits the inflammatory enzymes COX-1 and COX-2, decreases synthesis of pro-inflammatory prostaglandins	*Two RCTs.* ASA had a significant effect on symptom reduction. FEP and inpatients benefitted more
Celecoxib	Specific COX-2 inhibitor and inhibits conversion of arachidonic acid to prostaglandins	Five RCTs. Three were positive; two were negative. Effect size not significant
Minocycline	Inhibits nitric oxide synthase and pro-inflammatory enzyme 5-lipoxygenase	Seven RCTs. Modest gains in symptoms. Effect size not significant
Antioxidants, free radical scavengers, and nutrients (Sommer et al. 2014; Arroll et al. 2014; Joy et al. 2006; McGorry et al. 2017)		
N-Acetylcysteine	Reduces hydroxyl radicals; modulates synthesis and degradation of anti- and pro-inflammatory cytokines	Two RCTs. Modest benefit in negative symptoms over placebo
Vitamin C (l-ascorbic acid)	Reduce oxidative stress by breaking free radical chain formation	One RCT. Vitamin C was better than placebo in improving symptoms globally
Vitamin E (tocopherols and tocotrienols)	Free radical scavenger, stimulates antioxidant enzymes, enhances intracellular glutathione to maintain cell membrane integrity	11 RCTs in tardive dyskinesia patients. Overall, no significant benefit
Melatonin (N-acetyl-5-methoxy tryptamine)	Antioxidant and mitochondrial protection. Upregulates antioxidant enzymes and downregulates NO synthases and lipoxygenases	Two RCTs: improved sleep and mood. No specific antipsychotic effects
Cotinine	Positive allosteric modulator of nicotinic cholinergic receptors with anti-inflammatory action	No studies
Omega-3 PUFAs	Antioxidant, free radical scavenging. Modulates microglial activity in the expression of TNF-α, IL-6, NO synthase, and COX-2; inhibits peroxidation (antioxidant)	Eight trials with over 792 subjects. Only two were clearly positive. Low, non-sig effect size 0.09 (Sommer et al. 2014) Largest trial (McGorry et al. 2017) for prevention of psychosis was negative
L-Theanine	Antagonist of AMPA and Kainic acid receptors, weak agonist of NMDA receptors	One RCT. Benefit only for anxiety symptoms
Gluten-free diet	Avoids proteins such as gliadin and prolamines (wheat gluten, barley, and rye) that trigger destructive antibody formation	Equivocal results. No consistency in results across studies
Antipsychotic drugs (Maes et al. 1997; Müller et al. 2013; Bumb et al. 2015; McGorry et al. 2017; Baumeister et al. 2016; Potvin et al. 2008; Debnath and Venkatasubramanian 2013; Chan et al. 2011; de Witte et al. 2014; Chen et al. 2012; Diaz et al. 2010)		

(continued)

Table 1 (continued)

Agent	Anti-inflammatory actions	Clinical evidence
Antipsychotic drugs	• Diminish pro-inflammatory cytokines such as TNF-α, IL-1β, and IL-6 • Stimulate anti-inflammatory cytokines such as IL-4, IL-10, and IL-17 • Attenuate cell-mediated immune activation and oxidative and nitrosative stress • Target intra- and intercellular inflammation pathways: p38 MAPK, NFkB, and COX	Risperidone and olanzapine: one study, 180 FEP subjects. Both reduced IL-10 and IL1-RA. May explain benefit in negative symptoms and occurrence of metabolic side effects Clozapine: one study. Increased IL-18 (enhanced Type 1 response) Haloperidol: one study. Increased levels of CRP after 3 months but no difference from baseline after 12 months
Biologicals (Miller and Buckley 2016; Girgis et al. 2017)		
Tocilizumab	Biological agent to target specific cytokine Anti-IL-6 receptor antibody	One RCT. No benefit Likely due to not crossing blood-brain barrier
Peroxisome proliferator-activated receptors (PPARs) (Rolland et al. 2013)		
Rosiglitazone	Nuclear receptors regulating gene expression and intracellular anti-inflammatory actions	Rosiglitazone: one study with clozapine, 19 subjects, no benefit in negative symptoms or total symptom scores
Pioglitazone		Pioglitazone: one RCT with 40 subjects, improvement seen in negative symptoms and total scores
Neuroprotectors (Cox et al. 2015; Sommer et al. 2014)		
Davunetide	Neuroprotective protein, also reduces TNF-α	One study with two doses of davunetide. No significant benefit
Estrogens	Reduce microglia activation, TNF-α, and NO, to reduce antioxidative stress	Seven studies, $n = 261$, reduced positive symptoms (females) with strong effect size of 0.51
Herbals (Arroll et al. 2014)		
Medicinal herbs Ginger, turmeric, *Ginkgo biloba*	Inhibition of microglia-mediated neuro-inflammation. Decrease PGE2, IL-1β, and TNF-α by downregulation of COX-2, p38MAPK, and NFkB expression	Ginkgo: one RCT with 109 patients. Improved response in otherwise refractory patients

serotonin antagonists. In addition to the dopamine and serotonin actions, SGAs have a multitude of actions on other receptors such as alpha adrenergic, histaminic, and cholinergic. They may also have indirect effects on the glutamatergic receptors, especially the NMDA receptor. Regarding their behavioral effects, in summary, antipsychotics are more effective with acute positive psychotic symptoms such as hallucinations, delusions, and disorganized behavior, less effective with chronic positive symptoms, and ineffective with negative symptoms (asociality, avolition, blunted affect, etc.) and cognitive symptoms (impaired working memory, impaired executive functions, etc.). Clozapine has distinguished itself from other agents as having a superior efficacy in otherwise treatment-refractory patients with schizophrenia. FGAs are prone to cause side effects such as extrapyramidal symptoms, and SGAs tend to increase weight and may lead to hyperglycemia and the metabolic syndrome. Although infrequent, clozapine has additional toxicities associated with it such as agranulocytosis, seizures, and myocarditis (Kane and Correll 2010). Of interest here is the question if APDs have pro- or anti-inflammatory effects when used in the treatment of schizophrenia and how this may help us better understand their benefits and adverse effects and optimize their use.

2.3.2 Anti-inflammatory Actions of APDs

There are several lines of evidence to attribute anti-inflammatory actions to antipsychotic drugs (Müller et al. 2013; Bumb et al. 2015; Baumeister et al. 2016; Potvin et al. 2008; Debnath and Venkatasubramanian 2013; Chan et al. 2011; Chen et al. 2012; Diaz et al. 2010). Certain APDs are found to attenuate cell-mediated immune activation as well as oxidative and nitrosative stress. In particular, second-generation APDs appear to stimulate production of anti-inflammatory cytokines such as IL-4, IL-12, and IL-17. Thus, the deleterious effects of ongoing inflammation that has been proposed in schizophrenia might be reduced by these agents. In a large clinical study of this question with 180 patients by de Witte et al. (2014), 6 weeks of risperidone and olanzapine treatment decreased IL-10 levels which correlated with improvement in negative symptoms and cognitive functioning.

APDs have also been shown to diminish pro-inflammatory cytokines, such as TNF-alpha, IL-1beta, and IL-6 (Chan et al. 2011). A recent study using a SNP-based analysis of neuroactive pathways implicated PGE_2 as a mediator of the effects of risperidone, olanzapine, and quetiapine. Also, chronic administration of olanzapine or clozapine is shown to reduce PGE_2 concentration in the rat brain. Risperidone also reduced L-prostaglandin D2 synthase (PGDS)/15d-PGJ2/PPAR-γ, which is known to help maintain the inflammatory state, and COX, the enzyme that converts arachidonic acid into prostaglandins. Murine studies have revealed that many APDs target intracellular and intercellular signaling inflammation pathways including p38 mitogen-activated protein kinase (p38MAPK), nuclear factor kappa B (NFkB), inducible NO synthase, and m-PGES-1. Thus, we may conclude APDs might be attenuating the proposed over-activated inflammatory process in psychotic states (Cox et al. 2015; Debnath and Venkatasubramanian 2013).

A number of additional candidate biomarkers have also been reported to change in response to antipsychotic treatment, including S100B, prolactin, IL-2, insulin, leptin, interleukin-1 RA (receptor antagonist), IL-8, and interleukin-2 RA. Reduced sICAM-1 levels in schizophrenia have been found to increase even after short-term antipsychotic therapy, and expression of the ICAM-1 ligand leukocyte function antigen-1 (LFA-1) increases significantly with antipsychotics (Müller 2014; de Witte et al. 2014).

Another line of evidence of the helpful role of APDs in modulating dysfunctional inflammatory responses comes from the pro-immunological role of APDs in states where the immune response is blunted. Clozapine is thought to enhance the Type 1 response by elevating IL-18 levels which may explain clozapine's relatively better effects on negative symptoms compared to other APDs. In vitro studies show that the blunted IFN-γ production is normalized after treatment with antipsychotics, especially clozapine. This too suggests that clozapine enhances the blunted T1 response in schizophrenia.

A 2008 meta-analysis of 62 studies of peripheral immune alterations in schizophrenia patients compared with controls documents many of the above findings (Potvin et al. 2008). Thus, it may be said that certain APDs blunt or reduce the presumably over-activated Type 2 response in schizophrenia and possibly enhance the blunted T1 response. In contrast to SGAs, haloperidol, an FGA, does not seem to share the above anti-inflammatory effects. In a study comparing the effects of haloperidol and risperidone on dendritic cells, it was observed that only risperidone affected the cytokine and chemokine production of mature dendritic cells by inhibiting Th1 cytokines (IFN-γ) and increasing Th2 cytokines (IL-6, IL-8, TNF-α, and IL-10) (Chen et al. 2012). A 2010 study of 111 patients examined the effects of haloperidol, olanzapine, and risperidone on CRP levels after 3 months and 12 months of treatment, respectively (Diaz et al. 2010). Haloperidol treatment actually led to an increase in the levels of CRP after 3 months. However, there were no significant differences in CRP levels between baseline and 12 months, for any of the treatments, suggesting development of tolerance to the effects of APDs on CRP levels.

2.3.3 Adverse Effects of APDs

The metabolic syndrome with weight gain, hyperglycemia, dyslipidemia, and hypertension can be a disturbing adverse effect of SGAs. While mechanisms such as antihistaminic effects, adipose tissue metabolism, and humoral effects on the glucocorticoid system have been proposed, there is no consensus on any single mechanism to explain these adverse effects. Adverse immunomodulation may offer another viable mechanism to consider.

The pro-inflammatory cytokines sIL-2R and IL-12 which are increased at baseline in schizophrenia increase further with antipsychotic treatment. TNF-alpha another inflammatory immunochemical also has been shown to increase after risperidone and quetiapine treatment. Clozapine has been reported to induce mitochondrial damage and thus promote inflammation in insulin-responsive cell and obesity-associated cell types. In a large study of serum cytokine levels in 180 patients with schizophrenia, 6 weeks of risperidone and olanzapine resulted in reduced levels of the anti-inflammatory cytokines IL-10 and IL1-RA (de Witte et al. 2014). All the above findings might help explain the occurrence of the metabolic syndrome by SGAs through increase in select inflammatory processes.

Neutropenia is another side effect of APDs, especially certain SGAs. Although not always consequential, it could lead to more serious hematological conditions. The study by Chen et al. referred to above (Chen et al. 2012) on the effects of APDs on dendritic cell-mediated immunity revealed that the production of TNF-α by risperidone-treated mature dendritic cells accelerates apoptosis of neutrophils, leading to drug-induced neutropenia.

In summary, the above make a strong case for the effects of APD on immune processes in a beneficial way in schizophrenia as well as explain possible immunological mechanisms to explain certain side effects. However, there is considerable inconsistency in the immunological findings, and some studies report findings in the opposite direction (Miller et al. 2011; Baumeister et al. 2016; Potvin et al. 2008; Debnath and Venkatasubramanian 2013). These inconsistencies include:

- Adding antipsychotic medications to plasma drawn from healthy control subjects that was stimulated by toxic shock syndrome toxin, producing elevated cytokine levels had no effects on plasma IL-6 levels.
- An increase of TNF-α and TNF-α receptors has been seen during therapy with clozapine and other antipsychotics.
- Clozapine has been shown to suppress the stimulated production of IFN-γ (in contrast to much weaker effects by haloperidol) in cultured peripheral blood mononuclear cells.
- The meta-analyses by Miller et al. (2011) concluded cytokines levels seen in schizophrenia were independent of antipsychotic drugs.
- Patients who are psychiatrically resistant to APDs continued to exhibit high pro-inflammatory cytokine levels, indicating that such immune overactivity had not been restored by APDs (Potvin et al. 2008).
- Some studies have reported that the SGAs did not significantly alter the higher levels of chemokines and cytokines in FEP.
- In vitro studies of antipsychotics effects on inflammatory chemicals are not consistent either. For example, IL-6 upregulation is not consistently found, and other studies report decreased pro-inflammatory proteins IL-1 β and IFN-ɣ (Miller et al. 2011).

Thus, the findings do not fit neatly into an explanatory theory of APD immunological actions. The proposed normalization of the Type 1 and Type 2 immune responses by APDs needs to be shown to be independent of the psychotic state. These inconsistencies highlight the complexities of the immunological processes, the state of the science, and the importance of designing trials to carefully delineate disease states, disease progression, and lifestyle while studying immunological biomarkers (Cox et al. 2015; Miller et al. 2011; Müller et al. 2013; Bumb et al. 2015; Baumeister et al. 2016; Potvin et al. 2008; Debnath and Venkatasubramanian 2013; de Witte et al. 2014).

3 Biologicals

An exciting approach to the use of immunotherapy in schizophrenia is through biological agents that target specific cytokines. We are increasingly seeing such agents in connective tissue disorders. Since the biological agent is specifically designed to modulate a receptor, there is no ambiguity regarding the target of action. In schizophrenia, an open-label study of six patients given two injections of tocilizumab reported improvement in cognition (Miller and Buckley 2016). Tocilizumab is a humanized monoclonal antibody against the IL-6 receptor and is administered as a once-monthly intravenous injection. It is used in the treatment of rheumatoid arthritis and juvenile idiopathic arthritis. Clinical effects are usually seen after 1 week. As mentioned earlier IL-6 has been shown to be elevated in schizophrenia patients. This agent was tested in a recent DB RCT and yielded negative results (Girgis et al. 2017). In this trial, 36 clinically stable and moderately symptomatic patients with schizophrenia were given three monthly infusions of 8 mg/kg tocilizumab or placebo (normal saline). Outcome was assessed at 12 weeks using the Positive and Negative Symptom Scale (PANSS) and the MATRICS battery of neuropsychological tests. Disappointingly, there were no significant effects on either of the outcome measures. The study also measured baseline CRP and various cytokine levels (IFN-γ, TNF-α, IL-1b, IL-2, IL-6, IL-8, IL-10, IL-12, and IL-17a), but there were no correlations between the inflammatory markers and the clinical outcomes. During the study period IL-6 and IL-8 increased, and CRP decreased in the active treatment group as predicted. The authors concluded that since tocilizumab does not cross the blood-brain barrier, it might not have impacted brain function. Inclusion of patients regardless of evidence of inflammation (e.g., elevated CRP) could be another reason for non-response to the anti-inflammatory drug. Various other reasons were also considered to explain the negative study such as sample size and heterogeneity of subgroups. Another possibility was that increases in IL-6 observed in schizophrenia are not causal or its impact on psychopathology may be indirect and not amenable to directly targeted therapy such as tocilizumab. However, recent population-based longitudinal and Mendelian randomization studies support a potentially causal role for IL-6/IL-6R pathway in schizophrenia (Rolland et al. 2013; Khandaker et al. 2014, 2018). In future, RCTs of immunotherapies based on patients with elevated CRP levels might be a useful way forward. This approach has shown promising results in a trial of infliximab for depression (Hartwig et al. 2017).

Another small but promising study of immunotherapy albeit of a less specific compound reported that injection of IFNg1b given weekly as a subcutaneous injection to two patients with schizophrenia resulted in improvement of psychosis. The field of biological therapies in psychiatric disorders is nascent, and more studies are likely in the near future. IL-1b, IL-6, TNF-a, and IFN-g represent the most promising initial therapeutic targets for such studies.

4 Miscellaneous Agents

4.1 *Cotinine*

Cotinine is an alkaloid present in tobacco leaves and the main metabolite of nicotine. It has psychotropic actions including antipsychotic effects through its modulation of the serotonergic, cholinergic, and dopaminergic systems. It behaves as a positive allosteric modulator of the nicotinic cholinergic receptors resulting in anti-inflammatory effects which may explain its psychiatric effects. There are no schizophrenia trials as of yet with this agent.

4.2 *Davunetide*

Davunetide is a neuroprotective protein that also reduces TNF-α. It is being studied in Alzheimer dementia and supranuclear palsy. It was studied in one schizophrenia trial wherein two doses of the active drug (5 and 30 mg) were added on to existing antipsychotic medication for a period of 12 months. Twenty-two patients received the active drug and twenty patients received placebo. There was no benefit (Sommer et al. 2014).

4.3 *Pregnenolone*

Pregnenolone is an endogenous compound and is involved in the synthesis of several steroids. It is richly distributed in the brain and acts as a neurosteroid. It enhances synaptic communication and especially promotes NMDA and GABA receptor functions. It also promotes myelinization and thus may be neuroprotective (Cox et al. 2015). Two studies have reported that pregnenolone used as add-on therapy to antipsychotics significantly decreased negative symptoms, such as avolition, anhedonia, and blunted affect, and thus appears to be a promising agent for further studies.

4.4 *Estrogen*

Estrogens have multiple anti-inflammatory effects through microglial activation, reducing TNF-α and NO and thus reducing oxidative stress. Estrogens are also known to have modulating effects on neurotransmission, for example, dopamine. There are seven reported studies of ethinyl estradiol (six studies) and estrone (one study) as adjuncts in schizophrenia patients including one study of male patients and six studies with females. A patch 0.5 mg strength or 2 mg oral tablets were used. Duration of the studies varied from 2 to 8 weeks. Total sample size was 138 subjects receiving estrogens and 121 receiving placebo, in double-blind design. A meta-analysis showed a strong and significant effect size of 0.51 even after removing an outlier study (ES: 0.51). Adjunct estrogen significantly decreased positive and negative symptoms in women with chronic schizophrenia. Sommer et al. (2014) concluded that the results of estrogen addition to antipsychotic treatment seemed promising but cautioned that further investigation is needed to establish that it is the anti-inflammatory actions that result in the benefit, as estrogens have a broad spectrum of actions.

4.5 Erythropoietin

Erythropoietin (EPO) is a hematopoietic growth factor which appears to have anti-apoptotic, anti-inflammatory, antioxidant, neurotrophic, angiogenic, and synaptogenic activity. Recently, in one open-label study, EPO application was found to have a positive effect on cognition in schizophrenia patients.

4.6 Peroxisome Proliferator-Activated Receptors (PPARS)

PPARS are compounds with intracellular anti-inflammatory actions. They act as nuclear receptors working as transcription factors and regulate genomic expression. They also influence synaptic communication. The β type of PPAR is richly expressed in the brain. Currently PPARs are used in diabetes management for increasing insulin sensitivity. They may have a role in psychiatric disorders.

Glitazones such as pioglitazone and rosiglitazone are PPAR agonists and potent agonists of PPAR-γ. There are two studies, one with rosiglitazone as an adjunct to clozapine in schizophrenia patients which did not reveal any benefit and one DB RCT with 40 patients with schizophrenia given 30 mg/day of pioglitazone or placebo as an adjunct to risperidone for 8 weeks. In the latter trial, the pioglitazone group showed significantly more improvement in negative symptom scores and total psychopathology score (Raison et al. 2013).

4.7 Medicinal Herbs

Ginger, *turmeric*, and *Ginkgo biloba* have been found to have immunomodulatory effects possibly as antioxidants through inhibition of microglia-mediated neuro-inflammation. Products of ginger have been demonstrated to diminish production of PGE2, IL-1β, and TNF-α by downregulating COX-2, p38MAPK, and NFkB expression (Joy et al. 2006). In an RCT of 109 patients, extracts of *Ginkgo biloba* as an add-on to haloperidol were found to be beneficial over placebo in otherwise treatment-refractory schizophrenia patients (Joy et al. 2006).

4.8 Other Agents

Other ongoing clinical trials with anti-inflammatory agents as adjuncts to APDs include salsalate, fluvastatin, simvastatin, methotrexate, resveratrol, hydrocortisone, and ibuprofen.

5 Strategies for Consideration by Clinicians

As is often the case, practice precedes science and evidence. Currently, physicians often use NAC, omega-3 fatty acids, and L-methyl folate as adjuncts to antidepressant and antipsychotic medications, but there are no published reports of the outcomes of such clinical practice in schizophrenia. While science is not advanced enough to make definitive recommendations, based on the available literature so far, the following suggestions are provided as guidelines and in the spirit of helping our patients while hopefully not hurting them:

Practitioners should obtain a detailed clinical history, especially of inflammatory disorders such as gluten sensitivity, celiac disease, early life infections, and family history of immune disorders, and incorporate the information into the clinical formulation for a better understanding of the patient's disorder. The standard review of systems should include symptoms suggestive of deficiency of neuroprotective agents such as omega-3 PUFAs, vitamins C and E, folic acid, etc. This may include tendency toward frequent infections, skin and mucosal conditions, chronic fatigue, unexplained mood changes, etc.

Medical providers should routinely counsel patients on consuming a health balanced diet rich in nutrients reviewed above. More cautiously, they consider adding agents with some evidence of benefit to the standard antipsychotic therapeutic regimen after review of potential risks and benefits. Fortunately, the chances of any serious adverse events with such agents is low. Based on the current state of literature, aspirin, celecoxib, and NAC could be tried over 3–6 months with monitoring of outcome and continue/discontinue based on the results in the individual patient. Adding estrogen could be considered in female patients, especially postmenopausal women in collaboration with the gynecologist and after reviewing individual risk benefits. Physicians should keep abreast of the emerging studies and literature in the field of anti-inflammatory therapeutics including APDs with beneficial anti-inflammatory profiles and newer and more specific anti-inflammatory therapies and assess their relevance to their own patients and practices. We owe such diligence to our patients.

6 Opportunities, Challenges, and Future Directions

We have arrived at an exciting stage in the development of anti-inflammatory therapeutic approaches and agents for schizophrenia, often a devastating and disabling condition. The optimism is justified for the following reasons:

- Our understanding of the immune system and its complexities is significantly better in the last two decades.
- There is moderate convergence of findings suggesting immune abnormalities in schizophrenia, with studies spread over three decades.
- A theory of blunted T1 response and over-activated T2 response offers a credible model for large-scale empirical testing, with attention to some of the factors noted earlier.
- There appears to be some promise in trials with existing anti-inflammatory agents.
- There appears the possibility of optimizing the effects of APDs by understanding their pro- and anti-inflammatory effects, including efficacy with negative/cognitive symptoms, and in managing adverse effects.
- Biological agents such as monoclonal antibodies are arriving on the scene. Medical disorders such as rheumatoid arthritis, systemic lupus erythematosus, and diabetes mellitus that have some parallels with schizophrenia are benefitting from biological immune therapies. A role for similar biological therapies in schizophrenia can easily be envisaged, at least for a subgroup of patients and/or particular stages of the disease.
- There is progress in the field of personalized medicine, and identifying patients with high levels of specific biomarkers such IL-6, genetic polymorphisms of cytokines, etc. with customized immunotherapy is a possibility. For illustrative purposes we might suggest studies that target narrowly defined and homogenous subgroups, such as those with acute psychoses within 5 years of illness onset and similar socioeconomic backgrounds for measuring and targeting pro-inflammatory cytokines, and a different study with chronic patients with greater than 10 years of illness and predominantly negative symptoms for measuring and targeting anti-inflammatory, Type 1 response blunting markers and follow their levels as patients receive therapies aimed at improving negative symptoms.

The enthusiasm regarding immunotherapy however needs to be tempered. As reviewed earlier, the evidence is neither consistent across studies nor always in the expected direction. Future studies could add to our knowledge by including measurement of cytokine levels in serum and/or CSF at baseline, prodromal, early psychosis, remitted state, pre-relapse, chronic and deficit syndrome stages, etc. Other targets could include measurable functional targets such as white matter integrity and connectivity with diffusion tensor imaging and functional activity of specific neural circuits with fMRI. The personalized medicine approach combining known genomic, proteomic, and immune markers to characterize biological subgroups of patients and then treating with an agent with the potential to reverse a known abnormality in these markers is another promising direction in the future. Among individual APDs, clozapine should be investigated more thoroughly because of its unique clinical profile. Understanding whether its superiority or unique adverse effects arise from its immunological activity could be a key to more effective and safer drug development. The naturally occurring herbal antioxidants are another promising group of agents to be pursued further with large systematic trials. Lastly, the development of the MABs targeting biologically homogenous groups of patients with demonstrable cytokine abnormalities also holds much promise. While newer biological therapies will surely raise the issue of exorbitant costs as well as adverse effects, our patients deserve no less than the best therapies we can offer.

References

Arias I, Sorlozano A, Villegas E et al (2012) Infectious agents associated with schizophrenia: a meta-analysis. Schizophr Res 136:128–136

Arroll MA, Wilder L, Neil J (2014) Nutritional interventions for the adjunctive treatment of schizophrenia: a brief review. Nutr J 13:91

Baumeister D, Ciufolini S, Mondelli V (2016) Effects of psychotropic drugs on inflammation: consequence or mediator of therapeutic effects in psychiatric treatment? Psychopharmacology (Berl) 233(9):1575–1589

Bumb JM, Enning F, Leweke FM (2015) Drug repurposing and emerging adjunctive treatments for schizophrenia. Expert Opin Pharmacother 16(7):1049–1067

Chan MK, Guest PC, Levin Y et al (2011) Chapter 5 – Converging evidence of blood-based biomarkers for schizophrenia: an update. Int Rev Neurobiol 101:95–144

Chen ML, Tsai TC, Wang LK et al (2012) Risperidone modulates the cytokine and chemokine release of dendritic cells and induces TNF-α-directed cell apoptosis in neutrophils. Int Immunopharmacol 12(1):197–204

Cox D, Chan MK, Bahn S (2015) The potential of immune biomarkers to advance personalized medicine approaches for schizophrenia. J Nerv Ment Dis 203(5):393–399

de Witte L, Tomasik J, Schwarz E (2014) Cytokine alterations in first-episode schizophrenia patients before and after antipsychotic treatment. Schizophr Res 154(1–3):23–29

Deakin B, Suckling J, Barnes TRE et al (2018) The benefit of minocycline (Benemin) on negative symptoms of schizophrenia with recent onset psychosis. Lancet Psychiatry 5(11):885–894

Debnath M, Venkatasubramanian G (2013) Recent advances in psychoneuroimmunology relevant to schizophrenia therapeutics. Curr Opin Psychiatry 26(5):433–439

Diaz FJ, Pérez-Iglesias R, Mata I et al (2010) Possible effects of some antipsychotic drugs on C-reactive protein in a drug-naïve psychotic sample. Schizophr Res 121(1–3):207–212

Girgis RR, Kumar SS, Brown AS (2014) The cytokine model of schizophrenia: emerging therapeutic strategies. Biol Psychiatry 75(4):292–299

Girgis RR, Ciarleglio A, Choo T et al (2017) A randomized, double-blind, placebo-controlled clinical trial of tocilizumab, an interleukin-6 receptor antibody, for residual symptoms in schizophrenia. Neuropsychopharmacology 43(6):1317–1323

Hartwig FP, Borges MC, Horta BL, Bowden J et al (2017) Inflammatory biomarkers and risk of schizophrenia: a 2-sample Mendelian randomization study. JAMA Psychiat 74(12):1226–1233

Joy CB, Mumby-Croft R, Joy LA (2006) Polyunsaturated fatty acid supplementation for schizophrenia. Cochrane Database Syst Rev 19(3):CD001257

Kane JM, Correll CU (2010) Past and present progress in the pharmacologic treatment of schizophrenia. J Clin Psychiatry 71(9):1115–1124

Khandaker GM, Pearson RM, Zammit S et al (2014) Association of serum interleukin 6 and C-reactive protein in childhood with depression and psychosis in young adult life: a population-based longitudinal study. JAMA Psychiat 71(10):1121–1128

Khandaker GM, Zammit S, Burgess S et al (2018) Association between a functional interleukin 6 receptor genetic variant and risk of depression and psychosis in a population-based birth cohort. Brain Behav Immun 69:264–272

Maes M, Delange J, Ranjan R et al (1997) Acute phase proteins in schizophrenia, mania and major depression: modulation by psychotropic drugs. Psychiatry Res 66:1–11

McGorry PD, Nelson B, Markulev C et al (2017) Effect of ω-3 polyunsaturated fatty acids in young people at ultrahigh risk for psychotic disorders: the NEURAPRO randomized clinical trial. JAMA Psychiat 74(1):19–27. https://doi.org/10.1001/jamapsychiatry.2016.2902

Miller BJ, Buckley PF (2016) The case for adjunctive monoclonal antibody immunotherapy in schizophrenia. Psychiatr Clin N Am 39:187–198

Miller BJ, Buckley P, Seabolt W et al (2011) Meta-analysis of cytokine alterations in schizophrenia: clinical status and antipsychotic effects. Biol Psychiatry 70:663–671

Müller N (2014) Immunology of schizophrenia. Neuroimmunomodulation 21:109–116

Müller N, Myint AM, Krause D et al (2013) Anti-inflammatory treatment in schizophrenia. Prog Neuro-Psychopharmacol Biol Psychiatry 42:146–153

Nitta M, Kishimoto T, Muller N et al (2013) Adjunctive use of nonsteroidal anti-inflammatory drugs for schizophrenia: a meta-analytic investigation of randomized controlled trials. Schizophr Bull 6:1230–1241

Potvin S, Stip E, Sepehry AA et al (2008) Inflammatory cytokine alterations in schizophrenia: a systematic quantitative review. Biol Psychiatry 63(8):801–808

Raison CL, Rutherford RE, Woolwine BJ et al (2013) A randomized controlled trial of the tumor necrosis factor antagonist infliximab for treatment-resistant depression: the role of baseline inflammatory biomarkers. JAMA Psychiat 70(1):31–41

Rolland B, Deguil J, Jardri R et al (2013) Therapeutic prospects of PPARs in psychiatric disorders: a comprehensive review. Curr Drug Targets 14(7):724–732

Sommer IE, van Westrhenen R, Begemann MJ et al (2014) Efficacy of anti-inflammatory agents to improve symptoms in patients with schizophrenia: an update. Schizophr Bull 40(1):181–191

Torrey EF, Davis JM (2012) Adjunct treatments for schizophrenia and bipolar disorder: what to try when you are out of ideas. Clin Schizophr Relat Psychoses 5(4):208–216

The manufacturer's authorised representative in the EU is Springer Nature Customer Service Centre GmbH, Europaplatz 3, 69115 Heidelberg, Germany. If you have any concerns regarding our products, please contact ProductSafety@springernature.com

Printed and bound by CPI Group (UK) Ltd, Croydon, CR0 4YY
30/09/2026
02225458-0009